Factory Air:

Cool Cars in Cooler Comfort

An Illustrated History of Automotive Air-Conditioning

Volume 1: 1940 to 1942—The Innovative Years

Allen B. Simons

October 20, 2021

Copyright © 2021

All Rights Reserved

Author note:
Illustrations are the keynote of this historical work. Originating some 80 years ago, various images are of lower quality, yet serve as an important support for the narrative.

Acknowledgments

I must thank many friends and colleagues for their interest in this project, along with their belief that this compilation of Packards, Cadillacs, and Chryslers will serve as a meaningful research document on the inception of automotive factory air-conditioning.

I salute my Special Lady and wife, Candy, for her encouragement, comments (both good and otherwise), evaluations, and patience with this multi-year project that generated stacks of research publications and proof copies of each chapter, unpublished until now in this format.

A sincere thank you goes to Dwight Heinmuller, the Packard Guru of Information. Not only did he offer editing recommendations, but he also contributed an unprecedented amount of previously non-published Packard documents that enhance the Packard air-conditioning story.

Especially invaluable are the images sent to me, the invitations for photoshoots, and their legacy found in automotive publications of the 1940-era factory air-conditioned Packards, Cadillacs, and Chryslers.

Spotlighted automobiles and equipment feature extensive coverage.

1. West Peterson's 1940 Packard Custom Super-8 One Eighty, photos by West Peterson

2. Nelson Bates' 1940 Packard One Twenty (non-A/C equipped), photos by Allen B. Simons

3. Jim Hollingsworth's 1940 Packard One Eighty, photos by David W. Temple

4. Royce and Sheila Baier's 1941 Packard One Eighty (non-A/C equipped), photos by Charlie Kietzman

5. Sal and JoAnn Saiya's 1941 Packard One Eighty, photos by Dave Czirr, Dwight Heinmuller, and Jack Swaney

6. Cellarette available in 1941 Packard One Eighty, information and photos by Joe Block; Mitch Frumkin, Chicago Auto Show; Dwight Heinmuller; Andrew Hook via Steven Kelley; ownership by David E. McGahey, verified by Annette Simons Story; *San Francisco Examiner*

7. Jim Hollingsworth's 1941 Packard One Twenty, photos by David W. Temple

8. Terry Weiss' 1941 Packard Clipper Custom, photos by Allen B. Simons

9. General Douglas MacArthur's 1942 Packard Clipper Eight, story and photos by Jim Hollingsworth

10. Hyman Ltd. Classic Cars' 1942 Packard One Eighty photos

11. Dr. Richard Zeiger's 1941 Cadillac Fleetwood Series Sixty Special, photos by Ronald Verschoor

12. Doug Houston's 1941 Cadillac Series Sixty-Two, photos by Larry Edsall, *ClassicCars.com Journal*; David G. Helmer, Braun & Helmer Auction Service; Doug Houston; Richard Kughn; West Peterson; and Buck Varnon

13. Richard Kughn's 1941 Cadillac Fleetwood Series Seventy-Five, photos by Gene Dickirson, Richard Kughn, and Rex Roy

14. David O. Selznick's 1942 Cadillac Series Seventy-Five, written and photos by Roy A. Schneider

15. Unknown owner's 1942 Chrysler Crown Imperial Limousine, photos by Warren H. Erb and Online Imperial Club member Tony

16. T. C. Gleason's report and photos of the 1942 (?) Chrysler through the Society of Automotive Engineers (S.A.E)

A hearty thanks go to Amanda McFarlane, editor and proofreader. I enjoyed her surprise, interest, and reaction to the 1941 Packard Cellarette, "The car that makes ice cubes." In addition, her experienced, level-headed comments smoothed out difficult wording situations for a better read.

To the following additional contributors and sources of brochures, owner's and shop manuals, accessory listings, data books, service letters, magazines, newspapers, and web sites, I thank you.

Chapter 1, 1940 Packard Factory Air-Conditioning:

Jeff Steidle, Regress Press: 1940 Packard air-conditioning brochure; *ASHRAE Journal*, ashrae.org; Allen B. Simons Collection (ABS): *Popular Science, Popular Mechanics, The Syracuse Herald-American, Automotive Industries, Fortune, The New York Times,* new car cabin air filter; Dwight Heinmuller: Henney-Packard brochure, 1940 Packard Weather Conditioner brochure; James Hollingsworth: *Packard 1940: A Pivotal Year*; Howard Hanson, Thomas A. McPherson Collection: 1938 Henney Motor Company ambulance; *The History Channel*; Guscha, Kevin Waltman, The Collection of Fred and Carol Mauck, *PackardInfo.com*: *Air-Conditioning & Refrigeration News, The Austin American, Texas,* "Packard Promotional Pointers," *Saturday Evening Post*, 1940 "Packard Service Letter;" Romie Minor, Carla Reczek, Detroit Public Library, National Automotive History Collection (DPL NAHC): *Motor*, 1940 Packard Weather Conditioner demonstrator and dealership photos, 1940 Packard factory photos of air-conditioning compressor, *New York Journal and American*; Matt Hocker, David Reilly, and Chris Ritter, Antique Auto Club of America Library & Research Center (AACA Library): *1940 Senior Packards* brochure, 1940 Packard brochure images, The *Packard Data Book for 1941*; James J. Bradley, George Hamlin, Dwight Heinmuller, L. Morgan Yost: *Packard: A History of the Motor Car and the Company*; John Lawrence: *Automotive News Celebrating the 40th Anniversary–Packard*; Phil Gaffney: 1953 Oldsmobile photo.

iv

Chapter 2, 1941 Packard Factory Air-Conditioning:

Matt Hocker, David Reilly, and Chris Ritter, Antique Auto Club of America Library & Research Center (AACA Library): 1941 Packard brochure images, *Packard Data Book for 1941*, 1940 Packard factory photo, 1941 *Chrysler Air Refrigeration System* service manual; Benson Ford Research Center, The Henry Ford: 1941 Packard brochure; Joe Block: *Packard 1940 Data Book*, 1941 Packard Cellarette photos; Andrew Hook via Steven Kelley: Supplement to the *Packard Data Book for 1941*; Dave Czirr, David Flack, Kevin Waltman, *PackardInfo.com: Information on the 1941 Packard*, 1941 Packard correspondence, 1940 Packard Service Letter, 1941 Packard One Eighty photos; Dwight Heinmuller: 1940–1941 Packard correspondence and brochures, 1941 Packard air-conditioning service manual, 1941 Packard Cellarette photos, 1941 Packard One Eighty photos; Jeff Steidle, Regress Press: *Saturday Evening Post, 1941 Packard Clipper Data Book*; Wesley Boyer: *Automotive Industries; San Francisco Chronicle* article: San Francisco Library; Jack Swaney: 1941 Packard dealer album, 1941 Packard One Eighty photos; Mitch Frumkin, Historian/Archivist Chicago Auto Show: Auto show flyer advertising the 1941 Packard Cellarette; Royce and Sheila Baier: 1941 Packard One Eighty, Charlie Kietzman photos; Allen B. Simons Collection (ABS): *San Francisco Examiner, Saturday Evening Post*, 1941 Packard Clipper ad, 1941 Packard Clipper and Pan American World Airways Boeing 314 Clipper flying boat photo, 1941 Packard Clipper owned by Terry Weiss, photos by ABS; Paul Fluckiger: 1942 Packard photos; Howard Hanson: 1946 Packard Accessory book; Romie Minor, Carla Reczek, Detroit Public Library, National Automotive History Collection (DPL NAHC): 1941 Packard Clipper factory photo, 1941 Packard air-conditioning factory photo; West Peterson: 1940 Packard evaporator; Paul Ayres, Cadillac & LaSalle Club Museum and Research Center (CLCMRC): 1941 *Cadillac Air-Conditioning Manual*.

Chapter 3, 1942 Packard Factory Air-Conditioning:

Matt Hocker, David Reilly, and Chris Ritter, Antique Auto Club of America Library & Research Center (AACA Library): 1942 Packard Clippers brochure, *Information on 1942 Packard Cars*, 1942 *Packard Data Book*; Benson Ford Research Center, The Henry Ford: 1942 Packard Clippers brochure; Hyman Ltd. Classic Cars: 1942 Packard One Eighty; Dwight Heinmuller: *1942 Packard Data Book*; Kevin Waltman, *PackardInfo.com: 1942 Packard Promotional Pointers, 1942 Packard Service Letters, 1943 Packard Service Letters, 1944 Packard Service Counselor*; Walter McCall: *80 Years of Cadillac LaSalle*; Jeff Steidle, Regress Press: *Packard Senior Cars 1942* press kit; Allen B. Simons Collection (ABS): *Popular Science*; 1941 Packard brochure; 1941 Packard Clipper owned by Terry Weiss, photos by ABS; 1943 Packard War Bond ad; Bud Juneau, Editor, *The Cormorant*, Spring 1975: 1942 Packard Clipper Eight article, "Destined to Survive," written and photos by Jim Hollingsworth, owned by General Douglas MacArthur; Bud Wells; *olive-drab.com*; Hyman Ltd. Classic Cars: 1942 Packard One Eighty owner and photos; West Peterson: 1940 Packard One Eighty owner and photos; Romie Minor, Carla Reczek, Detroit Public Library, National Automotive History Collection (DPL NAHC): 1942 Packard Clipper factory assembly line photo; Jack Swaney: photo of 1941 Packard One Eighty owned by M/M Sal Saiya; Paul Fluckiger: 1942 Packard photos.

Chapter 4, 1941 Cadillac Factory Air-Conditioning:

Dr. Richard Zeiger and Ronald Verschoor: 1941 Cadillac Fleetwood Series Sixty Special photos; Walter M. P. McCall: *80 Years of Cadillac LaSalle*; Theodore F. MacManus: *Saturday Evening Post*; Allen B. Simons Collection (ABS): 1941 Cadillac brochure, 1941 Cadillac Hydra-Matic ad; *Cadillac Milestones 1902–1942*: 1941 Cadillac sales; Roy A. Schneider: *Cadillacs of the Forties*; *ASHRAE JOURNAL*: 1939 prototype trunk-mounted air conditioner; GM Heritage Center: *1941 Cadillac Data Book*; Warren Rauch, Cadillac & LaSalle Club member (CLC): *1941 Cadillac Data Book Supplement*; Paul Ayres, Cadillac & LaSalle Club Museum and Research Center (CLCMRC): 1941 *Cadillac Air-Conditioning Manual*; L. Morgan Yost: *Packard: A History of the Motor Car and the Company*;

v

Kathy Adelson, GM Media Archive: 1941 Cadillac air-conditioning owner's manual, *Regarding the Air-Conditioning System on Your Cadillac;* Rod Barclay: *Boy! That Air Feels Good!;* GM Heritage Center: 1941 Cadillac optional sunroof; Northwest Motor Company, Seattle, WA: 1941 Cadillac pricing guide; Yann Sanders: *The Original Cadillac Database, car-nection.com;* Bud Juneau, Editor of *The Cadillac-LaSalle Self Starter:* a tribute article to the 1941 Cadillac, contributed by Terry Wenger, Sr.; Doug Houston: 1941 Cadillac Series Sixty-Two photos, "Cadillac's Air Conditioner" article, *cadillaclasalleclub.org Technical/Authenticity Forum,* 1941 Cadillac air-conditioning information, Cadillac & LaSalle Club Photo Gallery; David G. Helmer, Braun & Helmer Auction Service: 1941 Cadillac Series Sixty-Two photo; West Peterson: 1941 Cadillac Series Sixty-Two photos; Larry Edsall, *ClassicCars.com Journal:* 1941 Cadillac Series Sixty-Two photo; Rex Roy, *Motor City Dream Garages:* 1941 Cadillac Fleetwood Series Seventy-Five photos.

Chapter 5, 1941 Chrysler Factory Air-Conditioning:

Allen B. Simons Collection (ABS): "Chrysler Air Refrigeration System" brochure, 1942 De Soto brochure; Chrysler Historical Collection: 1941 Chrysler photo; Warren H. Erb, Docent, AACA Museum, *WPC News:* "1941 Chrysler Air Refrigeration System" article and service manual; Online Imperial Club (OIC): 1940 Chrysler Crown Imperial ads, 1942 Chrysler Crown Imperial photos; L. Morgan Yost: *Packard: A History of the Motor Car and the Company;* James Hollingsworth: *Packard 1940: A Pivotal Year;* Bishop & Babcock Mfg. Co. Bulletin; Chrysler Historical Society: 1942 Chrysler Crown Imperial Limousine photo; *oldcarbrochures.com:* 1941 Chrysler Crown Imperial brochure; *Classiccars.com:* 1942 De Soto photo; Dave Duricy: *1942 De Soto Refrigerated Air-Conditioning!;* T. C. Gleason, Society of Automotive Engineers (S. A. E.): 1942 (?) Chrysler air-conditioning photos; M/M Sal Saiya, Dave Czirr: 1941 Packard One Eighty photos; Hyman Ltd. Classic Cars: 1942 Packard One Eighty photos; Doug Houston: 1941 Cadillac Series Sixty-Two photo; Dr. Richard Zeiger and Ronald Vershoor: 1941 Cadillac Fleetwood Series Sixty Special photo; West Peterson: 1940 Packard One Eighty photo; 1942 Packard Clipper owned by General Douglas MacArthur: James Hollingsworth photo; Paul Wnuk: 1953 Oldsmobile photos; GM Media Archive: 1953 Oldsmobile brochure; Phil Gaffney: 1953 Oldsmobile photo; Robert Ober: 1953 Cadillac Fleetwood Series 60 Special photo.

Finally, there is no better way to quantify the evolution of automotive air-conditioning than to research, explore, and document the efforts of those who chose to publish its information. I thank those contributors and deem their collective works as The Legacy of Literature.

Allen B. Simons

October 20, 2021

Table of Contents

Chapter 1: 1940 Packard Factory Air-Conditioning 1

Chapter 2: 1941 Packard Factory Air-Conditioning 52

Chapter 3: 1942 Packard Factory Air-Conditioning 110

Chapter 4: 1941 Cadillac Factory Air-Conditioning 147

Chapter 5: 1941 Chrysler Factory Air-Conditioning 205

Index 252

Preface

How hot was it?

A Houstonian's viewpoint from the 1950s

by Allen B. Simons

Imagine going to school in Houston, Texas in the late spring or the first months after school began. There were no air-conditioned classrooms to temper the 90°F+ heat and 70%–90% humidity levels. While working at our desks the sweat from our arms stuck our paper to our arms. When asked to turn our work in, we peeled it off and passed it in to the teacher. After the paper dried, the sweat left wrinkle marks and smeared fountain pen ink stains.

Houston was known for frequent rain showers. Because of that, the school bus windows remained shut all day. Entering a bus that sat in the sun all day with closed windows yielded temperatures inside, according to experts, of 140°F and up. As we sat down on the frying-pan heat of blistering plastic seats, we opened the windows to cool us off, only to suffer the blasts of 90°F heat in our faces on our way home.

Throughout the home, there was no respite from the steamy heat as window air conditioner units, much less central air systems, were virtually nonexistent.

For a ten-year-old boy, life grew brighter as he read a three-year-old 1953 *Popular Science* magazine article announcing automakers' planned production of factory-installed air-conditioning.

He dreamed that *someday* he would ride in a car and not sweat out his clothes before getting there…

His epiphany led the author to pursue, collect, and research brand-specific automotive air-conditioning literature to support and identify this important subject. Subsequently, this extensive collection is now used to capture and create a unique and fascinating story for you, appropriately titled:

FACTORY AIR:

COOL CARS IN COOLER COMFORT

An Illustrated History of Automotive Air-Conditioning

Volume 1: 1940 to 1942 – The Innovative Years

Foreword

by West Peterson, Editor, Antique Automobile, Official Publication of the Antique Automobile Club of America

Dozens of automotive books traverse the histories of Packard, Cadillac, and Chrysler, meaning that few subjects remain to be chronicled about these titan automakers, so much so that anything so far overlooked are highly specific topics, appealing mostly to automobile enthusiasts and researchers that seek out –and enjoy– obscurity. I commend those authors who take the time and expend the effort necessary to ingratiate such audiences, painstakingly researching and documenting precise topics, rather than directing their efforts entirely toward wider appeal with resulting higher book sales. Their works are the historical accounts that represent accurate completions of history, making them the most valuable, so it is fortuitous that Allen B. Simons, in *Factory Air: Cool Cars in Cooler Comfort*, has documented a significant and much-forgotten piece of automotive history. Few today would consider the purchase of a car for daily transportation without air conditioning, so Simons' work also appeals to a wide audience.

From the time I was first able to grasp a Brillo pad enabling me to scrub the white sidewall tires of my father's classic cars, I have always had a voracious appetite for knowledge about antique automobiles, including the heritage that surrounds them and their mechanical workings. Always determined to "crack the code" whenever I became confused by cars with styling cues, badging, or markings from multiple years, I always sought out answers. Whenever my father introduced me to someone who was a marque "expert," in my quest to understand and recognize styling and engineering evolution, both of which could often be quite subtle between model years, I always seized the opportunity to ask as many questions as their patience would allow.

My attraction to Packards with air-conditioning stretches all the way back to 1966 when, at only six years of age, I first laid eyes on a factory-air-equipped 1940 Packard 180 touring sedan. The car was displayed at a family friend's museum located in Brooten, Minnesota, and was an imposing sight to a youngster, appearing to me to be about two blocks long. It is that exact car which has occupied my garage since 2005, so when Allen first contacted me in the spring of 2016 seeking information about early Packard air conditioners, I was more than willing to help, starting with extensive photographic documentation of such a low mileage, predominately all-original example.

In the late 1960s, my father purchased a 1942 Packard Clipper Eight touring sedan with factory air-conditioning. I was intrigued to later read a story in the Packard Club's *The Packard Cormorant* about an identical car used during WWII by General of the Army Douglas MacArthur. Like my father's car, General MacArthur's car was equipped with factory air. Many years later, my brother restored a factory-air-equipped 1942 Packard 180 formal sedan– coincidentally, first sold in the spring of 1943 in Allen's hometown of Houston, Texas– so I am privileged to have also shared with Allen my family's first-hand experiences with early Packard factory-air.

Throughout Packard's first 40-plus years as an automaker, it accumulated an extensive list of well-known "firsts," and the offering of factory-installed air-conditioning in a production car was one of them. Eighty-plus years have elapsed since the introduction of what is arguably one of Packard's best-ever engineering innovation firsts, and because Cadillac and Chrysler offered the same systems only one year later, it is not surprising that information about prewar air-conditioning systems has become mostly forgotten, other than being an answer to a trivia question.

With original-source information about the automobile industry's first air-conditioning systems so obscure, it was gratifying for me to contribute to Allen's research efforts. During the years in which Allen's research and writing were ongoing, he would inform me of items he needed to obtain that were essential to his quest for completeness, such as high-quality illustrations of the "workings" of early air-conditioning systems, so I was more than happy to disassemble my '40 Packard's interior, enabling its components to be photographically documented "from the backside," showing how the car had been specially prepared for the a/c system. The photos I provided include first-time published images of the original air filters installed in the trunk-mounted evaporator, along with photos of the insulation originally installed under the upholstery and carpet.

In *Factory Air: Cool Cars in Cooler Comfort*, you'll also be privy to Allen's collection of Packard factory brochures and publications on the subject, which I regard as the world's most complete, all legibly presented and, in many cases, enlarged, enabling easy interpretation. Also, for the first time published, you will find documentation and several photographs of Packard's little-known "Cellarette," a sleek walnut cabinet that was inset into the back of the front seat of a 1941 Model 180 Touring Sedan. The cabinet housed a freezer that had ice-making capability and contained cocktail glasses, measuring glasses, and a sterling silver stirrer and corkscrew. In addition, other chapters presented the three surviving 1941 Cadillacs equipped with factory air, plus photography of the only factory-air-equipped 1942 Chrysler known to survive.

The materials reproduced in Allen's book are from original sources, meaning its contents can be used and treated as research material or documentation for wherever authentication of factory-correct originality is required, such as for the Antique Automobile Club of America's stringent judging standards. Thanks to Allen's efforts, early prewar air-conditioning systems are now fully researched, documented, and expertly presented in a manner that is useful– and interesting– to all. For that, there can be no higher praise.

Initially, Allen's book was intended to be a complete compendium about the topic through 1960; however, because so much information was compiled throughout his many years of research, this first edition covers only the prewar years up to 1942. Allen will continue his publishing efforts with three follow-up volumes: Volume #2 will cover the eight 1953 models that offered factory-air, Volume #3 will cover years 1954–1956 and the changeover to "up-front" air conditioning, and Volume #4 will cover years 1957–1960 with the completion of the up-front installations.

I commend you, Allen, for your ongoing enthusiastic commitment to the proficient research and documentation of the origins of automotive "factory air."

West Peterson

Chapter 1

1940 Packard Factory Air-Conditioning

The 1940 Packard brochure announced, "Packard presents a new miracle, the air-conditioned motor car." (Fig. 1)

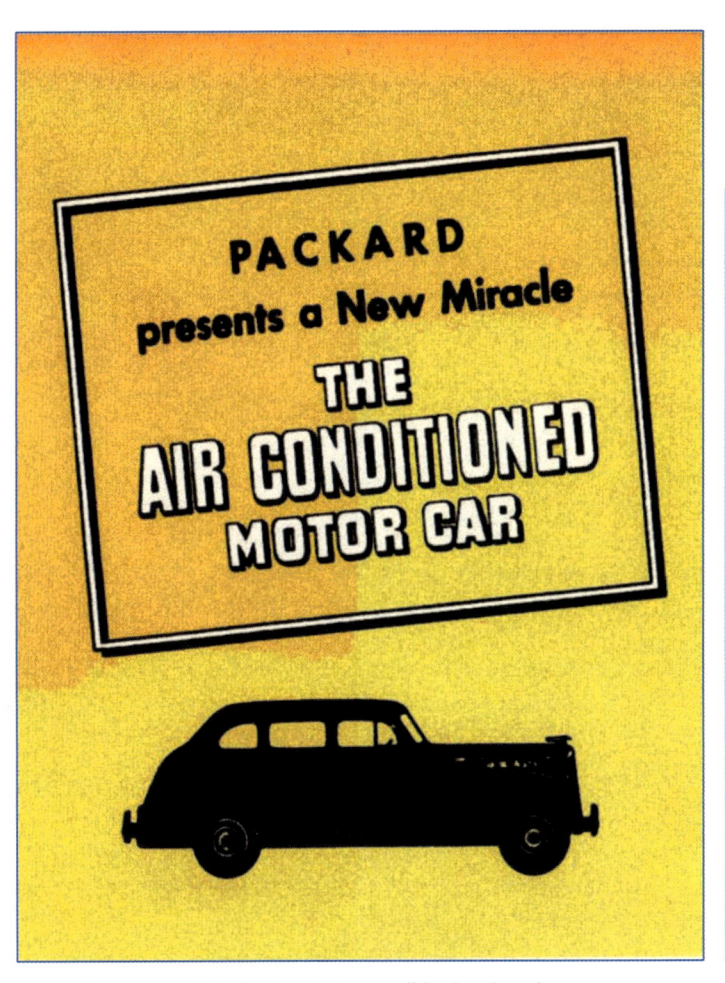

Figure 1. 1940 Packard air-conditioning brochure cover
Regress Press   Jeff Steidle

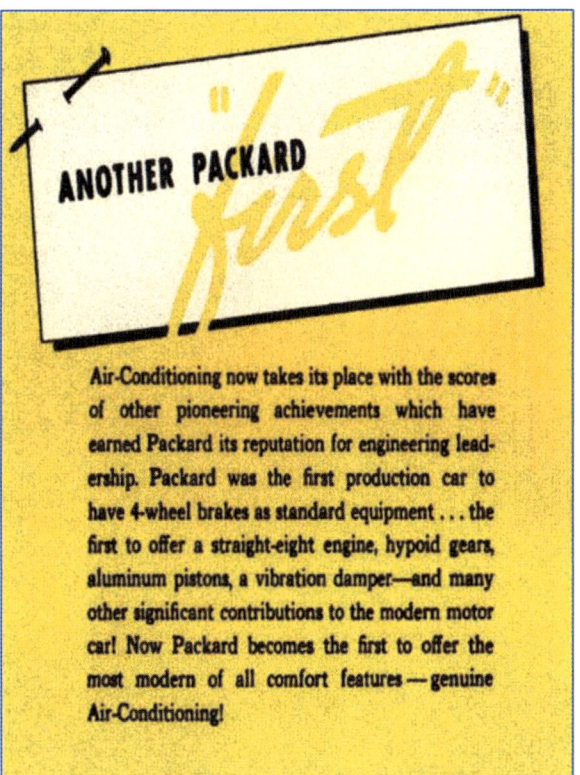

Figure 2. 1940 Packard air-conditioning brochure
Regress Press   Jeff Steidle

Packard, known for innovation and quality, advertised "Another Packard first,"

Air-conditioning now takes its place with the scores of other pioneering achievements which have earned Packard its reputation for engineering leadership…Now Packard becomes the first to offer the most modern of all comfort features—genuine air-conditioning! (Fig. 2)

The gold, black, and ivory-toned brochure continued with images and discussions of the benefits and comforts provided by its newly introduced accessory. Note how Packard's subtle use of blue-tinted interior coloration suggested coolness. (Fig. 3) The brochure headline stated:

"What an Air-Conditioned Packard Can Mean to You This Summer:"

Air-conditioning for motor cars is here! Yes, genuine *air-conditioning*—[italics in the original] that filters, dehumidifies, and *cools-by-refrigeration.* Packard offers it as a standard, extra cost, factory-installed accessory. Exclusive to Packard, you may have it in new 1940 closed models. *No other car offers anything like it.*

Six renderings illustrated the brochure's use of marketing hyperbole that encouraged passengers to:

1. TURN ON THE COLD!
2. COOL COMFORT FOR 6.
3. STEP OUT OF SUMMER HEAT into the delicious coolness of an Air-Conditioned Packard. [bold and italics in the original.]
4. SHUT OUT THE NOISE!
5. YOU'LL GET REFRIGERATED AIR.
6. HAY FEVER SUFFERERS REJOICE!

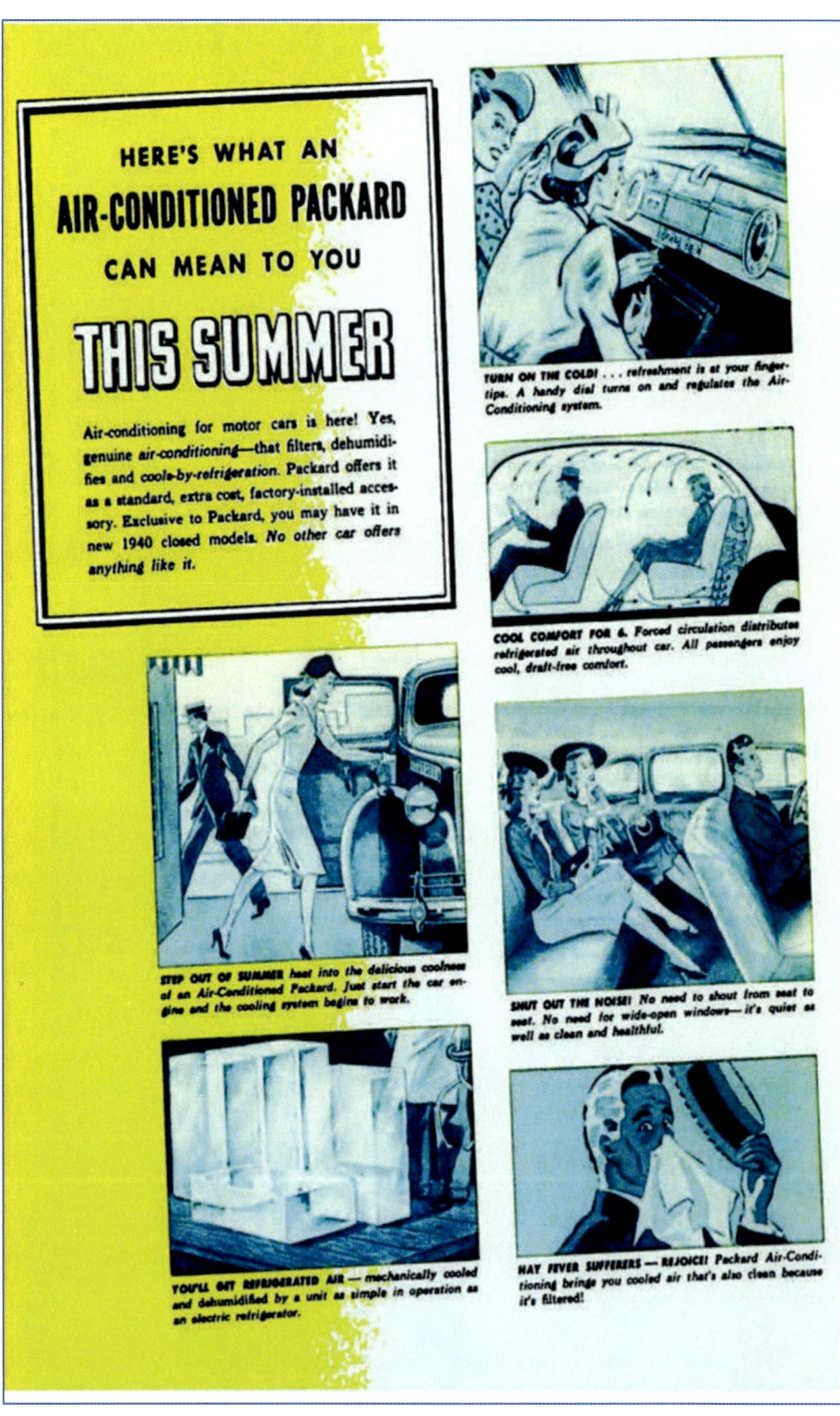

Figure 3. 1940 Packard air-conditioning brochure   Regress Press   Jeff Steidle

The brochure featured a two-page phantom illustration of the air-conditioning componentry installed on the Packard for 1940, author-colorized to show the conditioned air space (Fig. 4), followed by images of the compressor, the rear blower grille, and the trunk-mounted evaporator assembly. (Fig. 5)

It identified the condenser's locations in front of the radiator, a compressor that operated with a belt off the engine water pump pulley, the instrument panel four-speed control switch, Freon-12 receiver-reservoir, and the trunk-mounted evaporator assembly with the externally mounted expansion valve and drier. The evaporator housed the cooling coils and blower fan, with the blower grille on the package shelf behind the rear seat.

The conditioned air blew upwards along the headliner towards front seat passengers. In a circular airflow pattern, the conditioned air passed over the rear seat passengers and returned underneath a slightly elevated rear seat cushion, through air filters into the evaporator.

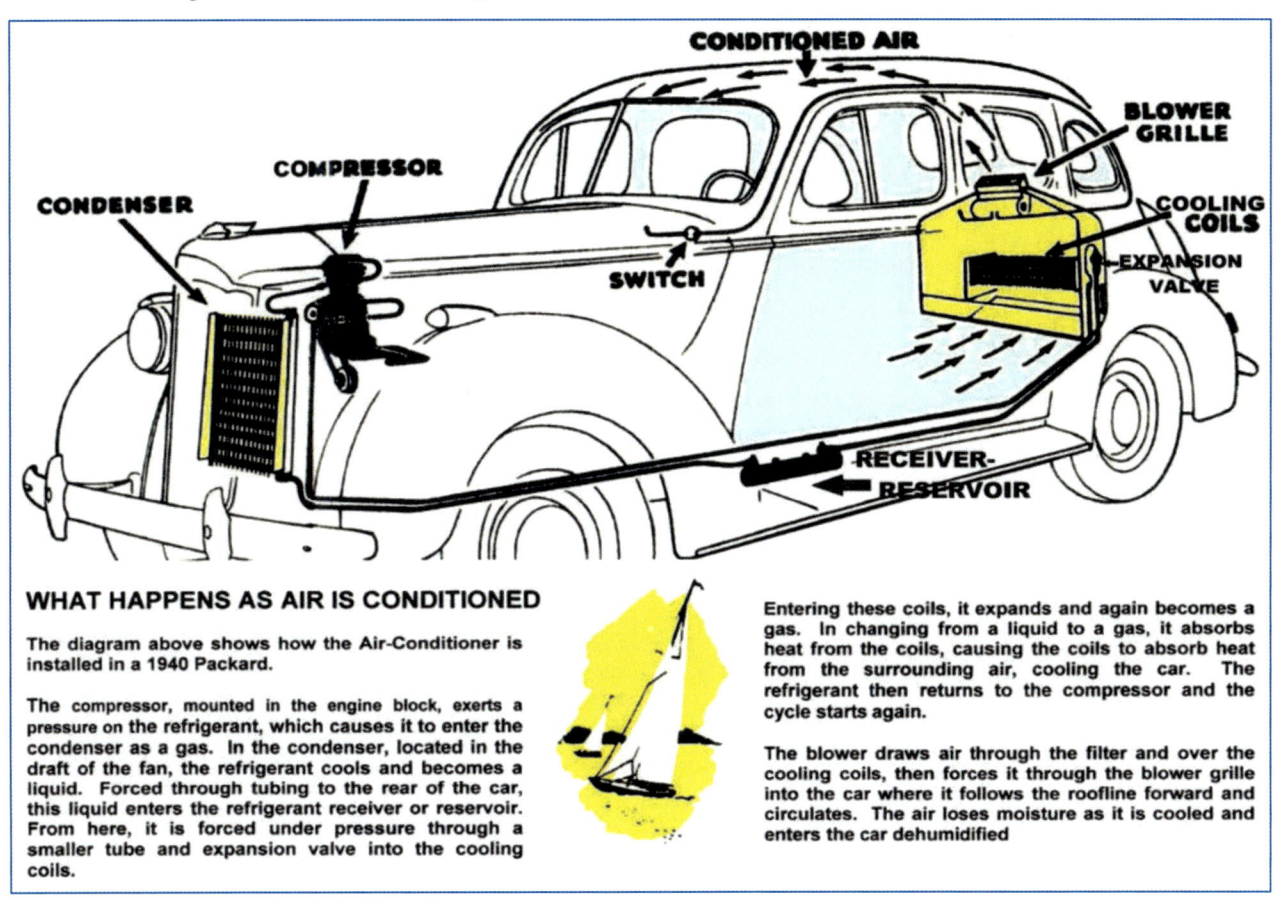

Figure 4. 1940 Packard air-conditioning brochure   Regress Press   Jeff Steidle

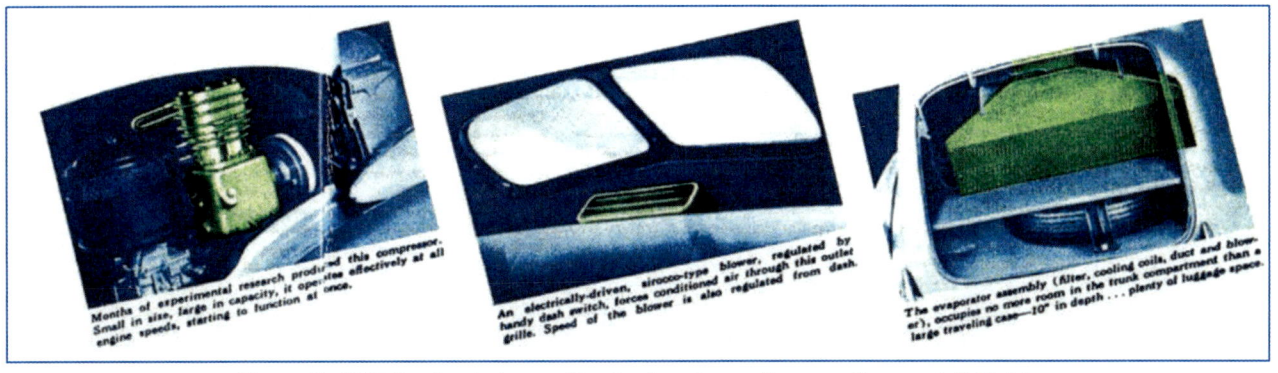

Figure 5. 1940 Packard air-conditioning brochure   Regress Press   Jeff Steidle

Referencing Figure 5, the air-conditioning compressor installed on the engine's passenger side served as the heart of the cooling system. Packard stated that its Servel-based compressor was "small in size, large in capacity, it operates effectively at all engine speeds, starting to function at once."

The under-dash-mounted four-speed paddle switch adjusted the "electrically driven, sirocco-type blower." Also known as a squirrel-cage blower, it circulated the cool air upward through the rear outlet grille, identified in the larger illustration as the "blower grille." Packard's design provided an adjustable louvered vent for airflow distribution.

The third highlighted component, the evaporator assembly, comprised the housing located forward in the trunk for the cooling coils and the blower fan. The brochure emphasized measurements of only ten inches deep, representing the size of a large suitcase.

"HOW PACKARD AIR-CONDITIONING WORKS" (Fig. 6) The brochure described the process of

> Genuine Air-Conditioning for motor cars!… It is not to be confused in any way with ordinary ventilating, 'fresh air,' or so-called 'conditioned air' systems.

It emphasized:

> Packard Air-Conditioning alone actually cools (by refrigeration), dehumidifies, and filters the air you breathe while driving.

The brochure continued with an explanation that its cooling process works the same as home refrigerators. The difference between a refrigerator and automotive air-conditioning lay in cooling, plus the added benefit of forced air circulation in the passenger compartment. It described how the electric blower distributed the cooled air and ensured "constantly circulating, cooled, dehumidified, filtered air" throughout the car interior and offered them the opportunity to "escape hot, sultry days."

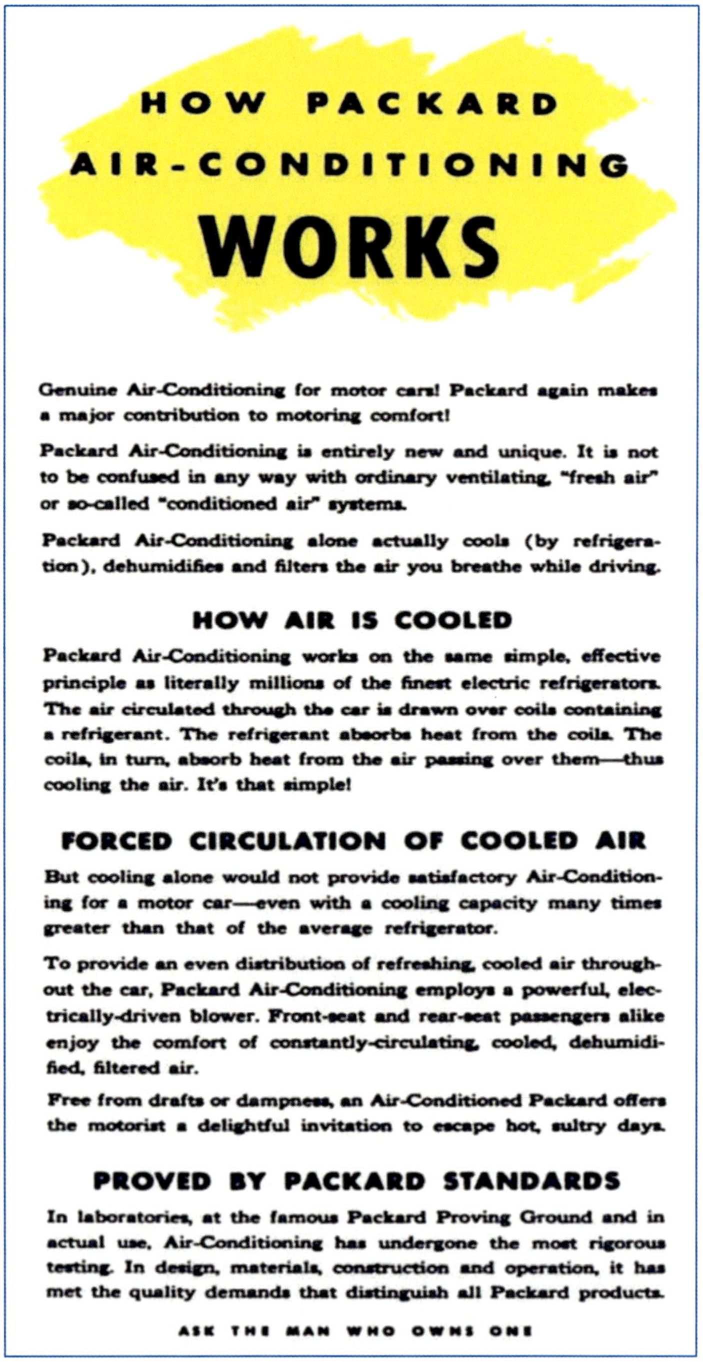
HOW PACKARD AIR-CONDITIONING WORKS

Genuine Air-Conditioning for motor cars! Packard again makes a major contribution to motoring comfort!

Packard Air-Conditioning is entirely new and unique. It is not to be confused in any way with ordinary ventilating, "fresh air" or so-called "conditioned air" systems.

Packard Air-Conditioning alone actually cools (by refrigeration), dehumidifies and filters the air you breathe while driving.

HOW AIR IS COOLED

Packard Air-Conditioning works on the same simple, effective principle as literally millions of the finest electric refrigerators. The air circulated through the car is drawn over coils containing a refrigerant. The refrigerant absorbs heat from the coils. The coils, in turn, absorb heat from the air passing over them—thus cooling the air. It's that simple!

FORCED CIRCULATION OF COOLED AIR

But cooling alone would not provide satisfactory Air-Conditioning for a motor car—even with a cooling capacity many times greater than that of the average refrigerator.

To provide an even distribution of refreshing, cooled air throughout the car, Packard Air-Conditioning employs a powerful, electrically-driven blower. Front-seat and rear-seat passengers alike enjoy the comfort of constantly-circulating, cooled, dehumidified, filtered air.

Free from drafts or dampness, an Air-Conditioned Packard offers the motorist a delightful invitation to escape hot, sultry days.

PROVED BY PACKARD STANDARDS

In laboratories, at the famous Packard Proving Ground and in actual use, Air-Conditioning has undergone the most rigorous testing. In design, materials, construction and operation, it has met the quality demands that distinguish all Packard products.

ASK THE MAN WHO OWNS ONE

Figure 6. 1940 Packard air-conditioning brochure
Regress Press Jeff Steidle

Packard supported its new air-conditioning accessory by testing "at the famous Packard Proving Grounds and in actual use." Further, it assured customers that the new system met all Packard standards for quality.

Earlier, several entrepreneurs exhibited their automotive cooling prototypes. Mohinder S. Bhatti, Ph.D., writing in the September 1999 *ASHRAE JOURNAL*, the American Society of Heating, Refrigerating and Air-conditioning Engineers, stated:

In 1930, C&C Kelvinator outfitted a customized 1930 Nash 8 (*Author note:* not a Cadillac as stated in the article), owned by a Houston Nash dealer, John Hamman Jr., with a 0.5-hp (0.37kW) Kelvinator refrigeration unit powered by a 1.5-hp (1.1 kW) gasoline engine.

Two flues on either side of the front seat took the air from a fan which circulated cool air throughout the passenger compartment. ...the unit looked like a trunk and fitted compactly onto the back of the car. (Figs. 7, 8)

Figure 7. 1930 Nash 8 with Kelvinator air-conditioning unit on rear, owned by Nash dealer, John Hamman, Jr. of Houston "RIDING IN COMFORT II," ASHRAE Journal, September, 1999 ©ASHRAE (www.ashrae.org)   Chuck Wilson

Figure 8. 1930 Nash 8 Kelvinator air-conditioning unit, owned by Nash dealer, John Hamman, Jr. of Houston "RIDING IN COMFORT II," ASHRAE Journal, September, 1999 ©ASHRAE (www.ashrae.org)

Charles Kettering chose Thomas Midgley Jr. to head research into new refrigerants. In 1928, Midgley and Kettering invented a nontoxic, nonflammable "miracle compound" called Freon® (Freon-12). The U.S. patent office approved the dichlorodifluoromethane formula, known as CFC, by the Frigidaire Division of General Motors. Equipped with the Freon-12 refrigerant, by 1935, sales of Frigidaire refrigerators totaled eight million.

Dr. Bhatti added:

> The development of the automotive air conditioner began in earnest in 1930 when General Motors Research Laboratories conceived the idea of the vapor compression system with R-12 (Freon-12) refrigerant. On September 23, 1932, the laboratory made a proposal to General Motors' management to develop such a system. The Cadillac Division displayed interest in the proposal. However, it was not until the summer of 1933 that work started.

The summer of 1933 marked the inception of Kinetic Chemicals, Inc., a joint venture between DuPont and General Motors. Automotive-based engineering calculations predicted the required cooling performance to be one-ton of cooling, 200 BTU per minute, 12,000 BTUs per hour. Research soon proved that a one-ton unit produced insufficient cooling performance in an automobile. Dr. Bhatti explained that the underestimated capacity resulted from two original hypotheses:

1. The engineers tested interior cooling with recirculated rather than freshly ventilated air.
2. Engineers projected the required temperature reduction to be no more than 10°F.

Dr. Bhatti continued:

> At that time, it was believed that if the difference between the outside air and the conditioned air exceeded 10°F, the occupant of the conditioned space could experience a thermal shock upon emerging into the outside air!

After years of experimentation, Cadillac produced a prototype trunk-mounted self-contained semi-constant speed drive air conditioner in 1939. (Fig. 9) Not satisfied, in 1941, Cadillac contracted with Bishop & Babcock Mfg. Co. of Cleveland, Ohio, the same air conditioner supplier as Packard used in 1940-1942. (Refer to the 1941 Cadillac chapter)

Figure 9. 1939 Cadillac prototype trunk-mounted, self-contained air conditioner installed in a 1938 Cadillac @ASHRAE JOURNAL, September 1999

Referencing Dr. Bhatti's comments about perceived mechanically created temperature differentials and potential physical effects, this author experienced a related situation when told about a prewar, residential central air-conditioning system in Houston, Texas.

Glenn McCarthy, famous Texas Wildcatter, and owner of Houston's legendary Shamrock (Hilton) Hotel, built a mansion on the outskirts of Houston circa 1940. His son, Glenn, Jr., said that their house featured all the modern conveniences, including an early residential central air-conditioning system installed over twenty years before it became commonplace.

When asked how comfortable the cool, dehumidified air from the early system was to live in, he replied that his mother would never turn on their air-conditioning system because she feared her family could contract polio! Although it was just an old wives' tale of that era, many people believed that chilled air temperatures were unnatural and could lead to severe illness, including polio.

Author note: Most Houstonians began purchasing window-mounted room residential air conditioners in the late-1950s. It was not until the early 1960s that new homes began installing the much more comfortable, expensive central air system. Houstonians rapidly retrofitted their homes, schools, offices, stores, and churches with central air, and by the mid-1960s, Houston's Chamber of Commerce promoted Houston as the "World's Most Air-conditioned City."

(Refer to the *Preface,* "How Hot Was It? A Houstonian's Viewpoint from the 1950s.")

In a November 1933 *Popular Science* article, "First Air-Conditioned Auto," an unnamed inventor demonstrated a self-contained air-conditioning system while driving a 1927 Packard Model 336 through the streets of New York City. He installed a sub-floor electric motor that ran a compressor system and fan that blew cold air upwards behind a closed limousine divider window. The air circulated in the passenger compartment and returned through a floor grille. (Fig. 10)

Figure 10. "First Air-Conditioned Auto," 1927 Packard Model 336
Popular Science, November 1933   ABS (Allen B. Simons Collection)

Known for its innovative technological advancements in the fine car field, Packard investigated an air-conditioning system for its passenger cars. It consulted Bishop & Babcock Mfg. Co., a Cleveland, Ohio producer of automotive thermostats, water heaters, forced air defroster motors, and a preliminary automotive air-conditioning developer.

Bishop & Babcock spent several years developing and perfecting automotive air-conditioning. In 1938, Edward L. Mayo patented the "Bishop & Babcock Weather Conditioner" that combined an air conditioner and heater based upon a 1936-design Servel commercial compressor. In late 1938, Packard contracted with them to supply its Weather Conditioner for use on Henney-Packard ambulances.

8 Chapter 1: 1940 Packard Factory Air-Conditioning

The April 1940 Henney *Program of Progress* brochure offered the "Hospital Type Henney-Packard Specially Equipped Ambulances." (Figs. 11, 12) The ambulances featured:

> …the most completely and scientifically equipped invalid cars ever built. Specifications were developed from the findings of an extensive survey conducted among leading hospitals, institutions, and safety councils.

The brochure noted equipment such as the "Genuine, mechanically refrigerated air-conditioning" and "Hot and cold running water" as features of their ambulance.

The bottom left image revealed the Packard Weather Conditioner evaporator installed behind the "compact steel cabinets." It closely resembled the evaporator image of the 1940 Packard One Eighty owned by West Peterson. (Fig. 82)

Figure 11. 1940 Henney Program of Progress brochure   Dwight Heinmuller

Figure 12. 1940 Henney Program of Progress brochure, pp. 12,13   Dwight Heinmuller

Henney, the standard-bearer of ambulances and funeral cars, advertised their doctor-requested, temperature-controlled air-conditioning system for their Packard-based chassis. (Fig. 13)

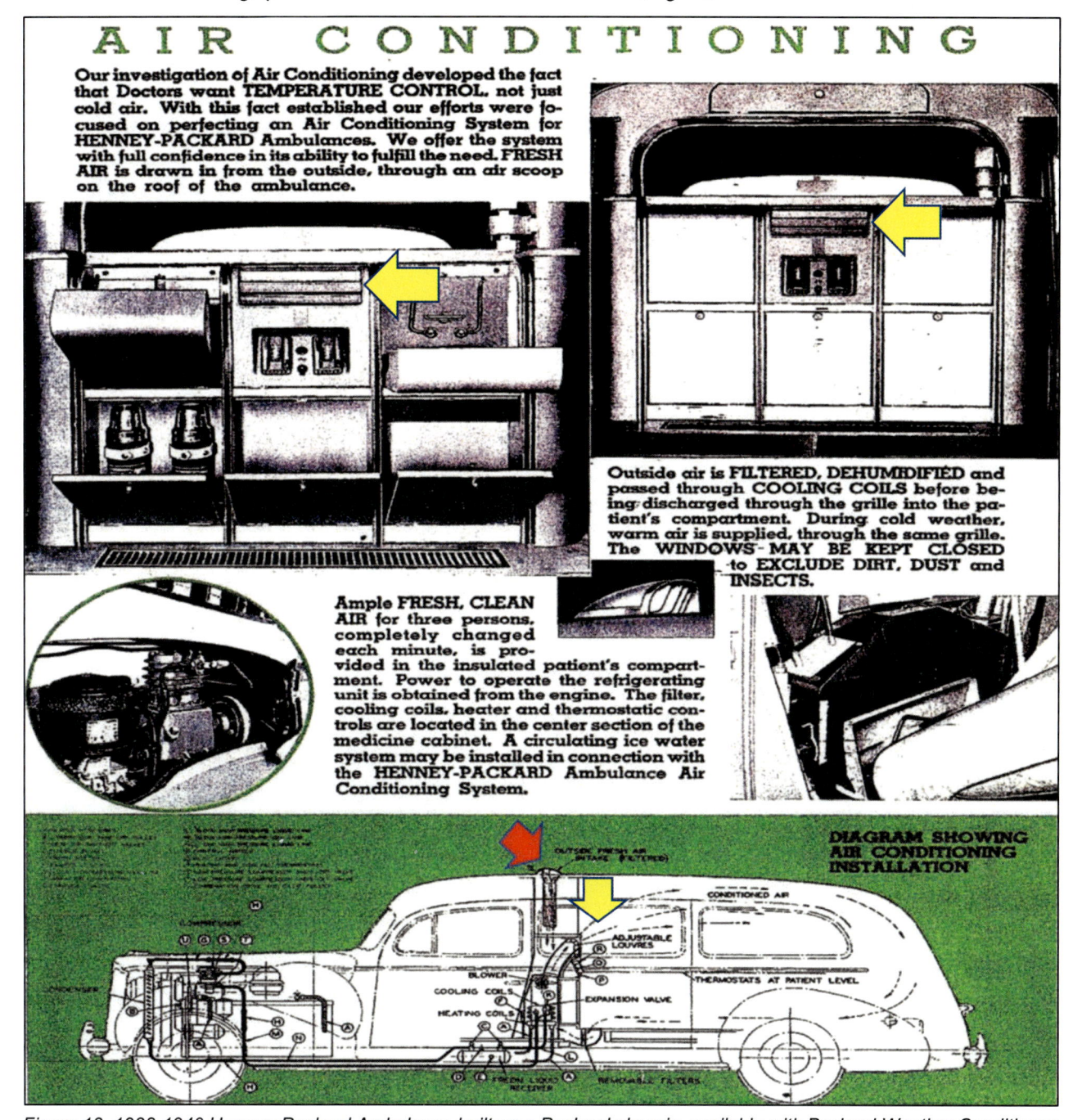

Figure 13. 1938-1940 Henney-Packard Ambulance built on a Packard chassis, available with Packard Weather Conditioner. Yellow arrows reveal cool air discharge vent louvers, Red arrow reveals fresh air intake on roof of ambulance
Packard 1940: A Pivotal Year   James Hollingsworth

The 1940 Henney-Packard ambulance schematic featured an equipment cabinet behind the front seat. The designers integrated the center medicine cabinet with the fan control, evaporator cooling coils, air filter, heater core, and the cooling register louvers, shown by yellow arrows. A red arrow showed an outside fresh air intake. The ambulance's air conditioner cabinet profile closely resembled that of the future trunk-mounted 1940 Packard Weather Conditioner.

A deluxe option offered a "circulating ice water system" when equipped with air-conditioning. Expected at a later offering, it continued, "Hot water is thermostatically controlled at 175°F."

Henney, in 1938, also contracted air-conditioning with the Trane Company to cool their ambulance. Trane featured a one-ton, rear, under-floor unit. According to information from the Thomas A. McPherson Collection, "one of the first of these air-conditioned ambulances was sold to the Kriedler Funeral Home, a long-time Henney customer in McAllen, Texas." (Fig. 14)

Life was good for Packard in 1939. Sales of its One Ten and One Twenty models thrived. According to the *Automotive News Celebrating the 40th Anniversary—Packard* edition,

Packard enjoyed an increase of 50.52 percent in production. Car shipments for (calendar year) 1939 totaled 76,366 units with a cash value of approximately $60M. Packard added 577 new dealers since September 1, 1939. The close of the year found the company with 1,896 distributer [sic] dealer outlets, achieving a new all-time high in Packard history. This dealer group does not include 450 foreign country distributers [s*ic*] operating under the Packard Motors Export Corp.

While writing *Packard 1940: A Pivotal Year*, the longtime Packard Club member James Hollingsworth (1930-2012) described several significant upgrades and innovations for the "Speed-Stream" styled 1940 Packard. (Fig. 15)

Introduced August 1, 1939, he reported the 1940 Packard's first sealed beam headlights. Equipped in the last year of Packard's freestanding headlight styling, the sealed beams provided illumination 50% brighter than before.

The powerful 160-hp eight-cylinder, 356-CID engine for the Senior One Sixty and One Eighty models replaced the twelve-cylinder, *Twelve*, effectively lowering maximum retail sales prices.

In Mr. Hollingsworth's opinion, however, the most noteworthy 1940 upgrade proved to be the introduction of Packard's Weather Conditioner accessory, a harbinger of comfortable hot weather driving. Packard Vice-President of Engineering, W. H. Graves, according to *Packard: A History of the Motor Car and the Company*, by Messrs. Martin, Bradley, et al., stated that it produced nearly 2,000 air-conditioning units from 1940 through their truncated 1942 sales season.

On February 9, 1942, Packard ended auto production and increased WWII war materiel manufacturing.

In conjunction with the Trane Company of LaCrosse, Wisconsin, The Henney Motor Company introduced the first mechanically air-conditioned ambulance in 1938. As seen in this cut-away view, a one-ton refrigeration unit utilizing Freon gas for cooling was installed under the rear compartment floor of a Henney-Packard Model 884. Powered by a 110-volt generator coupled to the engine and drawing air from outside the vehicle, this massive unit filtered, dehumidified, cooled and changed the interior air once every minute. One of the first of these air-conditioned ambulances was sold to the Kriedler Funeral Home, a long-time Henney customer in McAllen, Texas. *Thomas A. McPherson Collection*

Figure 14. Henney-Trane air-conditioned ambulance Thomas A. McPherson Collection Howard Hanson

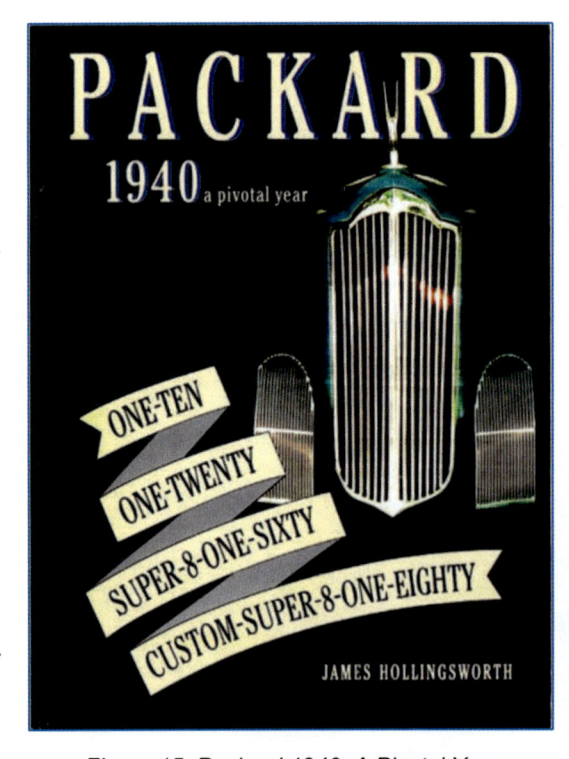

Figure 15. Packard 1940: A Pivotal Year book cover James Hollingsworth

In the fall of 1939, Packard displayed the first factory-installed automotive air conditioner at the 1940 Chicago Auto Show. *The History Channel*'s "This Day in History, November 4, 1939," observed:

On this day, the 40th National Automobile Show opened in Chicago, Illinois, with a cutting-edge development in automotive comfort on display: air-conditioning. A Packard prototype featured the expensive device, allowing the vehicle's occupants to travel in the comfort of a controlled environment, even on the most hot and humid summer day.

After the driver chose a desired temperature, the Packard air-conditioning system would cool or heat the air in the car to the designated level, and then dehumidify, filter, and circulate the cooled air to create a comfortable environment. The capacity of the air-conditioning unit was equivalent to 1.5-tons of melting ice in 24 hours, driven at highway speeds.

The 1940 Chicago Auto Show, November 4, 1939, premiered an automotive accessory that transcended all previously offered, comfort-based optional accessories. *The Syracuse Herald-American,* dated Sunday, November 5, 1939, printed the following announcement at the opening of the 1940 Chicago Auto Show: "New Packard Cooled with Filtered Air." (Fig. 16)

Mr. W. M. Packer of Packard Motor Car Company corrected the erroneous subtitle by adding:

> the refrigerating coils are located back of the rear seat in an air duct with heating coils in another compartment of the same duct.

New Packard Cooled With Filtered Air

Refrigeration System Is Installed Under the Rear Seat

The Packard Motor Car Company announces that it has the first automobile mechanical refrigeration air cooling system, a unit which also filters the air in a car and provides heat in winter. It is offered by Packard as a standard, factory installed, extra cost accessory.

Through use of the new device, air in an automobile body is cooled to the temperature desired by passengers, dehumidified, filtered and circulated through the car. A simple adjustment, easily made by a passenger, causes the system to circulate filtered warm air when desired.

Cooling is obtained by means of a refrigerating plant which is much the same in principle as that of any home mechanical refrigerating unit, according to W. M. Packer, vice president of distribution of the Packard Motor Car Company.

"The refrigerating coils," said Mr. Packer, "are located back of the rear seat in an air duct with heating coils in another compartment of the same duct. An electrically-driven fan forces air through either set of coils at the will of the passengers, the filter operating as air travels through the duct. It has been found in tests that this filter removes practically all dust from the air in the car. It also so largely removes pollen as to be helpful to hay fever victims.

"The treated air follows along the top of the car so that it does not blow directly on any of the passengers. It returns along the floor, passing under the rear seat to the coils for recooling, or reheating, and is then recirculated by the fan. Humidity is lowered by the operation of cooling the air. The compressor is mounted on top of the engine and is driven from the radiator cooling fan.

Figure 16. "New Packard Cooled with Filtered Air" The Syracuse Herald-American, November 5, 1939 ABS

12 Chapter 1: 1940 Packard Factory Air-Conditioning

Concurrent with the 1940 Chicago Auto Show, *Air-conditioning & Refrigeration News*, dated November 8, 1939, published an article titled, "Factory Installed Heating and Cooling System for Packard Sedan." Next to the phantom componentry image of the 1940 Packard, it added a ribbon award, declaring it as "The 1st Air-Conditioned Automobile." (Fig. 17)

The last sentence announced the Chicago auto show, unexpectedly coupled with the Oklahoma City auto show, as exhibition cities for Packard's 1940 Weather Conditioner.

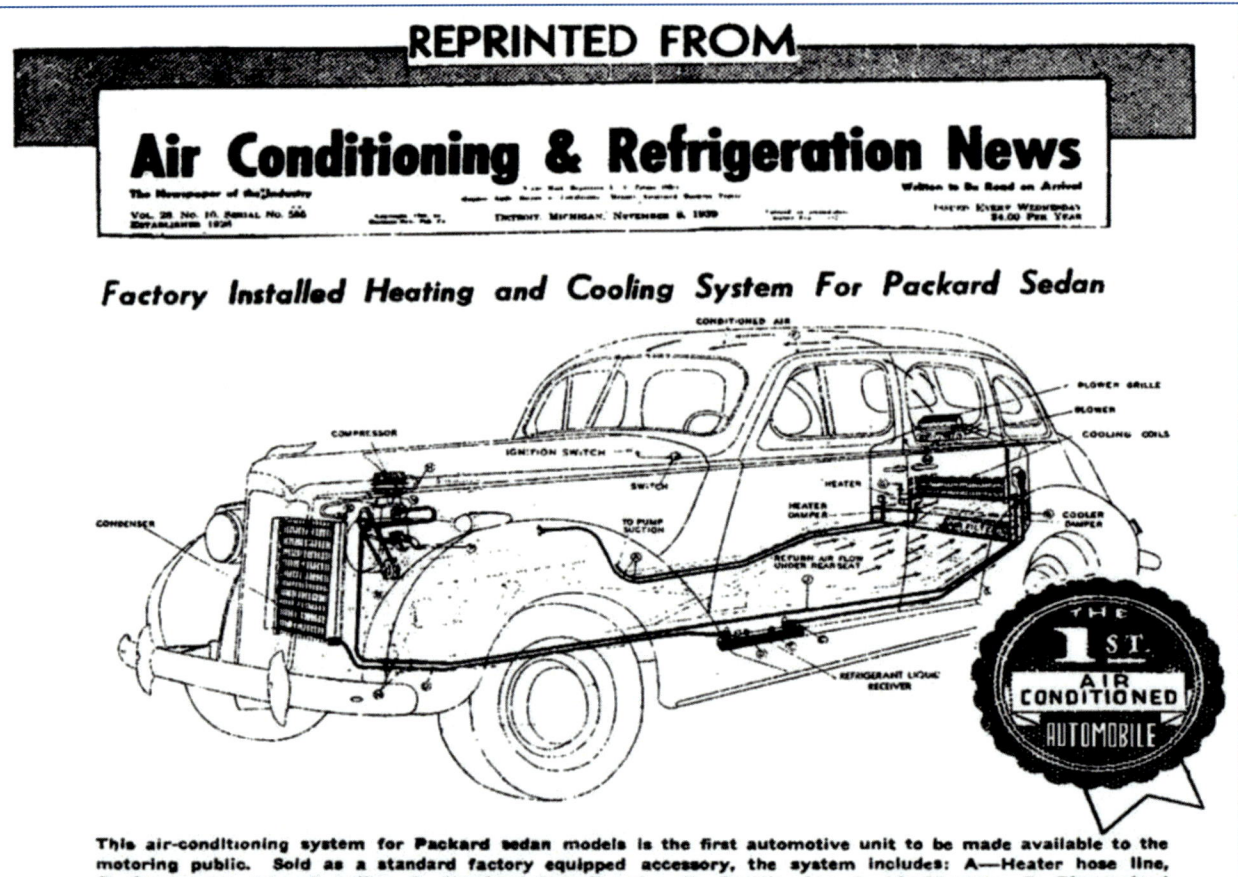

REPRINTED FROM

Air Conditioning & Refrigeration News

The Newspaper of the Industry — Written to Be Read on Arrival
Vol. 28 No. 10, Serial No. 586   Detroit, Michigan, November 8, 1939   Issued Every Wednesday $4.00 Per Year
ESTABLISHED 1926

Factory Installed Heating and Cooling System For Packard Sedan

This air-conditioning system for Packard sedan models is the first automotive unit to be made available to the motoring public. Sold as a standard factory equipped accessory, the system includes: A—Heater hose line, B—Compressor take-off pulley, C—Receiver shut-off valve, D—Fusible plug, E—Liquid tester, F—Blower lead wire, G—Damper control, H—½" dia. high-pressure gas line, J—⅝" dia. low-pressure gas line, K—Condenser shut-off valve, L—⅜" dia. high-pressure liquid line, M—Blower ground wire, N—Vibration eliminators, P—Extra length cylinder head studs, Q—½" dia. high-pressure liquid line.

Packard Introduces Car Cooling Unit As an Accessory

DETROIT —First announcement of an air-conditioning system using mechanical refrigeration, installed on passenger cars as a standard, factory built, extra-cost accessory, was made by the Packard Motor Car Co. here this week. Using a reciprocating compressor driven from a pulley on the fan belt shaft, the new system develops 1½ tons of refrigeration at 60 miles per hour, and 2 tons at 80 miles per hour.

The new conditioning system also includes provision for winter heating, and both the cooling and heating coils are located back of the rear seat in standard sedans. Air is drawn under the rear seat, over the coil, and then introduced to the car body at a point directly behind the heads of rear seat passengers. Air is deflected upward along the roof of the car by means of the fan. Control of the fan is by means of a rheostat on the dash. One hundred per cent recirculated air is used.

Refrigerant lines run from the compressor, which is mounted on the motor, to a condenser, mounted directly in front of the radiator. From this point the refrigerant goes to a receiver located underneath the body and thence to the low side coil behind the rear seat. Standard refrigerant connections are used, but refrigeration lines are mounted against the frame where they are not subject to twisting or vibration.

The conditioning unit is equipped with an air filter, which is said to remove the majority of dust and pollen from the air. Ventilation is obtained by using the standard wing ventilators on the front windows.

Installation price of the new conditioning system is expected to be approximately $275 including the cost of special insulation in the top and side walls of the sedan body. No systems will be sold or installed on cars, except in the regular course of factory production.

Change from summer to winter driving is accomplished by the regulation of two dampers located on the sides of the cooling unit mounted in the trunk compartment.

In announcing the new conditioning equipment, W. M. Packer, vice president of distribution of the Packard Motor Car Co., asserted that the new unit, which will be called the Packard Weather Conditioner, operates on the same principle as that of the home mechanical refrigerator.

Ralph M. Williams, service engineer for the Packard company, states that service on the air-conditioning systems will be handled by established dealers and distributors for Packard cars.

Cars equipped with the complete heating and cooling system are being exhibited at the Chicago and Oklahoma City automobile shows.

Figure 17. "Factory Installed Heating and Cooling System for Packard Sedan"
Air-Conditioning & Refrigeration News, November 8, 1939 PackardInfo.com   Kevin Waltman

From near the author's home in Central Texas, *The Austin American, Texas,* dated November 10, 1939, wrote, "Winter Auto Balmy as Spring, Summer Car Cools as Mountains, More than Engineers Dream." (Fig. 18)

Other than the United Press title and introductory comment, the article reported similar Packard marketing phrases in the *Automotive Industries* article below.

Figure 18. "Winter Auto Balmy as Spring..."
The Austin American, Texas, November 10, 1939
PackardInfo.com   Guscha

Automotive Industries wrote an article titled "Packard Offers Car Air-Conditioning System," dated November 15, 1939. It announced the Packard introduction of their mechanical refrigeration air cooling system as "a standard, factory-installed, extra-cost accessory. The unit also filters the air in a car and provides heat in winter." (Fig. 19)

Figure 19. "Packard Offers Car Air-Conditioning System"
Automotive Industries, November 15, 1939   ABS

*M*otor published two images of the newly introduced Packard air conditioner. (Fig. 20) The first image illustrated the two-cylinder Servel air-conditioning compressor, run by a belt off the engine water pump pulley and installed behind the radiator.

The second image featured a phantom rendering of the condenser and compressor in the engine bay, connected by copper tubing that carried the Freon-12 through a receiver tank, through tubing to the trunk-mounted evaporator assembly expansion valve and cooling coils. After passing over the cooling coils, a blower circulated the refrigerated air throughout the passenger cabin. It continued, "The cooling capacity of the refrigerating unit varies with engine speed, producing the equivalent of 1.5-tons of ice at 60 MPH."

The evaporator assembly served two purposes. An owner-controlled damper lever switched the cooling mode to the heating mode for winter operation. The radiator supplied hot water through hoses that connected to the heater core inside the evaporator. The fan-forced warmed air circulated throughout the interior, as did the refrigerated air.

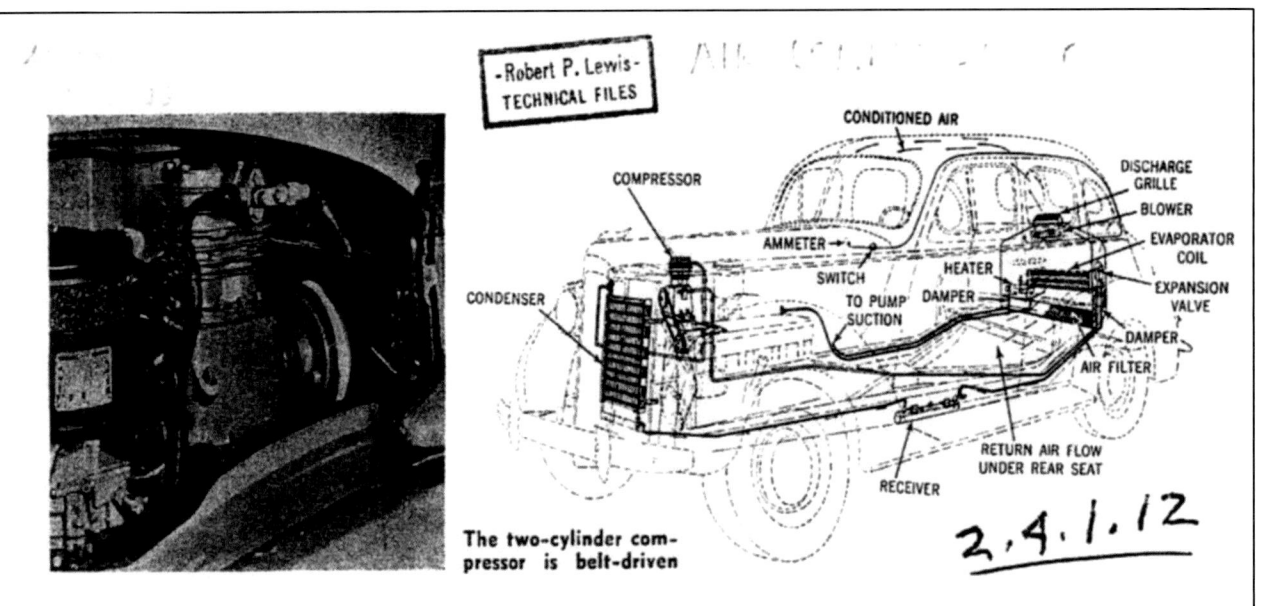

Air Conditioning for Packards . . . $274

OFFERED as optional equipment at $274 extra, Packard is the first to provide a complete air conditioning system consisting of a refrigerating unit for cooling and dehumidifying air in warm weather and a hot-water heater unit for warming the car in cold weather.

Both the heater and refrigerating coils are placed within a sheet metal chamber located at the back of the rear seat. Air enters through a filter at the bottom and is discharged by a blower at the top.

Air is cooled as it flows over the evaporator coils which are filled with a liquid which boils at a low temperature when the pressure is low. Low pressure is maintained by the suction of the compressor which draws off the vapor. The compressor raises the pressure of the vapor to several times its former value so that when it is cooled by flowing through the condenser it returns to liquid form. The liquid passes to a storage receiver and from thence to an expansion valve which permits the liquid to flow into the low-pressure evaporator coils without affecting pressure in the condenser. This system is the same in principle as that used in all electric refrigerators.

The cooling capacity of the refrigerating unit varies with engine speed, producing the equivalent of 1½ tons of ice at 60 mph. It is stated that the air filter largely removes pollen, a fact of interest to hay fever sufferers.

MOTOR for DECEMBER, 1939

Figure 20. "Air-Conditioning for Packards...$274" Motor, December 1939
DPL NAHC (Detroit Public Library National Automotive History Collection)   Romie Minor, Carla Reczek

Packard's dealer publication introduced its new 1940 Weather Conditioner in *Packard Promotional Pointers, Vol. VI, No. 6,* December 15, 1939. The article's headlines read, "Something New! Something Practical! Something Every Motorist Wants!" It confirmed the factory-installed optional equipment price of $274 that *Motor* described in its December 1939 issue and stated, "The Packard unit is an honest-to-goodness, all-year-round air-conditioning system." (Fig. 21a)

PACKARD PROMOTIONAL POINTERS

Vol. VI   December 15, 1939   No. 6

Something New!
Something Practical!
Something Every Motorist Wants!

AGAIN Packard pioneers the way. Again Packard adds to an already long list of "firsts." Again Packard gives its salesmen something to talk about and sell that no other salesman can offer today.

With its new *Weather - Conditioning* unit, Packard fills a long existing gap in motoring comfort. This new genuine air - conditioning system for Packard cars will be known and advertised as the Packard *Weather - Conditioner*. It is not a makeshift proposition. It is not a one-way ventilation contraption, nor just an under-seat heater arrangement, which are primarily units having their greatest appeals in winter. The Packard unit is an honest-to-goodness, all-year-round air-conditioning system — one that banishes perspiration just as effectively as it does goose pimples.

Think of the advantage this gives to you, a Packard salesman. Something new to talk about; something with a ready-made appeal; something every motor car owner would like to have in his next car.

And the price of the Packard *Weather-Conditioning* unit at $274 is reasonable. When it is considered that to equip one room with air-conditioning equipment it costs approximately $350, no one can object to a lower price for a genuine air-conditioning system in a motor car. And remember that this unit is a built-in proposition. It is not just an added accessory. In addition to the mechanism itself, a special system of insulation is required for roof, floor and sides.

At the present time the Packard air-conditioning unit is available only on sedans, but it is expected to be furnished on coupes in the near future.

Figure 21a. "Something New! Something Practical! Something Every Motorist Wants!"
Packard Promotional Pointers, Vol. VI, No. 6, December 15, 1939   PackardInfo.com   Kevin Waltman

The bulletin continued with the marketing demographics, advised a sense of urgency to sell this exclusive Packard accessory, and recommended a showroom demonstrator model for potential customers. (Fig. 21b)

• **PACKARD PROMOTIONAL POINTERS** •

The Market

While the desire for air-conditioning is almost a universal one, we believe the following classifications will be particularly interested:

1. Those who like to be "first" with anything new or novel. You'll find plenty of these people in every town.
2. Those who already have air-conditioning systems installed in homes or offices.
3. Those who use their cars for business and are in them the major part of the day, such as salesmen, doctors, etc.
4. Tourists who visit all parts of the country in all kinds of weather.
5. Elderly persons who are susceptible to ill effects of drafts, heat and cold.
6. Persons afflicted with hay fever caused by various pollens that are prevalent during certain seasons.
7. That big group which foregoes driving for pleasure in hot weather because of dislike for heat and road dust.
8. Those who have cars in rental service, such as cars for funeral trade, sight-seeing and taxi service. Air-conditioning gives these people a big advantage over competition not offering such equipment in their cars.
9. All owners of 1937-1938-1939 Senior Packards.

Make Hay While the Sun Shines

You will certainly be overlooking a good thing if you don't get into action on the Packard *Weather-Conditioner* at once. Other companies won't stand still and let Packard enjoy this field exclusively any longer than necessary. The time to strike is now.

Attached to this bulletin you will find an attractive folder which fully describes the new *Weather-Conditioner* and emphasizes its many comfort and health advantages.

Get to work now on a list of names covering people whom you think should know about this new development by Packard. Send them the folders and then after a day or two call them up and begin to sell with enthusiasm.

Demonstrate

A new Packard equipped with a *Weather-Conditioner* makes a wonderful demonstration, and every Distributer or Dealer should have a Packard car so equipped. To prepare a *Weather-Conditioned* Packard for demonstrating in the showroom requires but a simple hook-up with an electric motor, and to that end we will forward instructions within the next few days. We can think of nothing that will draw people to your showroom more quickly than one of these weather-maker cars.

Again Packard provides you with an interest-getting "first"—one which is going to have a decided appeal to the many people who think "there isn't much new in motor cars." The Packard *Weather-Conditioner* is something that motordom has been awaiting for a long time.

It drops opportunity right in your lap, so resolve to make the most of it.

SALES PROMOTION DEPARTMENT
PACKARD MOTOR CAR COMPANY

PRINTED IN U.S.A.

Figure 21b. "Something New! Something Practical! Something Every Motorist Wants!"
Packard Promotional Pointers, Vol. VI, No. 6, p. 2, December 15, 1939   PackardInfo.com   Kevin Waltman

Factory Air: Cool Cars in Cooler Comfort 17

Forecasting the typical Texas heat waves of the area, Packard's Assistant Sales Promotion Manager, N.C. Rogers wrote on January 16, 1940:

> Steadily increasing interest is being displayed in the new Packard Weather Conditioner, and this interest will be intensified as warmer weather approaches.
>
> To further the sales of cars equipped with the Weather Conditioner, we recommend that a car so equipped be placed on the showroom floor where it will be accessible at all times for a demonstration of the unit.
>
> To prepare a car for showroom demonstration requires the use of an electric motor for power purposes and necessary attaching brackets, etc. The electric motor is mounted under the bonnet out of sight, and the entire installation can be easily and quickly made by one of your servicemen.
>
> We will not supply the electric motor, which costs approximately $32. You can either purchase this locally or procure it on a rental basis. The motor should be one horsepower, frame #204 of a suitable voltage to operate on your lighting circuit.

Figure 22. 1940 Packard One Sixty Weather Conditioner demonstrator of "The New AIR-CONDITIONED Packard" (yellow arrow) in Dallas showroom DPL NAHC   Romie Minor, Carla Reczek

Figure 23. Packard Dallas customers entered past a showroom display of a 1940 Packard One Ten or One Twenty convertible DPL NAHC   Romie Minor, Carla Reczek

The 1940 Packard Weather Conditioner accessory proved welcome in the brutal Texas summer heat of the Dallas-Ft. Worth area, home of cattle, oil, and land executives wealthy enough to afford the expensive option. With an air-conditioned demonstrator appropriately displayed, per dealer bulletin of January 16, 1940 (Fig. 21b), it complemented the first-ever Packard air-conditioned showroom of Packard Dallas, which opened December 16, 1939. (Figs. 22, 23)

A small wall plaque in the Packard Dallas dealer's showroom advertised, "The New <u>AIR-CONDITIONED</u> Packard." (Fig. 22 and yellow arrow) The dealer recorded long showroom lines of curious potential customers who experienced cooled, conditioned air in the Packard One Sixty demonstrator.

18   Chapter 1: 1940 Packard Factory Air-Conditioning

Packard introduced its cars to the public primarily through brochures and magazine advertising. The 1940 Packard One Ten and One Twenty combination brochure cover illustrated a family looking through a Packard dealer's showroom window.

Shown here is a prime example of a restored eight-cylinder 1940 Packard One Twenty Deluxe Touring Sedan, Model 1801, Body number 1392DE in Centennial Blue, a CCCA (Classic Car Club of America) winner owned by Nelson Bates of the Ark-La-Tex Packard Club of Texas, a region of The Packard Club. (Figs. 24–30)

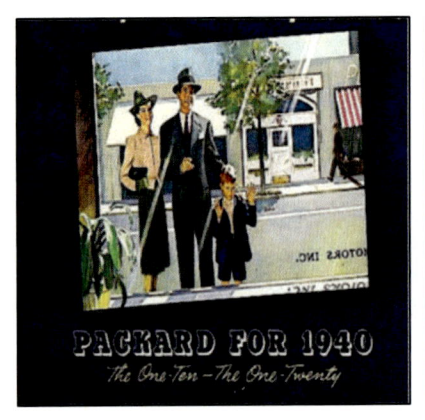

Figure 24. 1940 Packard One Ten and One Twenty brochure cover   ABS

Figure 25. 1940 Packard One Twenty Touring Sedan   ABS

Figure 26. 1940 Packard One Twenty with Author Allen B. Simons and Owner Nelson Bates   ABS photo

Figure 27. 1940 Packard One Twenty "Bonnet Louver Nosepiece Medallion"   Owner N. Bates   ABS photo

Figure 28. 1940 Packard One Twenty Deluxe Touring Sedan Bedford Tan pinstriped fabric interior   Owner N. Bates   ABS photo

Figure 29. 1940 Packard One Twenty Deluxe Touring Sedan in Centennial Blue with optional trunk rack and backup lamp   Owner N. Bates   ABS photo

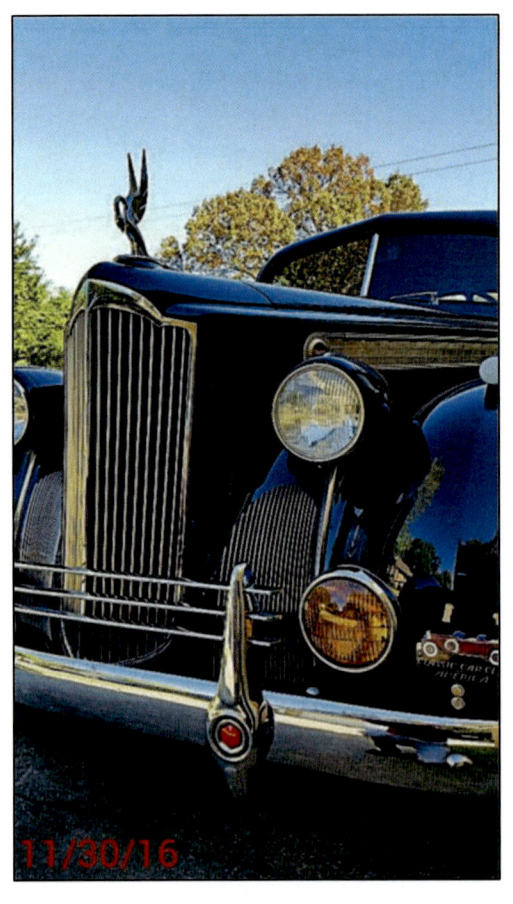

Figure 30. 1940 Packard One Twenty Deluxe Touring Sedan views of Cormorant, grille, and optional Packard Road and Fog Lamps   Owner N. Bates   ABS photo

"The End of an Era, Chapter Eighteen," written by L. Morgan Yost in the book *Packard: A History of the Motor Car and the Company,* offered a succinct description of the 1940 Packard Weather Conditioner with images and text. (Figs. 31–33)

Figure 31. 1940 Packard air-conditioning control under dash and air-conditioning compressor
Packard: A History of the Motor Car and the Company, p. 363   L. Morgan Yost

If the new Packards, excepting the Darrins, were not cool—in the street lingo sense of the word—on the outside, the company decided to make them so on the inside. "Cooled by Mechanical Refrigeration in Summer" read a footnote to an advertisement appearing in February 1940. There was a home-appliance ring to the phrase, but then Packard was the first manufacturer in the industry to offer air conditioning on an automobile, so the quandary about what to call it was understandable. The company decided on Weather-Conditioner. It was bulky. The evaporator and heater plenum chamber were combined into one and installed in the trunk compartment under the package shelf, with a continuously operating compressor mounted up front on the right side of the engine near the water inlet. Thus the faster a car was driven the cooler it became; there was no clutch or automatic disconnect and it was recommended the drive belt be removed when cool weather came. But this did not solve a problem. There was no thermostat, a three-speed switch on the dashboard controlling the fan which blew the cooled air forward out of the package-shelf grille directly onto the soon-to-be-stiff necks of rear-seat passengers. And, even with the fan turned off, the cold air from the coil would merely drop into the floor area by gravity. The price for all this was a reasonable $275—the frequently bruited thousand-dollar figure is a myth—but, still, keeping cool in the summer

Figure 32. 1940 Packard air-conditioning information
Packard: A History of the Motor Car and the Company,
p. 364   L. Morgan Yost

and not too warm in the winter was scarcely a can't-miss proposition, and customers generally gave the idea the bye, if they could. A snafu developed here as well. Special insulation for cars so equipped was provided at the factory but, although noted in Packard literature as "factory installed," the refrigerating unit itself was actually fitted into the car by Bishop & Babcock in Cleveland, and in theory vehicles ordered with it were shipped there, the unit was installed and then the cars were forwarded to the dealers. In practice, however, some cars showed up at dealerships with unordered air conditioning—and in order to rid themselves of the cars, dealers discounted the units substantially. Cooling the air was an expensive experiment all around for Packard, and although continued (but with separate heaters) as an option until World War II, it was dropped thereafter until 1953. Packard engineer W.H. Graves has estimated that about two thousand units were installed.

Figure 33. 1940 Packard air-conditioning information
Packard: A History of the Motor Car and the Company,
p. 365   L. Morgan Yost

20 Chapter 1: 1940 Packard Factory Air-Conditioning

A February 1940 *Fortune* magazine ad titled "Here is your hope!" displayed a top-of-the-line Packard One Eighty Club Sedan, "Cooled by Mechanical Refrigeration in Summer." (Fig. 34)

Here is your hope!

IF YOU'VE DREAMED OF A MIRACLE IN MOTOR CAR LUXURY —THIS IS YOUR CAR!

Illustrated—The Club Sedan, $2297* (white sidewall tires and Weather-Conditioner extra)

★ Cooled by Mechanical Refrigeration in Summer

—and heated by the same unit in winter—1940 Packards set a new standard in motor car comfort. By means of this latest Packard "first," the air inside your car is not only cooled or heated to the degree you may desire, but it is also filtered—cleaned of pollen and dust—and, in summer, dehumidified. This year-round accessory, the Weather-Conditioner, is available at extra cost, installed at the factory.

ASK THE MAN WHO OWNS ONE

PACKARD ONE-EIGHTY

$2228

AND UP, *delivered in Detroit. State taxes extra. Prices subject to change without notice.

SOCIALLY—AMERICA'S *FIRST* MOTOR CAR

Figure 34. 1940 Packard One Eighty Club Sedan advertisement, "Here is your hope!" "Cooled by Mechanical Refrigeration in Summer" Fortune, February 1940   ABS

The *Fortune* ad (Fig. 34) reminded readers that Packard continued to be "SOCIALLY— AMERICA'S FIRST MOTOR CAR."

Exclusive in its offering, Packard advertised a car that was "Cooled by Mechanical Refrigeration in Summer" and heated by the same unit in winter. Packard emphasized that the 1940 Weather Conditioner set a new standard for automotive comfort. It stated:

> By means of this latest Packard "first," the air inside of your car is not only cooled or heated to the degree you may desire, but it is also filtered—cleaned of pollen and dust—and, in summer, dehumidified.

Nash for 1940, in contrast to Packard, emphasized fan-forced ventilation and heating and suggested that ventilation alone would provide year-round comfort.

It advertised its "Weather-Eye" system, which offered "a full twelve months of glorious driving weather—not just the usual eight!" However, Nash did not address this claim to southern areas that suffered from heat and humidity several months each year. Their advertising hyperbole continued,

> There isn't any winter anymore—just spring, all year. But spring without dust, drafts, or bugs...It's "Weather Eye" magic...a completely new and revolutionary system of automatic heating and ventilating that literally makes your Nash more comfortable than a cozy room at home!

Packard in 1939 celebrated its 40th anniversary with an *Automotive News* publication in honor of its history, supported by vendor advertisements. Auspiciously timed, the magazine introduced Packard's Weather Conditioner with information and images of the passenger cooling componentry. (Fig. 35) It illustrated the engine bay air-conditioning compressor, the trunk-mounted evaporator, and the cool air grille behind the rear seat.

Importantly, *Automotive News* stated that Packard's Weather Conditioner "has proven its ability in proving ground tests to *lower the temperature inside a car 19 degrees below that of the outside air.*" (emphasis by author)

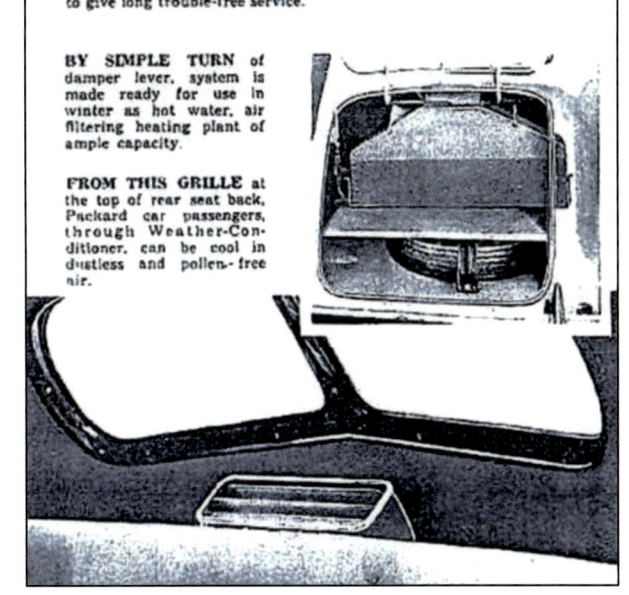

Figure 35. "Controlled Climate," Automotive News Celebrating the 40th Anniversary-Packard, February 5, 1940   John Lawrence

The *Automotive News* 40th Anniversary edition conferred congratulations to Packard from many of its suppliers, including its air conditioner manufacturer, The Bishop & Babcock Mfg. Co. of Cleveland, Ohio. (Fig. 36)

The cover of *Automotive News* captured the birth of a famous slogan, "ASK THE MAN WHO OWNS ONE," which "reflected an answer from the Packard president and owner, James Ward Packard, to a letter from a man in Pittsburgh in 1902." The story continued. His secretary asked,

> He wants to know about our carriage. What shall I tell him? Tell him I'll be down, said Packard, over his shoulder. Tell him—no, wait! I'll tell you what to do. Tell him... *just tell him to ask the man who owns one!* (Fig. 37)

During the Depression, Packard President M. M. Gilman realized that less expensive, high-quality models were the answer to its long-term corporate success. Packard introduced the 1935 One Twenty models, priced near $1,000, as its lowest-priced eight-cylinder car. In the 1935 model year, sales soared to 24,995. (Tables 1, 2, following page)

On February 5, 1940, the *Automotive News* article stated:

> so great was the public interest that ten thousand orders were on hand before a single One Twenty was ready for delivery. Public acceptance was instantaneous and phenomenal.

Later, in 1937, Packard introduced the *Six* models, a lower-priced six-cylinder car, with first model year sales of 65,400. (Tables 1, 2) Again, Gilman stressed Packard's quality and craftsmanship, this time in a lower 100-hp. model.

Automotive News reported, many critics declared the end of the prestigious Packard marque and complained it had severely undermined and diluted its upscale image and standing by pricing its models to match existing low to medium-priced brands. Perhaps so, however, Packard's 8,000 sales in 1934 increased exponentially to 98,000 in 1940 under President Macauley's marketing and manufacturing strategy, known as the "4-Year Plan."

The demise of prestigious names heavily influenced his decision: in 1931, Peerless; in 1934, Stutz and Franklin; in 1936, Cord and Auburn; in 1937, Duesenberg; in 1938, Pierce-Arrow; in 1940, Hupmobile. President Macauley's Packard brand succeeded where other brands failed because they did not offer lower-priced versions.

Figure 36. Bishop & Babcock, air-conditioning component supplier to the 1940 Packard, congratulatory ad Automotive News Celebrating the 40th Anniversary-Packard   John Lawrence

Figure 37. Packard's birth of a slogan in 1902 "—just tell him to ask the man who owns one!" cover for Automotive News Celebrating the 40th Anniversary-Packard, February 5, 1940   John Lawrence

Packard sales increased nearly four-fold from 1934 to 1935 and continued to do so until 1940. By 1937, the less expensive One Ten and One Twenty models made up over 94% of Packard sales. The six-cylinder One Ten models comprised 63% of the 1940 sales. Sales of the less expensive models eroded the exclusivity, panache, and prestige in past and future Senior Packard customers' eyes. (Tables 1, 2)

For 1941, Packard management introduced the sleek new Clipper Model to bolster its sales and rekindle its reputation.

PACKARD SALES INCREASE FROM LOWER PRICED MODELS
1934 TO 1940

	1934 SALES	1935 SALES	1936 SALES	1937 SALES	1938 SALES	1939 SALES	1940 SALES
110 MODELS, INTRO 1937				65,400	30,050	24,350	62,300
120 MODELS, INTRO 1935		24,995	55,042	50,100	22,624	17,647	28,138
SENIOR MODELS 160, 180	8,000	6,954	5,985	7,093	3,044	4,408	7,562
TOTAL SALES BY YEAR	8,000	31,949	61,027	122,593	55,718	46,405 (PLANTS CLOSED FOR UPDATING)	98,000

Table 1. Packard sales increase from lower priced models' 1934 to 1940
Packard: A History of The Motor Car and the Company   Dwight Heinmuller, George Hamlin, L. Morgan Yost, et al.

PACKARD LOWER-PRICED MODELS PERCENT OF SALES
1934 TO 1940

	1934	1935	1936	1937	1938	1939	1940
110 MODELS	0%	0%	0%	53%	54%	53%	63%
120 MODELS	0%	78%	90%	41%	41%	38%	29%
160, 180 MODELS	100%	22%	10%	6%	5%	9%	8%
LOWER PRICED MODELS PERCENT OF SALES	0%	78%	90%	94%	95%	91%	92%

Table 2. Packard lower priced models' percent of sales 1934 to 1940
Packard: A History of The Motor Car and the Company   Dwight Heinmuller, George Hamlin, L. Morgan Yost, et al.

Popular Mechanics magazine, March 1940 (Fig. 38), described Packard's new air-conditioning accessory in which,

> Air is cooled by a refrigerating plant, similar to that used in household refrigerators, and its moisture is condensed as it passes over coils that are defrosted automatically.

Author note: It is interesting to view how the writer described the system as a refrigerating plant that automatically defrosted the cooling coils. The descriptions appear elementary now, some 80 years later.

Erroneously stated, the compressor is not "driven by the radiator cooling fan," rather by "a "V" belt from a special pulley on the water pump shaft," according to the *Packard Data Book for 1941*. (Fig. 111)

POPULAR MECHANICS 391

Auto Air Conditioned for Summer and Winter

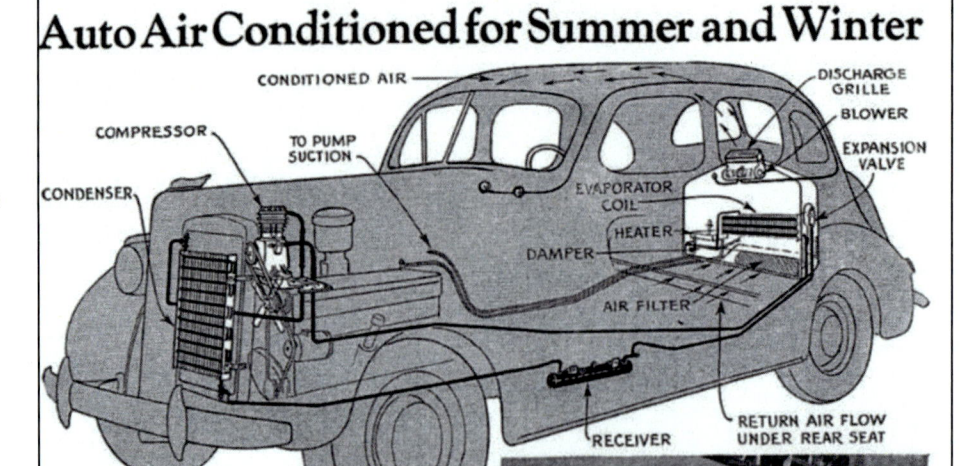

Air conditioning for automobiles, which includes cooling in summer and heating in winter, as well as filtering and dehumidifying, is being built into Packard sedans as extra equipment. Air is cooled by a refrigerating plant, similar to that used in household refrigerators, and its moisture is condensed as it passes over coils that are defrosted automatically. The refrigerator coils are located behind the rear seat in an air duct, which also contains the winter heater in a separate compartment, and two dampers easily reached through the trunk are opened and closed to select hot or cold air. Drawn into the duct by a fan, the speed of which can be varied by a switch on the instrument panel, air is filtered before it passes over the cooling coils, and the treated air is discharged through a grille above the rear seat toward the roof of the car, whence it is distributed naturally over the entire interior, entering the duct again under the rear seat. The compressor, mounted on top of the engine, is driven by the radiator cooling fan, and the refrigerant is admitted through an expansion valve to the coils. Its cooling capacity is rated at one and one-half tons of ice in twenty-four hours, at a car speed of sixty miles an hour.

Top, sketch shows installation of automobile air-conditioning equipment and how both warm and cool air are circulated. Bottom, compressor mounted over engine is run by auto fan

Figure 38. "Auto Air-Conditioned for Summer and Winter"
Popular Mechanics, March 1940   ABS

1940 Packard owners proudly regarded their innovative *FACTORY AIR* cars as,

COOL CARS IN

COOLER COMFORT

Factory Air: Cool Cars in Cooler Comfort 25

The New York Times announced on March 17, 1940, that a taxi distributor specially ordered New York's first 1940 Packard One Ten as a taxi model, equipped with the Weather Conditioner. (Fig. 39) The taxi featured air-conditioning, a six-cylinder, 245-CID, 100-hp. engine, 133-inch wheelbase, tan and blue leather seating, jump seats, heavy-duty clutch, and brakes. For 1941, Packard produced a dedicated six-cylinder taxi with air-conditioning, Chassis number T1900, Body T1462, and Type, New York.

Figure 39. 1940 Packard One Ten model taxi for New York City   The New York Times, March 17, 1940   ABS

James Hollingsworth, author of *Packard 1940: A Pivotal Year*, owned a 1940 Packard One Eighty Club sedan with the Packard Weather conditioner. (Figs. 40, 41) He reported some complaints by Senior Model Packard owners of longer wheelbase 138-inch and 148-inch models that the rear heater temperature proved tepid in severe winter circumstances.

The extra-length heater hoses from the radiator cooled more rapidly than in One Ten models with a 122-inch wheelbase and One Twenty models with a 127-inch wheelbase.

Figure 40. 1940 Packard One Eighty Club Sedan, Model One 1806, Body 1356, equipped with Packard Weather Conditioner. Displayed in $22 extra cost Miami Sand paint color.   Owner James Hollingsworth   David W. Temple Photo

Mr. Hollingsworth addressed the deleted heater core redesign in Packard air conditioners effective April 22, 1940:

> It is the author's guess that the time hot water circulated to the trunk compartment of a 138-inch or 148-inch wheelbase car in severe cold winter conditions, it would not effectively warm the car.

As an owner of one of 233 built, he continued,

> I own a One Eighty Club Sedan (127-inch wheelbase) with the early air-conditioning-heater combination, and the heater portion does adequately warm the car in mild Texas winter weather. (Figs. 40–43, and 34)

The Packard Weather Conditioner nameplate displayed instructions for "Summer Cooling" and "Winter Heating." (Fig. 42)

*Figure 41. 1940 Packard One Eighty Club Sedan, right rear view, Model 1806, Body 1356, equipped with Packard Weather Conditioner
Owner James Hollingsworth   David W. Temple Photo*

The air-conditioning blower grille featured adjustable louvers for directional airflow. (Fig. 43)

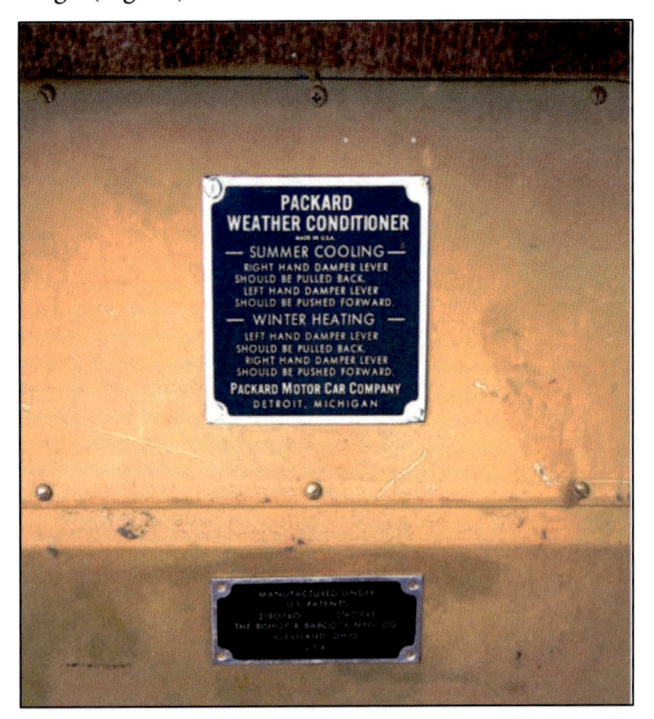

*Figure 42. 1940 Packard One Eighty Club Sedan displays Packard Weather Conditioner evaporator nameplate instructions for summer cooling and winter heating
Owner James Hollingsworth   David W. Temple Photo*

*Figure 43. 1940 Packard One Eighty Club Sedan Packard Weather Conditioner blower grille
Owner James Hollingsworth   David W. Temple Photo*

On May 10, 1940, W. M. Packer, Vice President of Distribution, wrote to dealers concerning the Weather Conditioner accessory: (Fig. 44)

Effective April 22, 1940, a memo addressed the following subjects: the Packard air-conditioning system sold separately from the heater, listed the Suggested Dealer Net Price of $206, and Suggested List Price, recommended under-seat type standard heater, advised length of time for delivery, restricted 1940 sales to specific models, and opened the air-conditioning availability to all lines for 1941.

Trade Letter No. T-3009
Dealers' TL-253

PACKARD MOTOR CAR COMPANY
DETROIT, MICHIGAN

To: PACKARD DISTRIBUTERS AND DEALERS   May 10, 1940

Subject: PACKARD AIR CONDITIONING EQUIPMENT

Gentlemen:

Effective April 22, Packard air conditioning equipment is supplied without the heater. Following are the prices for this equipment also effective April 22:

Suggested Dealer Net Price	Suggested List Price Installed
$206.00	$275.00

The heater to be sold with the air conditioning unit should be the standard heater (preferably the under seat type with defroster) best suited to the body type of the particular car.

This air conditioning equipment has been very well received by our field organization and by the public and, as you know, received widespread publicity. Many distributers and dealers have quickly recognized this opportunity for additional business; some have not; however, we have been compelled to reorder several times and are doing so again today. The volume of orders received in the last few weeks, added to the fact that it takes approximately six weeks for production because of time required to get this equipment from our supplier, points to a very definite shortage by the time hot weather arrives.

We are, of course, very much pleased with the reception given this unit during the first year that any automobile has been offered so equipped and we are glad to be able to tell you that we are making air-conditioned cars a very definite part of our program for the 1941 model year, at which time it will be available on the entire line. For the balance of the present model year, however, only the Packard One-Sixty and Packard One-Eighty are offered with air conditioning equipment.

In selling air-conditioned cars, it should be emphasized that Packard air conditioning actually cools, filters, dehumidifies and circulates the air. Packard is the only manufacturer offering a unit which cools the car through actual mechanical refrigeration.

Very truly yours,
PACKARD MOTOR CAR COMPANY

W. M. Packer
Vice President of Distribution

WMP:EB
(c)2800

APPENDIX 115
Figure 44. 1940 Packard letter sent from W. M. Packer to dealers May 10, 1940
Packard 1940: A Pivotal Year   James Hollingsworth

28   Chapter 1: 1940 Packard Factory Air-Conditioning

Packard published another dedicated air-conditioning brochure in the spring of 1940, titled: (Figs. 45–50)

"THE FIRST COMPLETELY WEATHER-CONDITIONED CAR"

This mailer-style fold-over brochure comprised six pages, appropriately colored in red and blue shades, suggesting the seasonal temperature-controlled ranges inside an air-conditioned Packard.

The brochure freely used marketing hyperbole: (Fig. 46)

1. *An Air-Cooled Car!* Packard's year-round "Weather-Conditioning system" cooled the car for summer driving.
2. *Quick Cooling!* Cooling functions immediately when the engine starts. For short and long trips, it is useful and enjoyable.
3. *Heat for Winter!* A simple damper change converts the cooling to heating during the winter. Air circulates from the same grille behind the seat, circulating the warm air throughout the car.
4. *Closed Window Comfort!* For summer and winter travel, the windows remain closed to outside dirt, dust, and insects. Filters remove pollen.
5. *Simple to Operate!* The Weather Conditioner operates by a handy switch under the dash. It regulates the blower speed to circulate either cooled or warmed air.

Figure 45. 1940 Packard Weather Conditioner brochure cover   Dwight Heinmuller

Figure 46. 1940 Packard Weather Conditioner brochure, p. 2   Dwight Heinmuller

Comprehensive and detailed, the brochure compared its system to a home refrigerator that circulated air over the refrigerant-filled coils, absorbing the heat and giving off refrigerated air. (Fig. 47)

Packard's unit differed in one respect:

Because the interior of a car is more spacious than the average refrigerator, it forces air circulation to ensure cooling for the whole car interior.

The cooling capacity and efficiency of the Weather Conditioner rate several times greater than the average household refrigerator.

The compressor, pressurized at 180 psi, pumped the hot gas to the condenser in front of the radiator.

The forward movement of air, aided by the fan, cooled the hot gas into a liquid. The liquid traveled through tubing into a reservoir, the receiver, and into an expansion valve where it expanded into the evaporator coils, absorbed heat, and returned as hot gas back to the compressor.

A squirrel-cage blower, known as the sirocco-type, blew cooled air through the blower grille behind the rear seat, followed the headliner to the front passengers, and then circulated the cool air over the rear passengers on its way back to the rear evaporator.

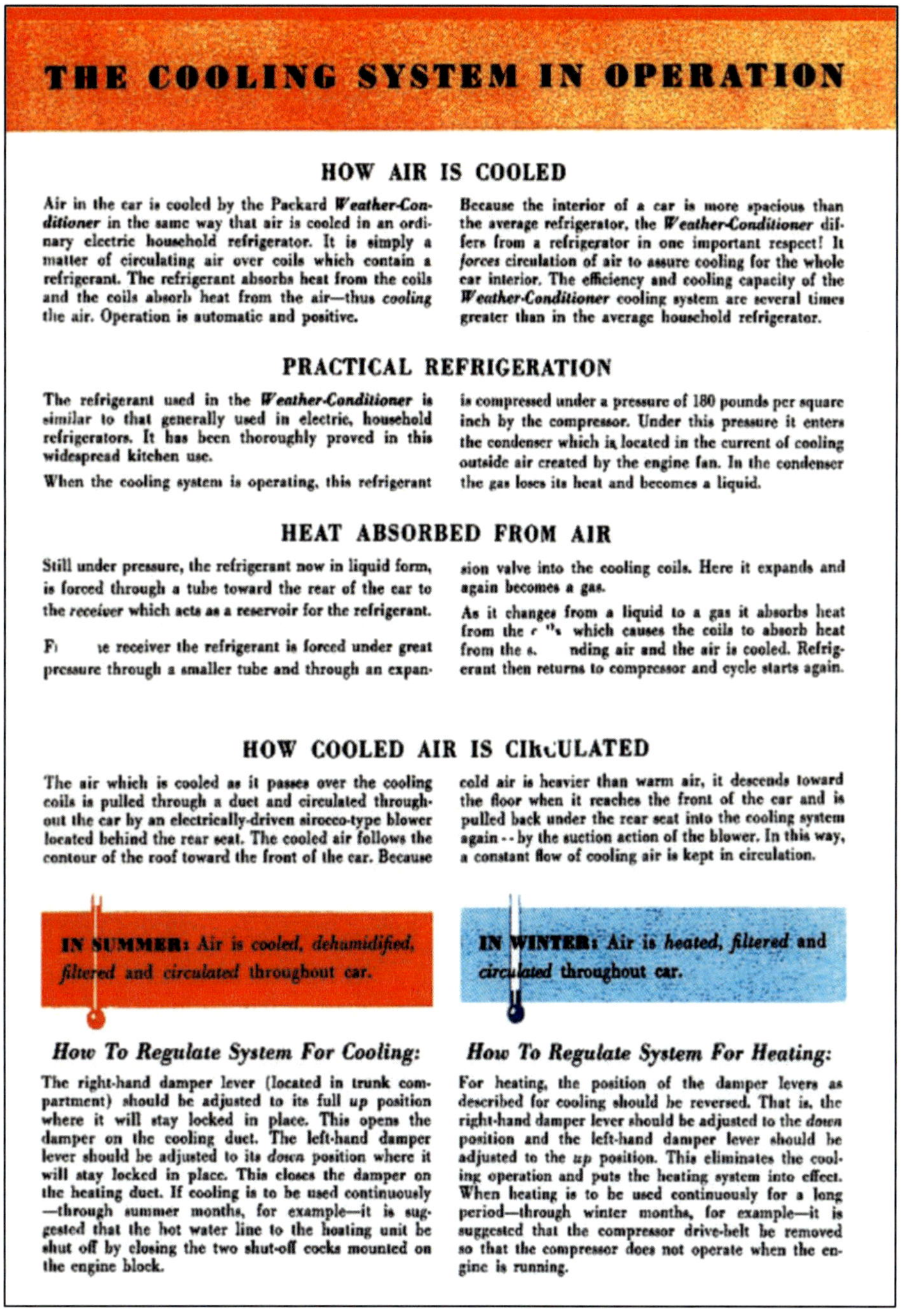

The brochure advised damper lever adjustments to select cooling or heating modes. It suggested shutting off the hot water supply in continuous summer operation and recommended air conditioner compressor drive-belt removal during winter heating to prevent compressor operation.

Figure 47. 1940 Packard Weather Conditioner brochure, p. 3   Dwight Heinmuller

30 Chapter 1: 1940 Packard Factory Air-Conditioning

Packard presented phantom componentry views of the Weather Conditioner. (Figs. 48, 49)

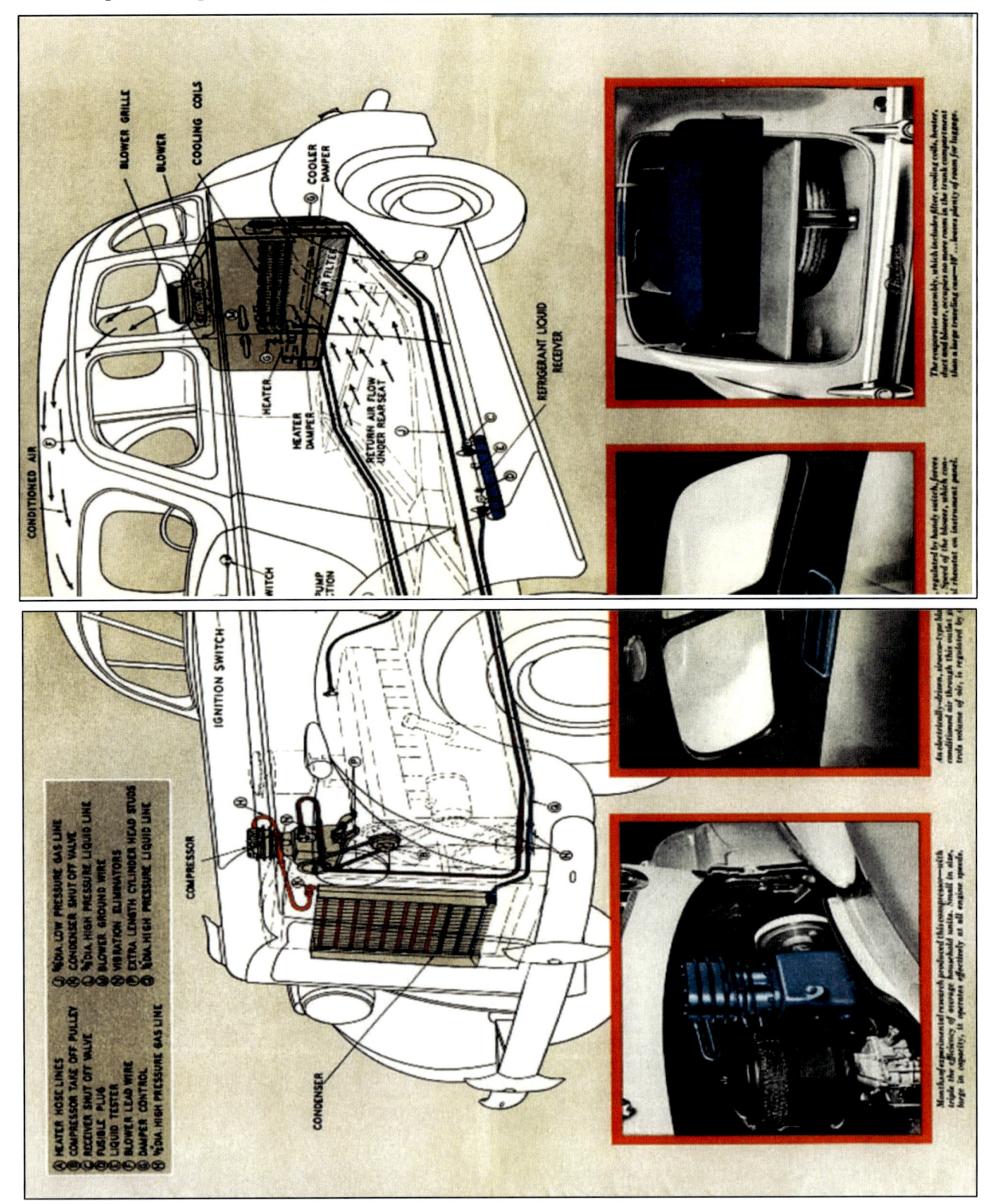

Figures 48, 49. 1940 Packard Weather Conditioner brochure, pp. 4, 5   Dwight Heinmuller

1. *Year-Around Comfort*—Provides heating and cooling. "Air is cooled by a mechanical refrigeration process." (Fig. 50)
2. *Cooling Principle*—same as home refrigerators.
3. *Cooling Effectiveness*—Cooling capacity of 1.5 to two tons of ice melted in 24 hours.
4. *Dehumidification*—Moisture removed while cooling.
5. *Heating Principle*—Hot water pipes lead into another area of the evaporator for winter heating.
6. *Filtering*—It filters air during heating and cooling.
7. *Simple Adjustment*—Damper control levers direct air to cooling or heating chambers. "During the winter season, the compressor may be disconnected," requiring belt removal by a mechanic.
8. *Insulation*—Special insulation added to the body.
9. *Factory Installation*—Factory installed; the system is available in all sedans.
10. *An Exclusive System*—"It is the *only* system now available as a standard accessory which provides refrigerated cooling as well as heating."

Figure 50. 1940 Packard Weather Conditioner brochure, p. 6   Dwight Heinmuller

ASK THE MAN WHO OWNS ONE

32 Chapter 1: 1940 Packard Factory Air-Conditioning

Subsequently published, the *1940 Packard Accessory Listing* identified April 22, 1940, as the effective date that limited its air-conditioning accessory sales to Packard Senior One Sixty and One Eighty models.

The date also marked the discontinuation of the year-round Packard Weather Conditioner moniker and the combination air conditioner-heater assembly. Instead, Packard renamed its system "Air conditioner," dropped the heater chamber, and sold the heater separately in four accessory versions:

1. Heater—standard front compartment $18.75
2. Heater—super deluxe with defroster   21.95
3. Heater—dual-stream (under the seat)   27.50
4. Heater—rear compartment   32.50

Figure 51. 1940 Senior Packards brochure cover   AACA Library (Antique Automobile Club of America Library & Research Center, Hershey, PA)

The *1940 Senior Packards* brochure (Fig. 51) promoted its innovative, exclusive air-conditioning system in the headline,

ENJOY COOL COMFORT THIS SUMMER WITH PACKARD AIR-CONDITIONING! (Fig. 52)

The newest of many Packard contributions to motoring comfort—*Air-Conditioning*–is filling a sensational demand as warm weather spreads over the country.

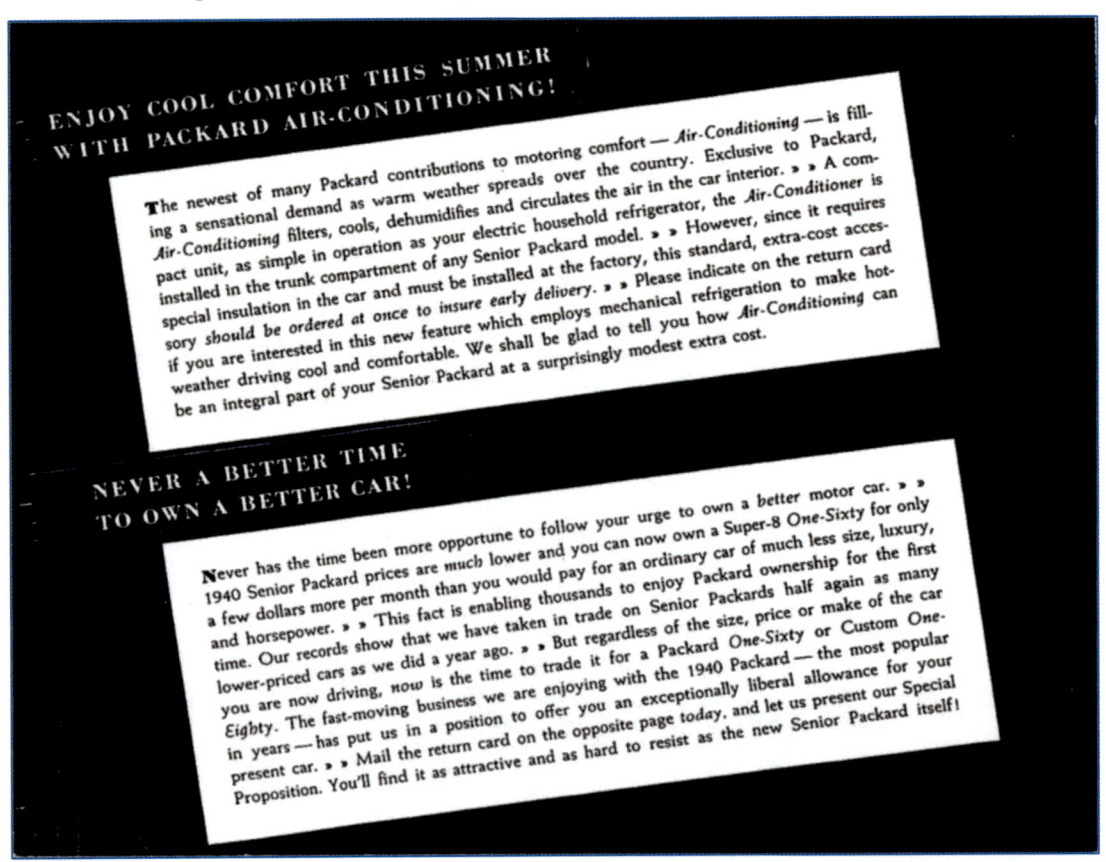

Figure 52. "Enjoy Cool Comfort This Summer with Packard Air-Conditioning!" 1940 Senior Packards brochure   AACA Library

Factory Air: Cool Cars in Cooler Comfort   33

The 1940 Packard prices ranged from its mid-priced $867 to $6300, according to *The Saturday Evening Post* "Air Cool-ditioned" Packard ad of August 24, 1940. (Fig. 53 and yellow arrow)

Author note: Refer to Figure 34: comparatively, a 1940 Packard One Eighty Club Sedan listed for $2228, priced far less than the $6300 custom-bodied cars cited in this advertisement.

Figure 53. "First with a Really Refrigerated Car!"   Saturday Evening Post, August 24, 1940
PackardInfo.com, from The Collection of Fred and Carol Mauck   Kevin Waltman

Spotlight on Packard

Presenting the 1940 Packard Custom Super-8 One Eighty, Seven-Passenger Touring Sedan, Equipped with Factory Air-Conditioning

Owned and Photographed by West Peterson

Figure 54. 1940 Packard One Eighty Cormorant hood ornament Owner and photo West Peterson

A glimpse of Packard's elegantly detailed hood ornament, the Cormorant, sets the tone for this luxury automobile's exceptional quality and impressive coachwork. (Fig. 54)

Inside, the eye focuses on inviting lavish appointments of custom-fitted, pinstriped wool broadcloth cushions, personal armrests with lighter and ashtray, center fold-down armrest, adjustable-length arm-loops, auxiliary seating, and elevated footrests.

Underfoot, fashionable, optional sheared mouton carpeting complements an inviting passenger lap robe of similar, luxurious quality.

Additionally, Packard's quiet living room on wheels provided a new level of passenger comfort. Refrigerated air flowed into the passenger cabin through the rear blower grille, signifying Packard's *exclusive* air-conditioning system. (emphasis by author) (Fig. 55)

Figure 55. 1940 Packard One Eighty Touring Sedan rear seating with one of two auxiliary seats shown. The air conditioner blower grille louvers aim cooled air upward in a circular airflow pattern   Owner and photo West Peterson

West Peterson recounted the history of his 18th Series 1940 Packard.

Originally purchased by a Kansas City, Missouri owner in July 1940, a small Minnesota museum acquired it in 1959. The museum displayed it for 45 years until Mr. Peterson's purchase in 2004.

The low mileage, 49,700-mile Packard's cowl-mounted serial plate, reflected it as the last production car of Model 1808, Body number 1351, yielding production number 283.

By extension, Packard installed the last air-conditioning system of the Model 1808 for 1940. Additional optional equipment included Econo-Drive overdrive, radio, dual heaters, dual sidemounts with mounted rear-view mirrors, mouton carpeting, and bumper guards. (Fig. 56)

Packard published a prestige hardbound brochure for the 1940 Packard One Eighty. (Figs. 57–59)

Figure 56. 1940 Packard One Eighty Seven-Passenger Touring Sedan, equipped with air-conditioning   Owner and photo West Peterson

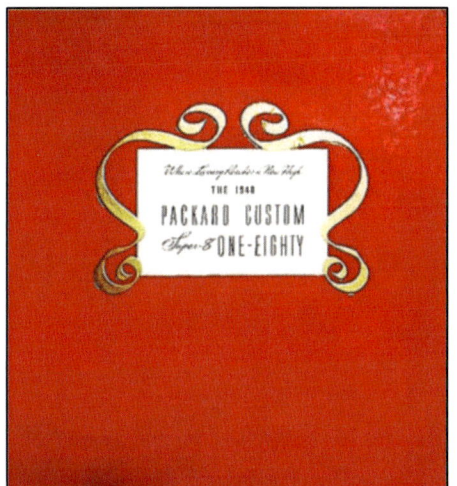

Figure 57. 1940 Packard Custom Super 8 One Eighty prestige hardbound brochure cover   West Peterson

Figure 58. 1940 Packard One Eighty Seven-Passenger Touring Sedan featured in hardbound prestige brochure   AACA Library

HOSPITALITY goes with you when this supremely luxurious car, with complete comfort for seven passengers, is at the disposal of you and your guests. The folding seats are deeply upholstered and far too comfort-giving to be considered mere auxiliary seating for brief trips or occasional use. It is a car that makes a week-end trip to the country or an extended family vacation tour thoroughly enjoyable from beginning to end. The beauty of its interior combines with that of its distinguished exterior to create an atmosphere of prestige and luxury unsurpassed in motoring.

Figure 59. 1940 Packard One Eighty Seven-Passenger Touring Sedan displayed an elegant interior for seven passengers on a 148-in wheelbase   AACA Library

36 Chapter 1: 1940 Packard Factory Air-Conditioning

Packard marketed the One Eighty series as its top-of-the-line models and equipped the interiors with premium appointments. The body, chassis, and systems featured state-of-the-art technology, backed by its legendary reputation for precise construction and exceptional craftsmanship. As an index of prestige and exclusivity, sales from legions of satisfied repeat and new customers consistently eclipsed Cadillac's next most prestigious American car. Customers believed Packard was "The Best." Not until the post-war years did Cadillac attain the top sales in the fine car field.

Mr. Peterson supplied multiple images of his 1940 Packard One Eighty that illustrated the air-conditioning system's specifics. In front, the Packard One Eighty air-conditioning condenser's horizontal tubing creates a checkered appearance behind the vertical grille louvers, similar to the 1940 Packard One Sixty air-conditioning demonstrator shown in the Packard Dallas showroom. (Figs. 60–62)

Figure 60. 1940 Packard One Sixty air-conditioning demonstrator grille appears cross-hatched due to condenser tubing behind it.   DPL NAHC Romie Minor, Carla Reczek

Figure 61. 1940 Packard One Eighty grille view   Owner and photo West Peterson

Figure 62. 1940 Packard One Eighty close-up view of air-conditioning condenser tubing behind grille   Owner and photo West Peterson

Figure 63. The cloisonné background found on the 1940 Packard One Eighty "Bonnet Louver Nosepiece Medallion" reflects careful attention to detail. Cormorant hood ornament lies in background   Owner and photo West Peterson

Grille views of Mr. Peterson's One Eighty allow a closer inspection of the air-conditioning condenser's checkered appearance.

Unfortunately, after removing the front clip during rejuvenation, using Mr. Peterson's term, the condenser remained installed between the grille and the radiator. Consequently, there are no separate images of his 1940 Packard condenser.

Outdoors, sunlight sparkled upon the red cloisonné artwork, the "Packard Super 8" inscription on the "Bonnet Louver Nosepiece Medallion," and, in the background, the Cormorant hood ornament. (Fig. 63)

Factory Air: Cool Cars in Cooler Comfort 37

Below the burled Carpathian Elm, faux wood-grained, ultra-high fashioned plastic "Tenite" dash, Packard installed the air-conditioning control panel on the right of the similarly grained steering wheel, with the heater control panel symmetrically located under the dash to the left of the steering wheel. (Figs. 64–67)

Figure 64. 1940 Packard One Eighty dash with air-conditioner "Temperature Control" panel on right of steering wheel and "Heat-Defrost" control panel on left of steering wheel   Owner and photo West Peterson

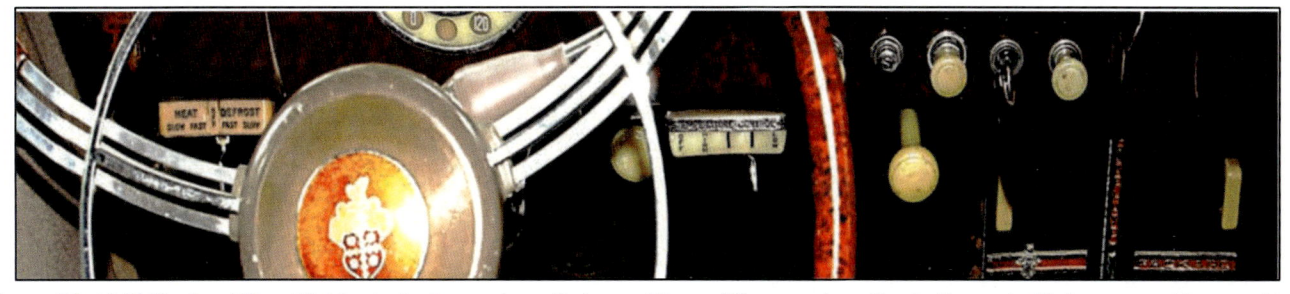

Figure 65. 1940 Packard One Eighty dash closer view of air-conditioner "Temperature Control" panel with four-speed fan markings on the right side of the steering wheel, the "Heat-Defrost" control panel on left   Owner and photo West Peterson

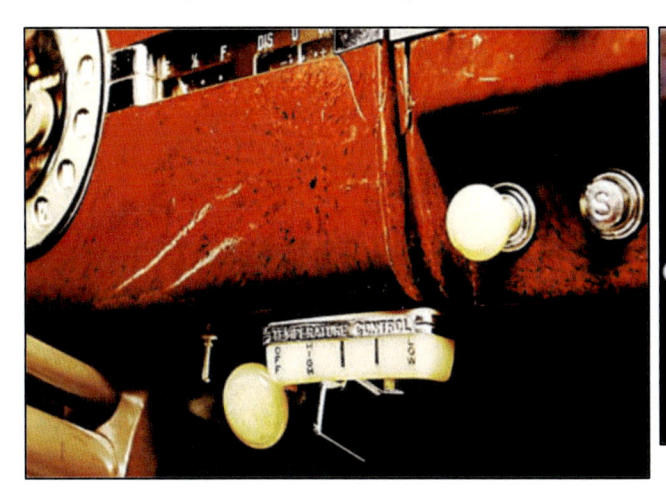

Figure 66. 1940 Packard One Eighty passenger side view of air-conditioning "Temperature Control" panel   Owner and photo West Peterson

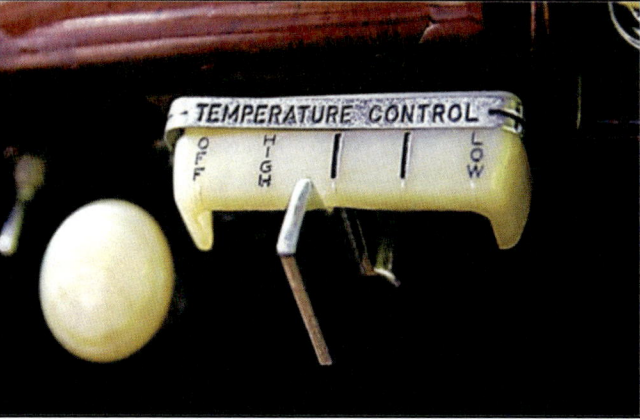

Figure 67. 1940 Packard One Eighty underside view of temperature control paddle switch on air-conditioning "Temperature Control" panel   Owner and photo West Peterson

Figure 68. Resplendent in black, twilight captures a left profile view of the 1940 Packard One Eighty Touring Sedan
Owner and photo West Peterson

For 1940, Packard offered the One Eighty models in three wheelbases of 127-inches, 138-inches, and 148-inches. The longer wheelbase 138-inch and 148-inch models featured foldable auxiliary seating for two additional passengers.

Mr. Peterson's 148-inch One Eighty, Seven-Passenger Touring Sedan, Model 1808, Body number 1351, had no divider window. (Figs. 68–71) It held two passengers in the front seat, two in the auxiliary seats, and three in the rear seat.

Figure 69. 1940 Packard One Eighty Touring Sedan view of two stored, foldable auxiliary seats and optional mouton carpeting
Owner and photo West Peterson

Figure 70. 1940 Packard One Eighty Touring Sedan view of opened and stored auxiliary seats, flanked by heater grilles
Owner and photo West Peterson

Figure 71. 1940 Packard One Eighty Touring Sedan interior view with one erect auxiliary seat and rear seating. Shown with the air-conditioning blower grille behind the rear seat and optional floor grille for the "Rear Compartment Hot Water Heater"
Owner and photo West Peterson

Factory Air: Cool Cars in Cooler Comfort 39

The Packard One Eighty Limousine factory photo, Model 1808, Body number 1350, illustrated optional sheared lamb's wool mouton carpeting. (Fig. 72) Mr. Peterson's Touring Sedan also featured the optional mouton carpeting and Packard lap robe, which added another level of sophistication to the luxurious air-conditioned rear cabin. (Fig. 73)

Figure 72. 1940 Packard One Eighty Limousine, model 1808, Body number 1350, factory illustration of optional mouton carpeting
Packard 1940: A Pivotal Year
James Hollingsworth

Figure 73. 1940 Packard One Eighty view of erect auxiliary seat, optional mouton carpeting, Packard accessory custom lap robe, and air-conditioning blower grille positioned behind rear passenger seat
Owner and photo West Peterson

Colorful brochure illustrations promoted the quiet, refrigerated passenger comfort and dehumidified, fan-forced air circulation inside the 1940 Packard passenger cabin. As a bonus to hay fever sufferers, the "Air Cool-ditioned" 1940 Packard offered clean, filtered air for its occupants. (Figs. 74, 75, 53, and yellow arrows)

Quiet, forced circulation! The fresh, clean, refrigerated air is circulated to every part of the car by a fan—concealed and quiet. Here is how it works: air follows along the top of the car, returns to the floor, under the rear seat to the cooling coils—where it is re-cooled and re-circulated. The comfort of an Air Cool-ditioned Packard is a revelation beyond what you have ever known in motoring.

Figure 74. 1940 Packard air-conditioning brochure
PackardInfo.com, from the collection of Fred and Carol Mauck
Kevin Waltman

Supplies clean, filtered air! If you can't get away during the hay fever season, your next best bet is an Air Cool-ditioned Packard. For in its spacious, comfortable interior you'll breathe *clean, fresh, filtered* air.

Figure 75. 1940 Packard air-conditioning brochure
PackardInfo.com, from the collection of Fred and Carol Mauck
Kevin Waltman

Packard technicians expected factory authorized training before servicing the innovative and sophisticated new Weather Conditioner. Instead, they received no formal factory training. No known reasoning explains why Packard's *Packard 1940 Data Book* did not include the air-conditioning information, nor did it publish a service manual.

Instead, Packard waited and published a *Packard Service Letter*, dated March 1, 1940. It stated, no technician should attempt to service the Weather Conditioner unless he "had special instruction and training… *If service is required, a good refrigeration man should be called in.*" (emphasis by author) (Fig. 76 and yellow arrow)

WEATHER CONDITIONED CARS

We are now shipping cars equipped with the new "Packard Weather Conditioner."

All Packard servicemen should know about these installations so that they can take care of the touring owner, as well as any which are in their own town.

Unless you have had special instruction and training, no attempt should be made to service the Weather Conditioner. The cooling system operates on the same principle as the ordinary electric household refrigerator. If service is required, a good refrigeration man should be called in. He will have the necessary gauges and equipment to enable him to quickly make any necessary adjustments. The service manual placed in the glove compartment of each Weather Conditioned car or the illustration of the installation, will enable him to locate the various parts and points of adjustment.

The refrigerant which provides the cooling effect circulates in the cooling system under high pressure. When working on a car equipped with the Packard Weather Conditioner, do not disconnect any of the piping or open or close any of the valves. To do so may permit the gas to escape and the cooling system will fail to operate.

If it should be necessary to remove the cylinder head of a Weather Conditioned car, *do not disconnect the compressor tubing.* The compressor should be unbolted from the base, a rope tied around the body of the cylinder and the compressor swung around out of the way and hung from the radiator tie rod. Be careful not to twist or strain the flexible connections. This might cause a leak which will permit the refrigerant to escape.

You already have the "Packard Service Parts List, Air Conditioning System." All parts should be ordered from the Packard Parts Department.

Numbers in the Parts Book preceded by an asterisk are especially designed for cars equipped with the Weather Conditioning system and differ from the corresponding standard parts used on cars without Weather Conditioning.

When ordering cylinder head studs, fan pulleys and radiator parts for a car equipped with Weather Conditioning, check the Weather Conditioning parts list first and if any parts required are listed there, order accordingly.

Report to the Factory Service Department immediately any case of complaint or mechanical difficulty which comes to your attention.

Figure 76. 1940 Packard Service Letter, March 1, 1940 PackardInfo.com Kevin Waltman

It also stated that a service manual (was) placed in the glove compartment of each Packard Weather Conditioned car. Unfortunately, records yield no such copy.

In the 1941 model year, Packard published air-conditioning information in their *Packard Data Book for 1941*, dated September 1940. It published the 11-page *1941 Packard Air-Conditioning Operation, Service, and Owner's Instruction* manual on April 1, 1941. (Refer to the 1941 Packard chapter)

Packard and the Bishop & Babcock Mfg. Co. of Cleveland, Ohio, combined mechanical refrigeration principles to air-condition the 1940 Packard automobiles. Previously discussed in 1938 and the 1940 brochure, the Henney-Packard ambulance optionally featured Packard's Weather Conditioner. (Figs. 11–13)

According to *Cleveland, the Making of a City* by W. G. Rose, Bishop & Babcock's heating and ventilating division began in 1915, automotive thermostats began in 1926, and the automotive hot water-based heater-defroster production began in 1935.

After several years of developing air-conditioning for the passenger compartment of an automobile, Bishop & Babcock introduced the first practical cooling equipment in 1939, selling it to the Packard Motor Car Company.

The article did not consider the Henney ambulance company installation in 1938 as the "passenger compartment of an automobile."

The 1940 Packard air conditioner system used the slightly modified 1936-based Servel commercial, two-cylinder compressor supplied by Bishop & Babcock. (Figs. 77–79, 93)

These four images revealed the passenger-side compressor placement on the engine block, including a close-up of the unusual U-shaped bracket attached around and between the cooling fins on the upper portion of Mr. Peterson's compressor.

The anti-vibration bracket stabilized the high and low-pressure tubing lines. Mr. Peterson stated he has never seen another bracket used in this manner and speculated that it could have served as a prototype; yet, there is no evidence of use on other 1940 or subsequent 1941 Packard air-conditioning compressors.

Figure 77. 1940 Packard One Eighty Servel air-conditioning compressor with unusual U-shaped bolt supporting the high- and low-pressure tubing Owner and photo West Peterson

Figure 78. 1940 Packard factory photo of driver side view of air-conditioning compressor. Caption reads, "Close-up view of air-conditioning unit in PACKARD car." Typed on back, "Packard accessories—Air-conditioning—Pre WWII, air-conditioned car." Stamped on back, March 19, 1940 DPL NAHC Romie Minor, Carla Reczek

Figure 79. 1940 Packard factory photo of passenger side air-conditioning compressor. Label on unit reads, "Important! Do Not Remove oil plug or add oil unless system is to be recharged. Use only special compressor oil supplied by Packard Motor Car Co." Typed on back, "Packard accessories—Air-conditioning—Pre WWII, air-conditioned car." Stamped on back, March 22, 1940 DPL NAHC Romie Minor, Carla Reczek

42 Chapter 1: 1940 Packard Factory Air-Conditioning

Copper tubing connected the condenser mounted in front of the radiator to the receiver-coolant reservoir underneath the car, next to the frame where it stored liquid Freon-12. The tubing continued and carried the Freon-12 into the expansion valve's trunk location, attached to the evaporator that housed the cooling coils inside. (Figs. 80–84)

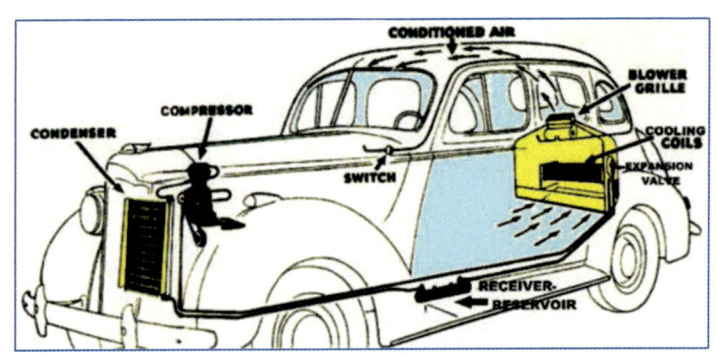

Figure 80. 1940 Packard air-conditioning brochure Regress Press   Jeff Steidle

Figure 81. 1940 Packard One Eighty air-conditioner receiver-reservoir and tubing located under driver's side floor pan   Owner and photo West Peterson

Built after the cut-off date of April 22, 1940, the July 1940 build date of Mr. Peterson's evaporator eliminated the heater core assembly, yet it maintained the previous Weather Conditioner identification.

Figure 82. 1940 Packard One Eighty air-conditioning trunk-mounted evaporator displays nameplates and expansion valve

Figure 83. 1940 Packard One Eighty Weather Conditioner nameplate on evaporator without heater in assembly   Owner and photo West Peterson

The nameplate on the evaporator contrasts with Mr. Hollingsworth's nameplate for cooling and heating instructions. (Figs. 83, 42)

Underneath the inspection plate, the original brown primer color contrasts with the evaporator's color. The rounded shape of the blower housing appeared above the brush-like bristles, or heat sinks, of the cooling coils. (Fig. 84)

Figure 84. 1940 Packard One Eighty evaporator inspection plate removed. This close-up view revealed the rounded air-conditioning blower placement and evaporator cooling coils shaped like brush bristles, or heat sinks.   Owner and photo West Peterson

A further study focuses upon the method of air circulation. The examination of the 1940 Packard One Eighty package tray, compared with a factory air-conditioned 1953 Oldsmobile Ninety-Eight package tray, revealed significantly different installations. (Figs. 85, 86)

Figure 85. 1940 Packard One Eighty blower grille in package tray   Owner and photo West Peterson

Figure 86. 1953 Oldsmobile Ninety-Eight package tray with one of two cool air ducts, return air grilles, and rear radio speaker grille   Owner and photo Phil Gaffney

The 1940 Packard One Eighty package tray solely contained the air conditioner's louvered blower grille. The 1953 Oldsmobile Ninety-Eight package tray incorporated two clear plastic cool air ducts (one not shown) that lead into the headliner for individual passenger cooling registers, two return air grilles, and a rear radio speaker.

There are no return air grilles on the 1940 Packard package tray. Earlier, a passage stated,

In a circular airflow pattern, the conditioned air passed over the rear seat passengers and returned underneath a slightly elevated rear seat cushion, through air filters into the evaporator cabinet. (Referenced above Fig. 4)

The return airflow space between the cushion and the floor measured about an inch. (Fig. 87)

Author note: Viewed in this original setting, the owner-added antimacassar embroidery lace, coupled with mouton carpeting, reflected a prewar, kinder, gentler era.

Figure 87. 1940 Packard One Eighty air-conditioning return airflow space between rear seat cushion and floor. Rear compartment heater grille in floor   Owner and photo West Peterson

44 Chapter 1: 1940 Packard Factory Air-Conditioning

Mr. Peterson (Fig. 88) removed the rear seat cushion and labeled specific areas. (Figs. 89, 90)

Figure 88. West Peterson in his 1940 Packard One Eighty   West Peterson photo

Figure 89. 1940 Packard One Eighty rear seat cushion removed for inspection
Owner and photo West Peterson

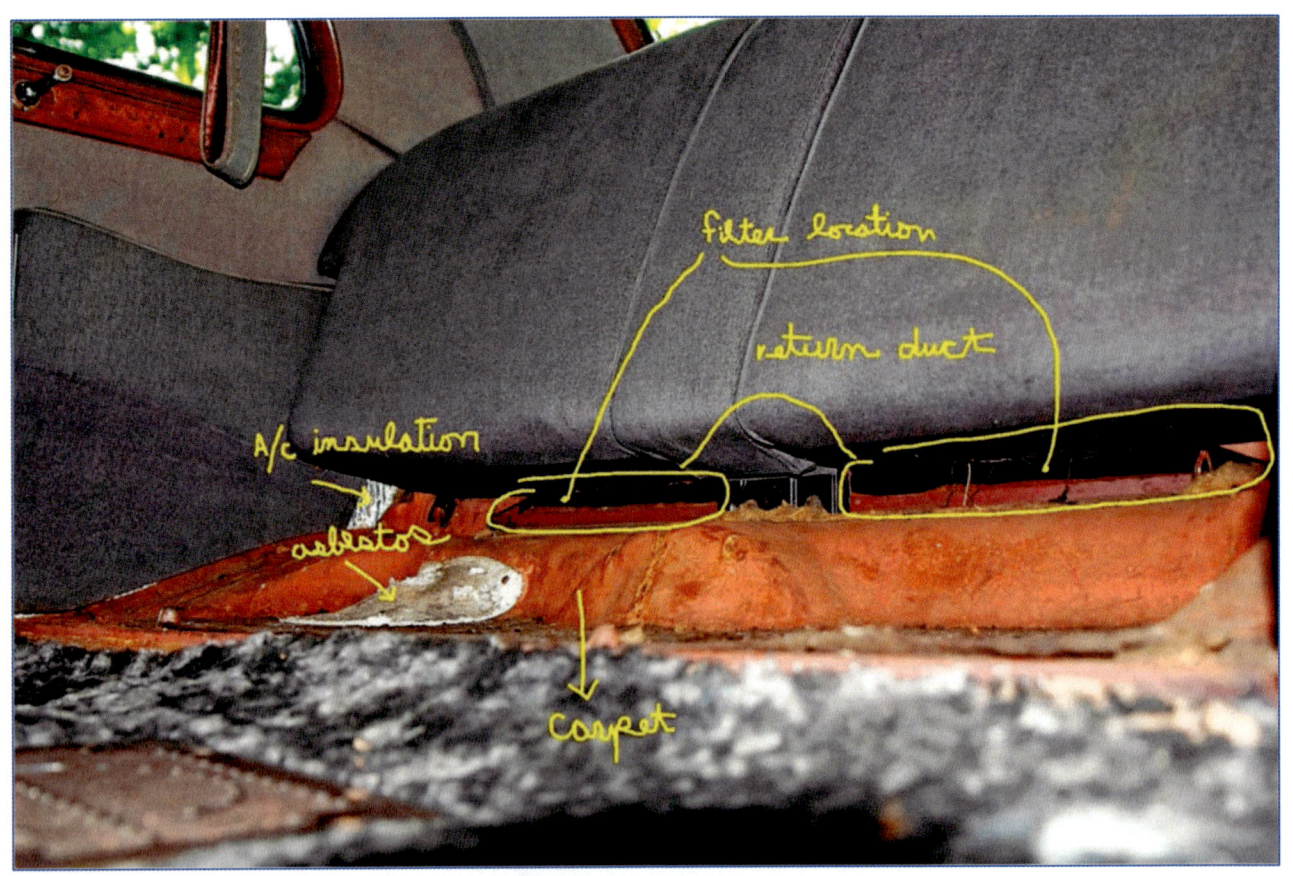

Figure 90. 1940 Packard One Eighty identification of carpet, air filter locations, return ducts, foil air-conditioning insulation, and asbestos area   Owner and photo West Peterson

The four-speed blower fan in the trunk-mounted evaporator drew the return air under the rear seat cushion through two disposable air filters located inside the return air ducts and upward through the trunk-mounted evaporator cooling coils. (Figs. 97, 98)

The cabin view photos with the rear seat cushion removed, and one with the front seat cushion removed illustrated Packard's insulation for air-conditioned cars.

After removal of the front seat cushion and carpeting, Mr. Peterson exposed the foil-backed, corrugated cardboard insulation added at the Packard factory during assembly, before shipping the car to Bishop & Babcock for installation of the air-conditioning componentry: (Fig. 91)

Figure 91. 1940 Packard One Eighty insulation installed in air-conditioned cars during assembly, viewed on front floor, kick panel, and underneath front seat frame   Owner and photo West Peterson

Black painted insulation also appeared on the engine cowl and firewall. (Figs. 92, 93)

Figure 92. 1940 Packard One Eighty close-up of factory-applied insulation added to cowl and firewall of air-conditioned body   Owner and photo West Peterson

Figure 93. 1940 Packard One Eighty cowl, firewall insulation, and the Servel air-conditioning compressor   Owner and photo West Peterson

46 Chapter 1: 1940 Packard Factory Air-Conditioning

While the *Packard 1940 Data Book* did not cover the air-conditioning insulation description, the *Packard Data Book for 1941* supplied the pertinent reference information: (Fig. 94)

> **Insulation**—In every Packard equipped with Air Conditioning special insulation is used in the roof, sides, and floor to reduce the amount of heat leakage into the car from sun heat on the roof and side panels, and from exhaust heat of the engine under the car floor. This special insulation aids materially in maintaining desired interior temperatures.

Figure 94. The Packard Data Book for 1941, p. 78   AACA Library

A closer examination revealed the composition of the insulation. Rather than "foil-backed corrugated Fiberglas," described in *Packard 1940: A Pivotal Year*, Mr. Peterson determined it is foil-backed corrugated cardboard, similar to cardboard box composition (Figs. 95, 96)

Figure 95. 1940 Packard One Eighty foil-backed corrugated cardboard used as body insulation in factory air-conditioned cars   Owner and photo West Peterson

Figure 96. 1940 Packard One Eighty close-up view of foil-backed corrugated cardboard body insulation
Owner and photo West Peterson

Particularly noteworthy, Mr. Peterson photographed the removable, air-conditioning disposable air filters. (Figs. 97, 98)

Figure 97. 1940 Packard One Eighty air-conditioning disposable air filters   Owner and photo West Peterson

Figure 98. 1940 Packard One Eighty close-up view of a Detroit Air Filter air-conditioning disposable air filter made by Detroit Lubricator Company, Detroit, Michigan   Owner and photo West Peterson

The fragile cardboard frames connected paper conduits fashioned in a cellular pattern similar to a beehive design for Packard's conditioned airflow. The airflow, under closer observation, passed through openings in the filter patterns.

Post-war and contemporary cabin air filters differ significantly from the primitive Bishop & Babcock air-conditioning system air filters supplied to Packard, Cadillac, and Chrysler in the 1940-1942 era.

The reintroduction of factory air in 1953 placed return air grilles on the package tray. (Fig. 86) The fans recirculated cabin air through grilles placed above mesh-type filters before discharging refrigerated air into the cabin.

By the mid-1950s, compact evaporator assemblies progressed, thus allowing engine bay installation. This development led to the "up-front style," under hood, air-conditioning component placement. Fans circulated airflow over the evaporator coils and discharged the refrigerated air through registers or nozzles in the dash.

Engineers depended on air filtration through the dehumidification of water vapor from the air as it passed over the evaporator cooling coils rather than by previously used air filters. The condensate on the cooling coils captured pollen and dust particles and discharged them down to the ground through the evaporator drain hose.

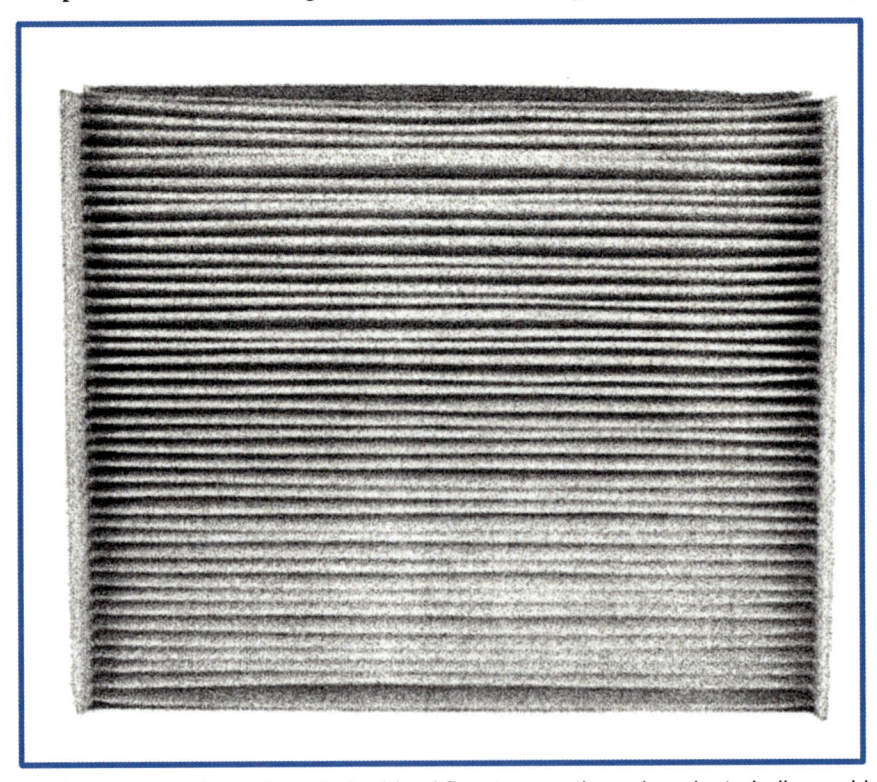

Figure 99. New car manufacturers require recirculated cabin airflow to pass through a pleated, disposable cabin air filter   ABS

Currently, carmakers still rely on condensate capture to remove pollen and dust in the air-conditioner's fresh air mode. To improve the cabin air quality, however, carmakers now equip cars with disposable cabin air filters. Air must now pass through a disposable, pleated paper cabin air filter in the recirculate mode. (Fig. 99) Some carmakers now offer charcoal-impregnated cabin air filters, citing even better cabin air quality.

Factory Air: Cool Cars in Cooler Comfort   49

The 1940 Packard Weather Conditioner system's introductory year ended with an air-conditioning summer road test on a New York State, New England, and New Jersey tour by Mr. R. M. Vandivert. Driving a Packard identical to Mr. Peterson's, he published the article, "New Air-Cooled Packard Foils Humidity on Tour," in the *New York Journal and American*, July 24, 1940. (Figs. 100, 101)

New Air-Cooled Packard Foils Humidity on Tour

New Device Like Home Refrigerator

By R. M. VANDIVERT

"Tempering the wind to the shorn lamb," often referred to as one of the great phenomena provided by an all-wise providence, has its parallel in motor car luxury made available in current Packard enclosed sedan models.

Packard engineers, profiting by the advances in house refrigeration and home air conditioning, have applied these developments to motor car comfort and created weather conditioning to meet all seasons and all outside temperatures.

Recently the writer took a four-day trip through New York State,

The seven-passenger Packard super-eight, equipped with air-cooling equipment, which was used on a recent test trip through New York and New England.

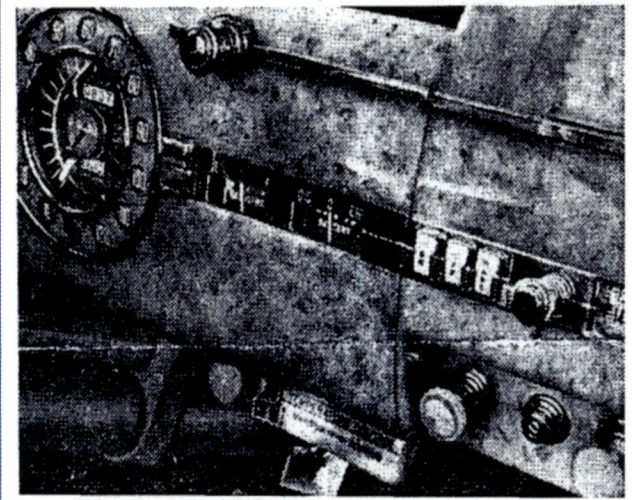

This simple switch, incorporated in the instrument panel, governs the speed of the blower which controls the volume of air. This is regulated by a dial rheostat.

The compressor, small in size but large in capacity, operates effectively at all engine speeds.

Figure 100. "New Air-Cooled Packard Foils Humidity on Tour"
New York Journal and American, July 24, 1940, Part 1 of 2   DPL NAHC   Romie Minor, Carla Reczek

New England and New Jersey, during which the variable weather provided an opportunity to give the "weather-conditioner" a thorough try-out. Not only did we ride in air-cooled comfort during two humid unpleasantly warm days, but we also encountered a period of cold, clammy rain when slightly warmed, dehumidified air added greatly to the creature comfort of "down east" touring.

SIMPLE SYSTEM.

The transition from air-cooled comfort to equal comfort with warmed air is simplicity itself. The same air circulatory system provides car cooling or car heating, as one may desire, by a simple adjustment of damper controls located in the trunk compartment of the car, an operation requiring no particular skill and but a few seconds of time.

For hot weather driving the car is cooled by mechanical refrigeration. The cooling principal is the same as that used in household refrigerators. It is simply a matter of circulating air over coils which contain a refrigerant. The refrigerant absorbs heat from the coils and a coils absorb heat from the air, thus cooling the air, which is dehumidified as it is cooled.

The capacity of the Packard cooling system at high touring speeds is rated as equivalent to the cooling effect of placing one and one half to two tons of ice in the car to be melted in twenty-four hours.

The evaporator assembly, which includes filter, cooling coils, heater duct and blower, is located at rear of the trunk compartment leaving plenty of room for ordinary travel luggage.

WORKS WITH ENGINE.

The cooling system functions emmediately when the car engine is started. It is regulated by a simple switch attached to the instrument panel within easy reach of the driver, which controls the amount of cooled air desired. With the windows closed, the air effectively filtered to remove dust or pollen as it passes through the air-conditioner, and regulated by the speed of the blower, passes through a duct located behind the rear seat, circulates along the roof to the front of the car and returns to the rear along the car floor. This flow of air effectually results in the permeation of conditioned air throughout the entire car.

To heat the car, the dampers are changed and air heated for circulation by passing through a score in which hot water is circulated.

COMFORTABLE RIDE.

The demonstrator we drove was a Packard Super-eight seven passenger sedan, big, roomy, luxurious but at the same time easy to handle at touring speeds on the highway and in traffic. During the trip we drove ultra-modern highways and back-country roads and all were negotiated with a maximum of comfort. Even in steep mountainous country in the Berkshires, the Catskills, the Ramapo and other ranges we crossed, the car took everything in high without effort.

In spite of the fact that the cooling unit was kept in almost constant use in much of this mountain travel, there was no perceptible loss of motive power in handling the car nor was there any noticeable extra consumption of gasoline. Considering the size of the car, the roads traveled, and the touring speeds maintained, the consumption of fuel was remarkably low during the entire trip.

Figure 101. "New Air-Cooled Packard Foils Humidity on Tour"
New York Journal and American, July 24, 1940, Part 2 of 2   DPL NAHC   Romie Minor, Carla Reczek

Mr. Vandivert's summer road tour followed hilly and mountainous areas and climates that changed, allowing enough time to evaluate the newly introduced Weather Conditioner's operation in a 1940 Packard One Eighty seven-passenger Touring Sedan. He stated:

> Not only did we ride in air-cooled comfort during two humid, unpleasantly warm days, but we also encountered a period of cold, clammy rain when slightly warmed, dehumidified air added greatly to the creature comfort of "down east" touring.

He discussed the system's mechanical operation, the control on the instrument panel, air circulation, filtration, and the damper changes necessary to switch from cooling to heating. The Packard powerfully navigated hills and mountains while remaining in high gear. He continued, "there was no perceptible loss of motive power in handling the car, nor was there any noticeable extra consumption of gasoline."

Author note: A visionary in the 1940s would not have imagined the popularity of "factory air." According to the Automotive Air-conditioning Association, Inc. figures, Detroit automakers sold 29,000 factory-installed units in the reintroduction year of 1953. Sales increased in 1958 to 198,000, in 1961 sales more than doubled to 438,000, and sales exceeded one million units for the first time in 1963.

Packard Pricing—Then and Now

The CPI (Consumer Price Index Inflation Calculator) lists January 1940 equivalent prices:

The 1940 Packard Weather Conditioner combination air conditioner-heat system list price:			$274
The CPI lists a January 2021 equivalent price:			$5,156
1940 Packard One Eighty 7-Passenger Touring Sedan:	$2,254	2021 equivalent price:	$42,417
1940 Packard One Eighty Club Sedan:	$2,248	"	$42,304
1940 Packard One Twenty Touring Sedan:	$1,166	"	$21,942

Packard advertised that it held title to many automotive "firsts." Accordingly, the air-conditioning brochure on the first page of this chapter stated:

"ANOTHER PACKARD 'FIRST,'" Air-Conditioning now takes its place with the scores of other pioneering achievements which have earned Packard its reputation for engineering leadership…Now Packard becomes the first to offer the most modern of all comfort features—genuine Air-Conditioning!

Packard air-conditioning passengers in 1940 experienced the first refrigerated and dehumidified climate control, the first effective relief from airborne pollen and dust, and the first dedicated system that insulated the cabin against extreme temperatures.

Packard's *Innovative Years* produced the first system in 1940 for passengers who selected

FACTORY AIR:

COOL CARS IN

COOLER COMFORT.

Chapter 2

1941 Packard Factory Air-Conditioning

P ackard announced,

the Class of '41...smarter, sleeker, more luxurious

...successor to a great success!

That is the enviable position of the 1941 Packard, for the car that preceded it proved one of the most popular in all of Packard history. Packard increased its fast-growing family of nearly half a million loyal owners by some 20% in a single year!

Their sales claim proved valid for its best sales year ever:

        1940 PACKARD   SALES TOTALED 98,000
        1941 PACKARD   SALES TOTALED 72,855
        1942 PACKARD   SALES TOTALED 33,776*

*Limited 1942 production through February 9, 1942, per WWII regulations.

The *Packard One Ten and One Twenty for 1941* prestige brochure cover sported a One Ten Touring Sedan.

In the first-year offering of one of seven new "Multi-Tone" color combinations, Packard displayed the Silver French Gray Metallic over Barola Blue Metallic. (Fig. 102)

Figure 102. Packard One Ten and One Twenty for 1941 brochure cover   AACA Library

~ 52 ~

Factory Air: Cool Cars in Cooler Comfort 53

For 1941, all bodies were entirely new; subtle styling changes freshened the front end yet retained "That Packard Look." The brochure image displayed a multi-tone color selection of Silver French Gray Metallic over Laguna Maroon Metallic. (Figs. 103–106)

The front lines of beauty!...distinctly new but still distinctively Packard! Headlamps faired into fenders and surmounted by streamlined parking lamps add to rakish, clean-cut appearance-- enlivened by sparkling cooling grilles.

Figure 103. 1941 PACKARD grille with faired-in headlamps inset into fenders   Benson Ford Research Center, The Henry Ford

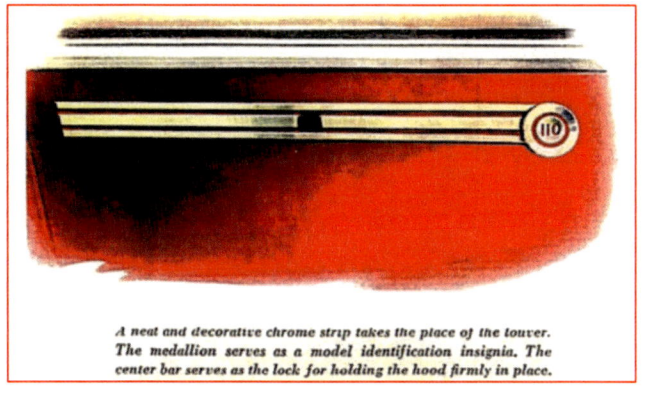

A neat and decorative chrome strip takes the place of the louver. The medallion serves as a model identification insignia. The center bar serves as the lock for holding the hood firmly in place.

Figure 104. 1941 Packard replaced louvers on the hood with a chromed strip and a model number inside a medallion   Benson Ford Research Center, The Henry Ford

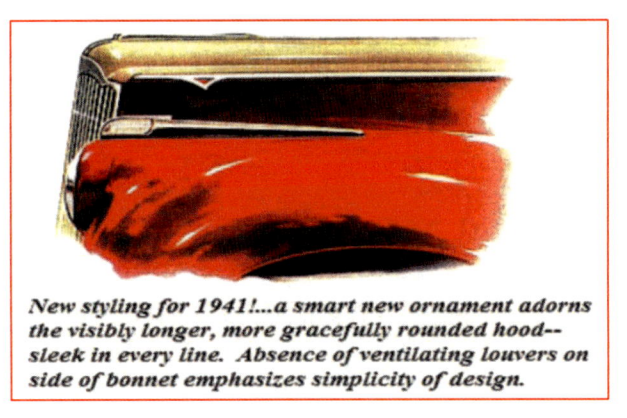

New styling for 1941!...a smart new ornament adorns the visibly longer, more gracefully rounded hood-- sleek in every line. Absence of ventilating louvers on side of bonnet emphasizes simplicity of design.

Figure 105. 1941 Packard new hood ornament, rounded hood and chrome strips replaced louvers   Benson Ford Research Center, The Henry Ford

Suggestive of the fleet performance that awaits you now in this Packard are sparkling speedline stripes on fenders. These give emphasis to the increased length and sleekness of the 1941 One-Ten

Figure 106. 1941 Packard "sparkling speedline stripes" accent rear fender   Benson Ford Research Center, The Henry Ford

The 1941 brochure promoted Packard Air-Conditioning for the second year. (Figs. 107, 108)

the Class of '41

... for the most welcome comfort improvement in years!

PACKARD AIR CONDITIONING

No longer need summer heat affect your motoring comfort. For in any 1941 Packard you may have genuine *Air Conditioning* that refrigerates, filters, dehumidifies and circulates the air you breathe. An extra, of course, but in the opinion of many, a feature worth a fortune when the mercury starts to roar. Do not confuse it with ordinary "fresh air" ventilating or heating systems.

Figure 107. 1941 Packard brochure   AACA Library

54   Chapter 2: 1941 Packard Factory Air-Conditioning

the Class of '41   **TURN ON THE COLD!**

Figure 108. 1941 Packard brochure illustrated passenger's air-conditioning benefits   AACA Library

Packard promoted its passenger air-conditioning based upon six primary benefits: (Fig. 108)

1. *Turn on the cold!* This motto underscored the establishment of the first exclusive air-conditioning system for motorcars. The driver adjusted the control on the dash while her friend awaited the cool air.
2. *Takes no more room than a large suitcase!* The 10-inch-deep evaporator compared with the size of a large suitcase, showing ample room for luggage.
3. *Packard AIR-CONDITIONING has a cooling capacity...* Packard rephrased its advertised cooling capacity rating from 1940 to 1941, stating it "has a cooling capacity equivalent to the melting of 2.5-tons of ice every 24 hours. Cools the entire car at high speeds or low." The *Packard Data Book for 1941*, page 78, stated a two-ton capacity at sixty mph. (Fig. 114)

Author note: For a "ton" cooling rating comparison that a customer would understand, a typical ranch-style home requires two and one half to three-tons of air-conditioning. The powerful Packard system produced nearly as much cooling as a typical home required. However, according to a Packard expert, it exaggerated the claim that it cooled sufficiently at low speeds.

4. *An oasis for 6!* The Packard brochure illustrated the cool air circulation within the car by "a powerful electric blower." It emphasized that it produced no drafts, dampness, or dryness.
5. *Where to go for hay fever?* The system filtered out pollen and dust for hay fever sufferers.
6. *No dust–No noise–No insects!* The air conditioner permitted dust-free, insect-free, quiet driving.

Packard introduced the first automotive mechanical refrigeration at the 1940 Chicago Auto Show in November 1939.

Unusual for Packard, the *Packard 1940 Data Book*, dated September 1939, did not print any product data in the Weather Conditioner's initial year, nor did it update it with supplementary data during the year. (Fig. 109)

The unavailable yet essential product information reflected poorly upon management because of the need for reference material. Dealership salespeople depended upon factory-based printed literature for more in-depth discussions with potential customers. Such a situation proved especially true when introducing a technologically important and expensive innovation, such as the first factory air-conditioning.

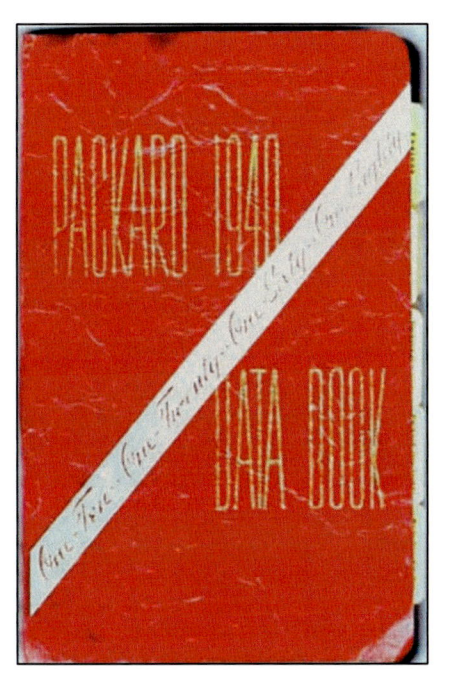

Figure 109. Packard 1940 Data Book cover   Joe Block

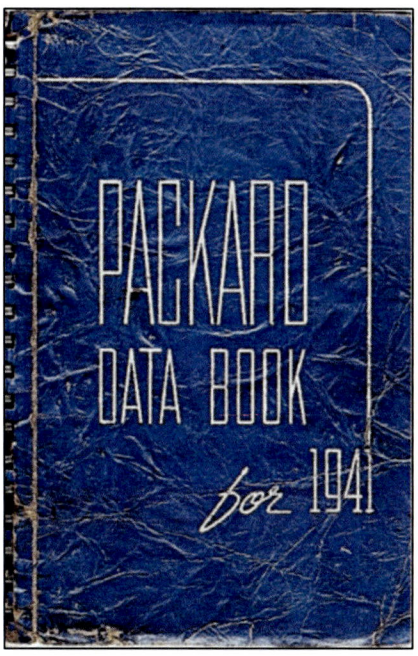

Figure 110. Packard Data Book for 1941 cover   Andrew Hook via Steven Kelley

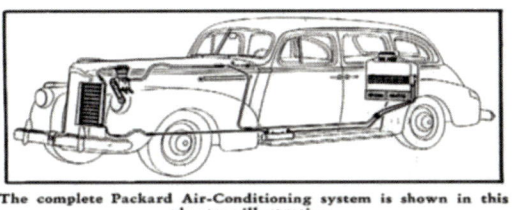

Figure 111. Packard Data Book for 1941, p. 75 AACA Library

Later, the *Packard Data Book for 1941*, dated September 1940, printed air-conditioning specifics for the 1940 and 1941 models on pages 75-80. (Figs. 110–116)

The *Packard Data Book for 1941* referenced the 1940 Packard air-conditioning introduction as another "First" for American motorists. (Fig. 111)

It emphasized, "Packard Air-Conditioning involves cooling by mechanical refrigeration," rather than just other manufacturers' descriptions of normal ventilation inside the car.

It stated that it "operates on the same…efficient principal…in millions of the finest electric household refrigerators." The Packard owner adjusts a switch on the instrument panel that "regulates the speed of the blower…and the volume of cooled and filtered air circulated in the body of the car."

The System in Operation: A phantom illustration of a 1941 Packard Touring Sedan illustrated the air-conditioning system components: a compressor mounted on the passenger side engine, driven by a pulley and "V" belt off the water pump, connected to a condenser in front of the radiator. (Fig. 111)

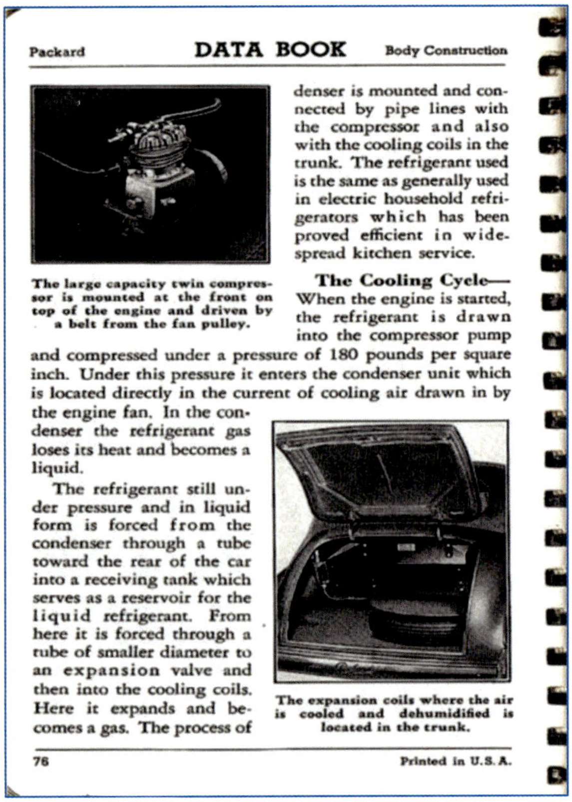

Figure 112. Packard Data Book for 1941, p. 76 AACA Library

Figure 113. Packard Data Book for 1941, p. 77 AACA Library

The condenser connected to both the compressor and the evaporator cooling coils by copper tubing "pipelines." It used a refrigerant (Freon-12) typically used in household refrigerators. (Fig. 112)

The Cooling Cycle: After starting the engine, the compressor drew refrigerant, compressed it to 180 psi, and pumped it into the condenser in front of the radiator. "Cooling air, drawn in by the engine fan," cooled the pressurized Freon to a liquid state and passed it through copper tubing to a reservoir tank known as a receiver. The pressurized liquid Freon passed on to an expansion valve attached to the evaporator cooling coils inside the trunk-mounted evaporator.

The expansion valve expanded the gaseous Freon as the cooling coils absorbed heat from the passenger cabin and cooled the air. The heated gaseous refrigerant returned to the compressor to restart the cooling cycle.

Author note: Packard oddly described the evaporator cooling coils as the "expansion coils where the air is cooled and dehumidified..." (Fig. 112, lower image)

Cool Air Circulation: An electrical "sirocco-type blower," known as a squirrel-cage blower, drew the passenger cabin air over the evaporator cooling coils and circulated cooled air throughout the car. (Fig. 113)

Packard stated its system differed from household refrigerators of the 1940s because "Cool air is *forced* to circulate throughout the whole car interior."

The new term, "cool air outlet," changed from the 1940 term, "blower grille," with an added 1941 louver adjustment handle, blew cooled air upward as it followed the headliner. It stated, "Because the cold air is heavier than warm air, it descends toward the floor near the front. It is then pulled back under the rear seat and into the cooling system again by the suction action of the blower." The blower circulated the cooled air.

Dehumidification: Moisture-laden warm air passed over the cooling coils, condensed, and drained its water to the ground. The dehumidified air blew out the cool air outlet. Another reference to the 1940s-era stated, "The cooling unit defrosts automatically."

Air Filtration: Warm air entered under the rear seat through filters discussed in the 1940 Packard chapter.

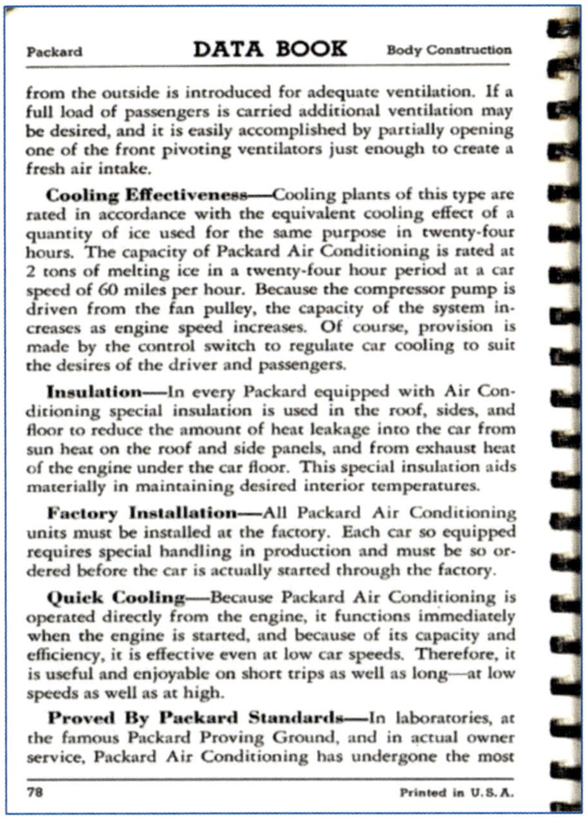

Packard **DATA BOOK** Body Construction

from the outside is introduced for adequate ventilation. If a full load of passengers is carried additional ventilation may be desired, and it is easily accomplished by partially opening one of the front pivoting ventilators just enough to create a fresh air intake.

Cooling Effectiveness—Cooling plants of this type are rated in accordance with the equivalent cooling effect of a quantity of ice used for the same purpose in twenty-four hours. The capacity of Packard Air Conditioning is rated at 2 tons of melting ice in a twenty-four hour period at a car speed of 60 miles per hour. Because the compressor pump is driven from the fan pulley, the capacity of the system increases as engine speed increases. Of course, provision is made by the control switch to regulate car cooling to suit the desires of the driver and passengers.

Insulation—In every Packard equipped with Air Conditioning special insulation is used in the roof, sides, and floor to reduce the amount of heat leakage into the car from sun heat on the roof and side panels, and from exhaust heat of the engine under the car floor. This special insulation aids materially in maintaining desired interior temperatures.

Factory Installation—All Packard Air Conditioning units must be installed at the factory. Each car so equipped requires special handling in production and must be so ordered before the car is actually started through the factory.

Quick Cooling—Because Packard Air Conditioning is operated directly from the engine, it functions immediately when the engine is started, and because of its capacity and efficiency, it is effective even at low car speeds. Therefore, it is useful and enjoyable on short trips as well as long—at low speeds as well as at high.

Proved By Packard Standards—In laboratories, at the famous Packard Proving Ground, and in actual owner service, Packard Air Conditioning has undergone the most

78 Printed in U.S.A.

Figure 114. Packard Data Book for 1941, p. 78
AACA Library

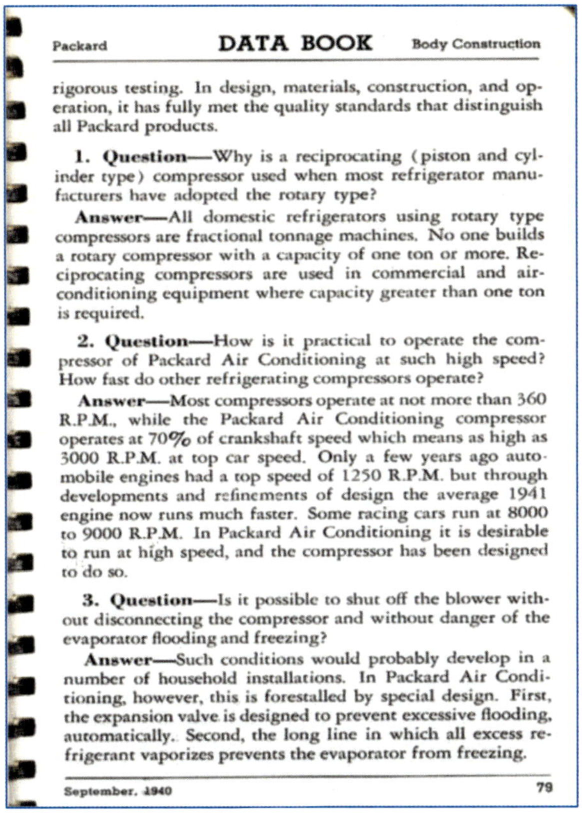

Packard **DATA BOOK** Body Construction

rigorous testing. In design, materials, construction, and operation, it has fully met the quality standards that distinguish all Packard products.

1. Question—Why is a reciprocating (piston and cylinder type) compressor used when most refrigerator manufacturers have adopted the rotary type?

Answer—All domestic refrigerators using rotary type compressors are fractional tonnage machines. No one builds a rotary compressor with a capacity of one ton or more. Reciprocating compressors are used in commercial and air-conditioning equipment where capacity greater than one ton is required.

2. Question—How is it practical to operate the compressor of Packard Air Conditioning at such high speed? How fast do other refrigerating compressors operate?

Answer—Most compressors operate at not more than 360 R.P.M., while the Packard Air Conditioning compressor operates at 70% of crankshaft speed which means as high as 3000 R.P.M. at top car speed. Only a few years ago automobile engines had a top speed of 1250 R.P.M. but through developments and refinements of design the average 1941 engine now runs much faster. Some racing cars run at 8000 to 9000 R.P.M. In Packard Air Conditioning it is desirable to run at high speed, and the compressor has been designed to do so.

3. Question—Is it possible to shut off the blower without disconnecting the compressor and without danger of the evaporator flooding and freezing?

Answer—Such conditions would probably develop in a number of household installations. In Packard Air Conditioning, however, this is forestalled by special design. First, the expansion valve is designed to prevent excessive flooding, automatically. Second, the long line in which all excess refrigerant vaporizes prevents the evaporator from freezing.

September, 1940 79

Figure 115. Packard Data Book for 1941, p. 79
AACA Library

Packard observed that adequate ventilation resulted from body leaks; however, a "front pivoting ventilator," or wind wing, provided extra ventilation when fully occupied.

Cooling Effectiveness: Cooling rates quantified the amount of ice melted over twenty-four hours. "The capacity of Packard Air-Conditioning is rated at 2-tons of melting ice in a twenty-four-hour period at a car speed of 60 miles per hour." The cooling capacity increased with engine speed; the driver regulated the cabin temperature by the fan speeds. (Fig. 114)

Insulation: "Special insulation is used in the roof, sides, and floor" to reduce heat incursion from the sun and engine. (Refer to 1940 Packard chapter insulation and, here, Figs. 16-19 for standard insulation data)

Factory Installation: Special-ordered air-conditioned Packards required additional insulation during assembly. Obfuscated, Bishop & Babcock installed the systems in Cleveland, Ohio.

Quick Cooling: The Packard air conditioner cooled immediately when the engine started and claimed that its capacity proved "effective even at low car speeds."

Author note: Although the control panel offered an "off" position, the compressor ran continuously, potentially leading to a mechanical breakdown far sooner than the expected design lifespan.

Far into the future, the reintroduced 1953 air-conditioning technology remained similar to the prewar systems. The 1954 Cadillac, Oldsmobile, Buick, Pontiac, and Nash cars and the 1955 Packard cars achieved a technological milestone by introducing an electromagnetic clutch that disengaged the compressor when the air-conditioning turned off, extending its compressor service life and saving gasoline.

Proved by Packard Standards: Laboratories, proving grounds, and designs ensured Packard's quality standards.

Question 1 Answer: Packard selected a reciprocating compressor rather than a rotary type because it produced more than a one-ton cooling capacity. (Fig. 115)

Question 2 Answer: Packard's compressor operated at 70% of crankshaft speed, up to a speed of 3,000 rpm.

Question 3 Answer: While "off," the expansion valve prevented flooding, and the low vaporization temperature prevented the evaporator from freezing.

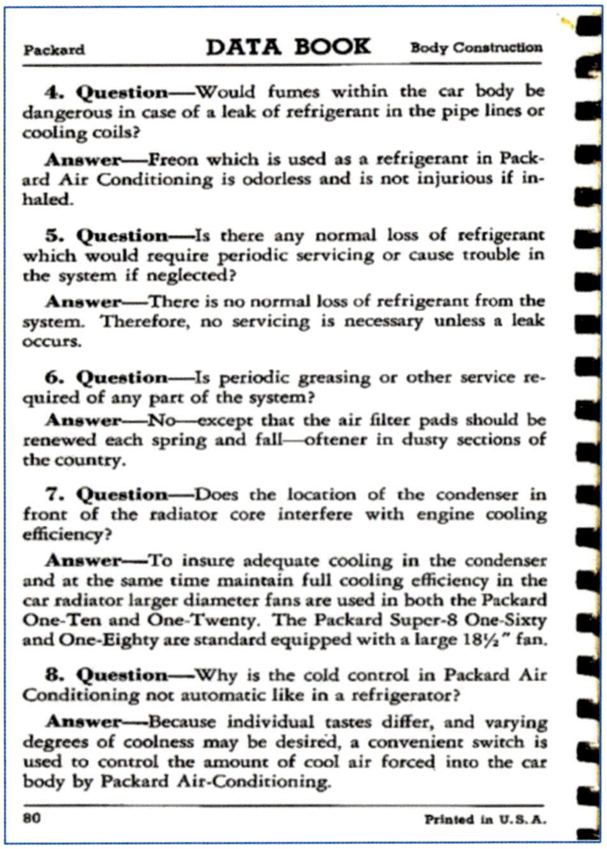

Packard **DATA BOOK** Body Construction

4. Question—Would fumes within the car body be dangerous in case of a leak of refrigerant in the pipe lines or cooling coils?

Answer—Freon which is used as a refrigerant in Packard Air Conditioning is odorless and is not injurious if inhaled.

5. Question—Is there any normal loss of refrigerant which would require periodic servicing or cause trouble in the system if neglected?

Answer—There is no normal loss of refrigerant from the system. Therefore, no servicing is necessary unless a leak occurs.

6. Question—Is periodic greasing or other service required of any part of the system?

Answer—No—except that the air filter pads should be renewed each spring and fall—oftener in dusty sections of the country.

7. Question—Does the location of the condenser in front of the radiator core interfere with engine cooling efficiency?

Answer—To insure adequate cooling in the condenser and at the same time maintain full cooling efficiency in the car radiator larger diameter fans are used in both the Packard One-Ten and One-Twenty. The Packard Super-8 One-Sixty and One-Eighty are standard equipped with a large 18½" fan.

8. Question—Why is the cold control in Packard Air Conditioning not automatic like in a refrigerator?

Answer—Because individual tastes differ, and varying degrees of coolness may be desired, a convenient switch is used to control the amount of cool air forced into the car body by Packard Air-Conditioning.

80 Printed in U.S.A.

Figure 116. Packard Data Book for 1941, p. 80
AACA Library

Question 4 Answer: A Freon leak inside the car was odorless and non-injurious. (Fig. 116)

Question 5 Answer: Service would only be necessary on the system if a refrigerant leak occurred.

Question 6 Answer: The system required no periodic service, except for changing the air-conditioning air filters twice yearly.

Question 7 Answer: The Packard One Sixty and One Eighty Series featured a standard eighteen and one-half-inch fan. The One Ten and the One Twenty Series cars with air-conditioning installed the same fan to adequately cool their radiators.

Question 8 Answer: Packard controlled the interior temperature by the driver selection of different fan speeds as needed, rather than using an automatic setting as in refrigerators.

Author note: The 1964 Cadillac introduced *Climate Control* air-conditioning technology that maintained a dialed cabin temperature number, summer or winter.

the Class of '41

1941 Packard owners regarded their *FACTORY AIR*, specially equipped cars as

COOL CARS IN

COOLER COMFORT

Separately, the *Packard Data Book for 1941* identified the extensively applied standard insulation by description and illustration. (Figs. 117–120)

After many years, it completed an insulation system that stifled body noises while protecting passengers from outside heat and cold. (Fig. 118) The insulation process comprised thirteen distinctive materials placed strategically throughout the car.

Roof: A "special material" cemented under the roof insulated the cabin from exterior heat and cold. A steel spring bow held up the insulation and prevented a drumming effect.

Panels: Rear quarter panels required insulation with the same material as the roof, while the housings over each wheel had a sprayed-on plastic insulating compound.

Doors: The door panel's semi-flat surfaces used another spray-on, a "thick asphaltic compound which never hardens or deteriorates."

Cowl: Heavy jute covered the cowl's top and sides. Under-dash areas required a full inch of a "special" material, covered by a "long-wearing, scuff-proof leatherette insulating board." Floor areas of the toe boards prevented heat and noise incursion by two thick layers of different insulations. (Fig. 118)

Floors: Outside noise, road noise, heat, and cold conditions required double insulations of sound-deadening asphalt-impregnated felt, covered with a heavy jute padding layer cemented under the carpet or mat.

Trunk: The trunk insulation prevented incursions of noise, temperature extremes, and moisture. The insulation process attached asphalt-impregnated felt to the trunk lid, while the sidewalls and floor received a heavy coating of another sprayed-on material.

After assembly, a high-pressure water bath tested each Packard body. In the event of a leak and subsequent repairs, another water bath confirmed its water resistance.

Figure 117. Packard Data Book for 1941, p. 68
AACA Library

Figure 118. Packard Data Book for 1941, p. 69
AACA Library

omprehensive body insulation applications included the underside of the roof (Figs. 119); the top and sides of the cowl (Area 2, not the first (3) typo listed in Fig. 19); front and rear doors (Area 3); the inner surface of doors (Area 4); the floor panels (Area 5); rear quarter panels (Area 6); trunk interior (Area 7); the underside of the trunk lid (Area 8).

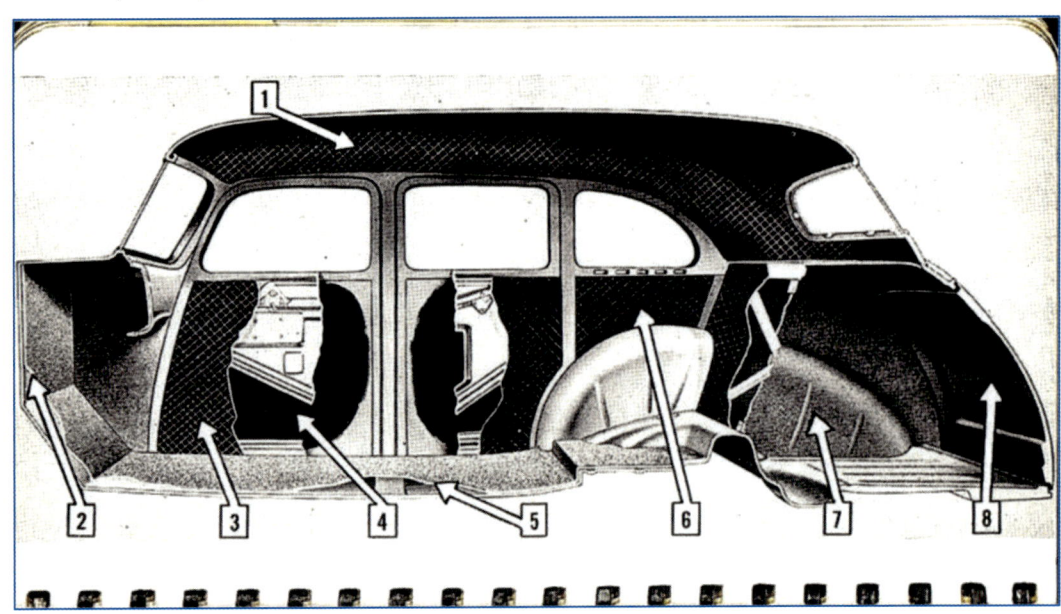

Figure 119. Packard Data Book for 1941, p. 70
AACA Library

COMPREHENSIVE BODY INSULATION

(1) A dense, sound deadening layer of special material permanently cemented against the under side of the steel roof insulates against sound, heat and cold.

(3) The top and sides of the cowl are lined with a thick pad of heavy jute, the dash is insulated with a full inch of special material and the toe board is covered with two layers of different insulating materials to keep out engine sound and heat.

(3) A panel of special insulating board is used beneath the trim material on both the front and rear doors.

(4) The inner surface of the outside panel of all doors is heavily sprayed with a thick viscous asphaltic compound that never hardens or deteriorates.

(5) Noise which might originate in the floor panels is eliminated and excess heat and cold are kept out of the body by a sound-deadening asphalt felt pad plus a heavy layer of jute padding cemented to the under side of the carpet or rubber mat.

(6) Rear quarter panels are insulated with a dense layer of insulating material similar to that used in the roof.

(7) The interior of the trunk is heavily insulated by a heavy coating of special sprayed-on compound.

(8) The under surface of the lid is covered with a thick layer of asphalt impregnated felt and a water seal of hollow rubber effectively seals the trunk compartment against rain and dust.

Figure 120. Packard Data Book for 1941, p. 71
AACA Library

Author note: Besides the standard body insulation, the specially ordered air-conditioned Packard also received foil-backed, corrugated paper insulation for the roof, sides, and floors during assembly.

Refer to the 1940 Packard chapter for images of the insulation's appearance and locations in the car.

Factory Air: Cool Cars in Cooler Comfort 61

The *Information on the 1941 Packard,* September 16, 1940, published the new model information and accessory pricing. The suggested price for the "Air-conditioning, without heater (All models except convertibles and station wagons)," listed for $275. (Figs. 121, 122)

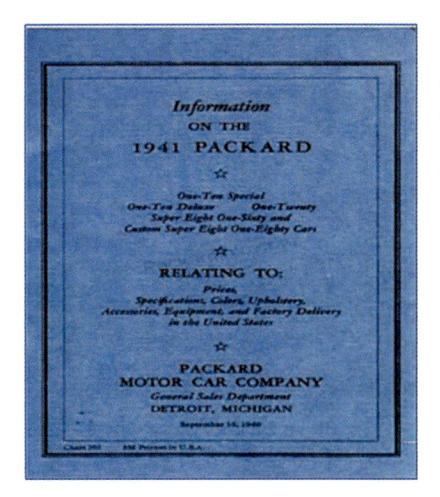

Figure 121. Information on the 1941 Packard, September 16, 1940 PackardInfo.com   David Flack

Figure 122. Packard accessory price list for 1941, September 16, 1940 Dwight Heinmuller

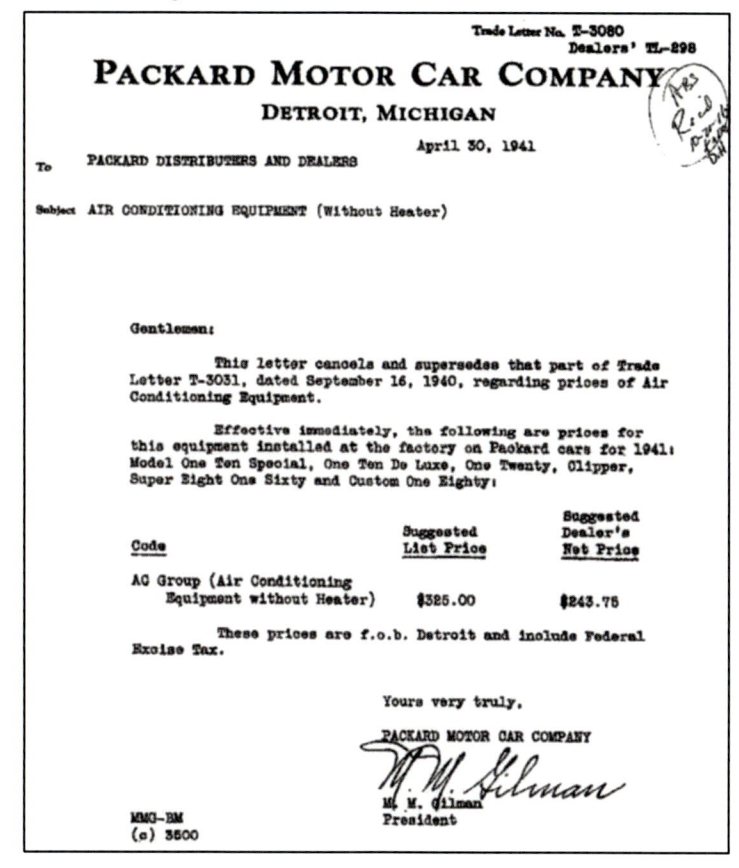

Figure 123. 1941 Packard air-conditioning price increase letter, April 30, 1941   Dwight Heinmuller

President M. M. Gilman's letter of April 30, 1941, raised the 1941 air-conditioning Suggested List Price for the "AC Group (Air-Conditioning Equipment without Heater)" to $325 with a Suggested Dealer's Net Price of $243.75. (Fig. 123)

The CPI indicates the inflation rate in current equivalent dollars for the following air-conditioning accessories:

1940 Packard air conditioner price, including heater	November 1939	$274	January 2021	$5,119
1940 Packard air conditioner price, heater deleted	April 1940	$275	January 2021	$5,138
1941 Packard air conditioner price, heater deleted	September 1940	$275	January 2021	$5,138
1941 Packard air conditioner price, heater deleted	April 1941	$325	January 2021	$5,945
1942 Packard air conditioner price, heater deleted	January 1942	$325	January 2021	$5,414

Author note: The $274, $275, or $325 air-conditioning accessory price appears inconsequential until viewed in 2021 equivalent dollars. One can empathize with a customer who purchased a new 1941 Packard Clipper

62   Chapter 2: 1941 Packard Factory Air-Conditioning

for $1375, for example, and could not stretch his budget to add another $325 to the purchase price. It would have added 24% to the purchase price, the equivalent of $5,945 in today's dollars. Headlines announced,

"Brimming with Beauty Bursting with News!"

The following month, in October 1940, Packard placed advertising in *The Saturday Evening Post, Collier's, Life, Newsweek,* and *Time,* with a full-color two-page ad for its restyled, newest model, the Packard One Ten Deluxe Sedan. (Figs. 124, 125)

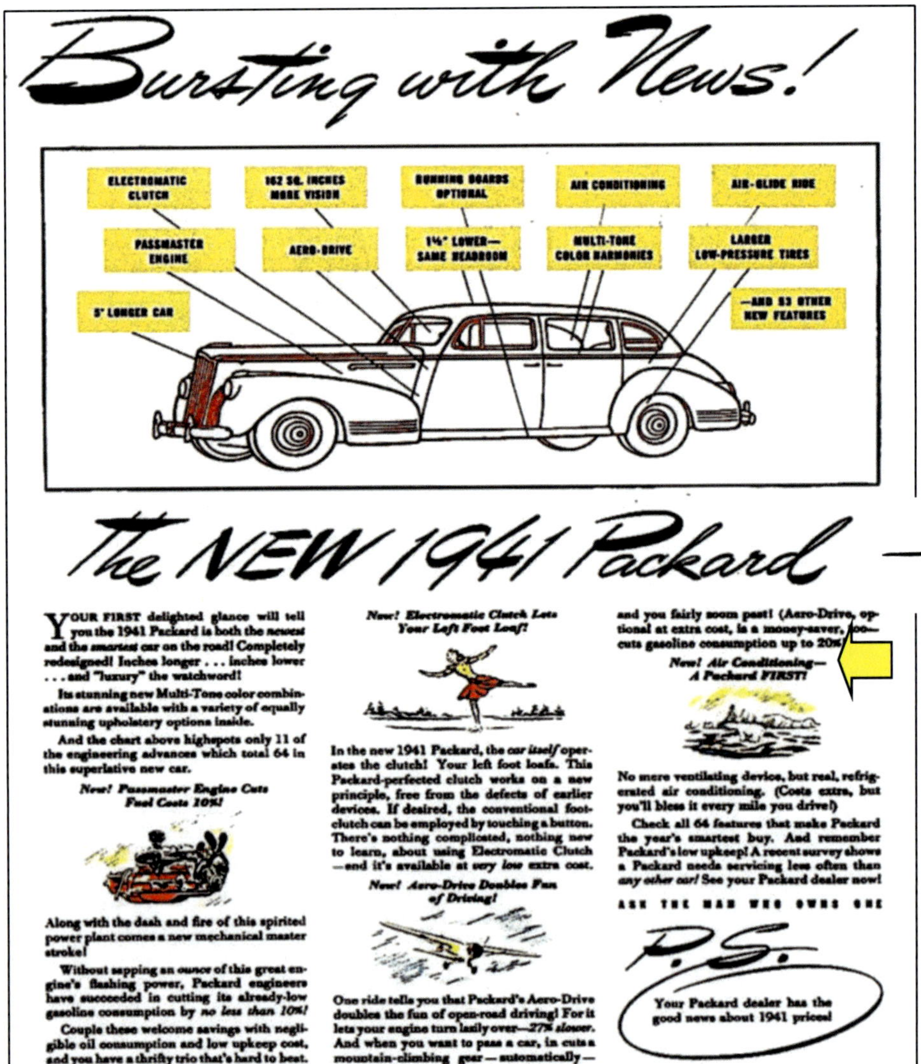

Figure 124. 1941 Packard One Ten Deluxe Sedan ad
Saturday Evening Post, October 1940
Regress Press   Jeff Steidle

The Packard ad displayed 12 of its 64 new selling points for the new 1941 models. Its slogan, "the Class of '41," represented a play on the word "class." (Fig. 125)

Hidden in the advertising copy, displayed almost as an afterthought, the ad promoted Packard air-conditioning for 1941. (Figs. 125 and yellow arrow, 126)

Figure 125. 1941 Packard One Ten Deluxe Sedan ad
Saturday Evening Post, October 1940   Regress Press   Jeff Steidle

Factory Air: Cool Cars in Cooler Comfort 63

An enlarged view of the Packard ad announced:
New! Air-Conditioning—A Packard First!
that featured a polar bear scene, symbolizing the "real, refrigerated air-conditioning" interior climate for its 1941 model. (Fig. 126)

At the bottom of the ad, the famous Packard slogan appeared:

ASK THE MAN WHO OWNS ONE.

Automotive Industries, October 1, 1940, announced the new 1941 Packard information: (Fig. 127)

Figure 126. 1941 Packard ad for air-conditioning
Saturday Evening Post, October 1940
Regress Press   Jeff Steidle

> **298**
>
> **P**ACKARD this year lays emphasis on new exterior and interior decorative treatment of its cars, including such items as two-tone color combinations, a wide variety of trim combinations in contrasting color and texture of upholstery cloth, or contrasting leather edging or colored leather piping; chrome belt moldings and window reveals, and chrome and plastic instrument panels.

Figure 127. 1941 Packard color and upholstery selections   Automotive Industries   Wesley Boyer

The article continued with a description of Packard's air-conditioning: (Fig. 128)

> Full mechanical refrigeration, first announced last year, is now available at extra cost on every model in the line. It has been simplified by removing the heater from the rear compartment, thus providing a simple cooling installation for regions where heating is not required. For operation in colder regions, the car can be fitted with the now conventional under-seat heater or the less-expensive dash-mounted heater.

Figure 128. 1941 Packard air-conditioning availability now in every model   Automotive Industries   Wesley Boyer

64 Chapter 2: 1941 Packard Factory Air-Conditioning

For 1941, Packard continued its air-conditioning evaporator placement in the forward area of the trunk. (Fig. 129)

The *Automotive Industries* article October 1, 1940, also published views of a 1941 Packard One Twenty Touring Sedan and one with a phantom view of Packard's air-conditioning componentry. (Figs. 130, 131)

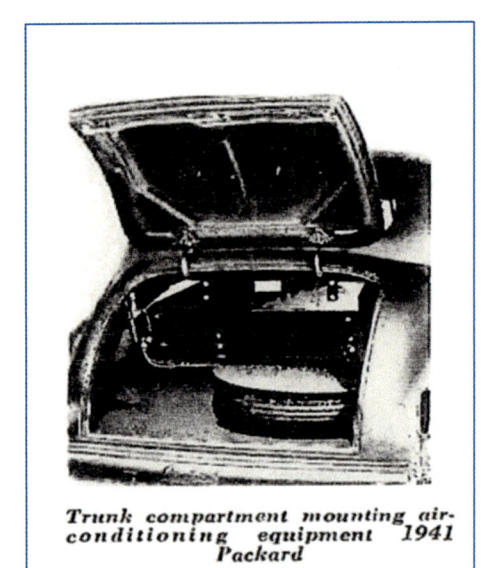

Figure 129. 1941 Packard air-conditioning evaporator in trunk
Automotive Industries   Wesley Boyer

Figure 130. 1941 Packard One Twenty Touring Sedan
Automotive Industries   Wesley Boyer

Figure 131. 1941 Packard air-conditioning componentry placement
Automotive Industries   Wesley Boyer

The *Automotive Industries* article of November 15, 1939, supplied an image of the air-conditioning compressor used in the 1940-1942 Packard. (Fig. 132)

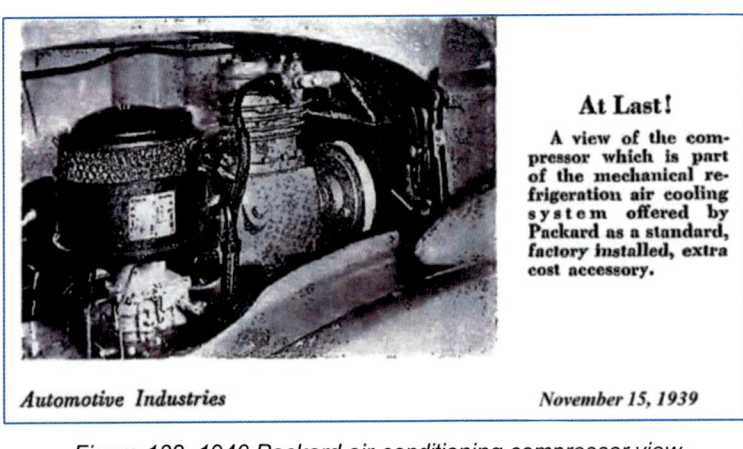

Figure 132. 1940 Packard air-conditioning compressor view
Automotive Industries   Wesley Boyer

The *Packard News Service* released information about their air-conditioning system: (Fig. 133)

PACKARD NEWS SERVICE

One of the great comfort innovations of the decade is the Packard air conditioner, introduced to the public for the first time in 1940. It found instant acceptance beyond the anticipation of Packard designers who introduced it without undue fanfare.

Offered again this year, as optional equipment, the mechanically refrigerated cooling system has undergone definite improvements in design and cooling effectiveness.

As the first authentic refrigerating unit of its kind offered for passenger car comfort, some idea of its utility may be gained from the rating, and the method of rating, its capacity. Since the Packard air-conditioner operates on the identical principle of popular household refrigerators, Packard designers have adopted a similar method of capacity rating. It is judged in accordance with the cooling effectiveness of a like quantity of natural ice used for the same purpose in a twenty-four-hour period.

Under this rating the Packard air conditioner is capable of producing $1\frac{1}{2}$ to 2 tons of ice, melting in each twenty-four-hour period at an engine speed of sixty miles per hour.

The system requires no mechanical knowledge to operate. A separate compressor mounted on the engine and driven from the fan belt provides power to the refrigerating coils and unit in the rear compartment of the car. Adjustable controls are provided on a simple switch located on the instrument panel. Quick, effective and continuous cooling is possible under all conditions.

* * * * *

Figure 133. 1941 Packard News Service announcement of Packard air conditioner
PackardInfo.com   Kevin Waltman

66   Chapter 2: 1941 Packard Factory Air-Conditioning

November 1940 marked the presidential election race between the two-time president, Franklin D. Roosevelt (FDR), and Republican candidate Wendell Wilke. The flamboyant San Francisco Packard dealer, Earle C. Anthony, projected the current political theme into an advertisement for the 1941 Packard One Ten Deluxe Touring Sedan.

It read, *Packard—A clean sweep in every state!* It also noted, *Real Air-Conditioning—A Packard First!* (Fig. 134 and yellow arrow)

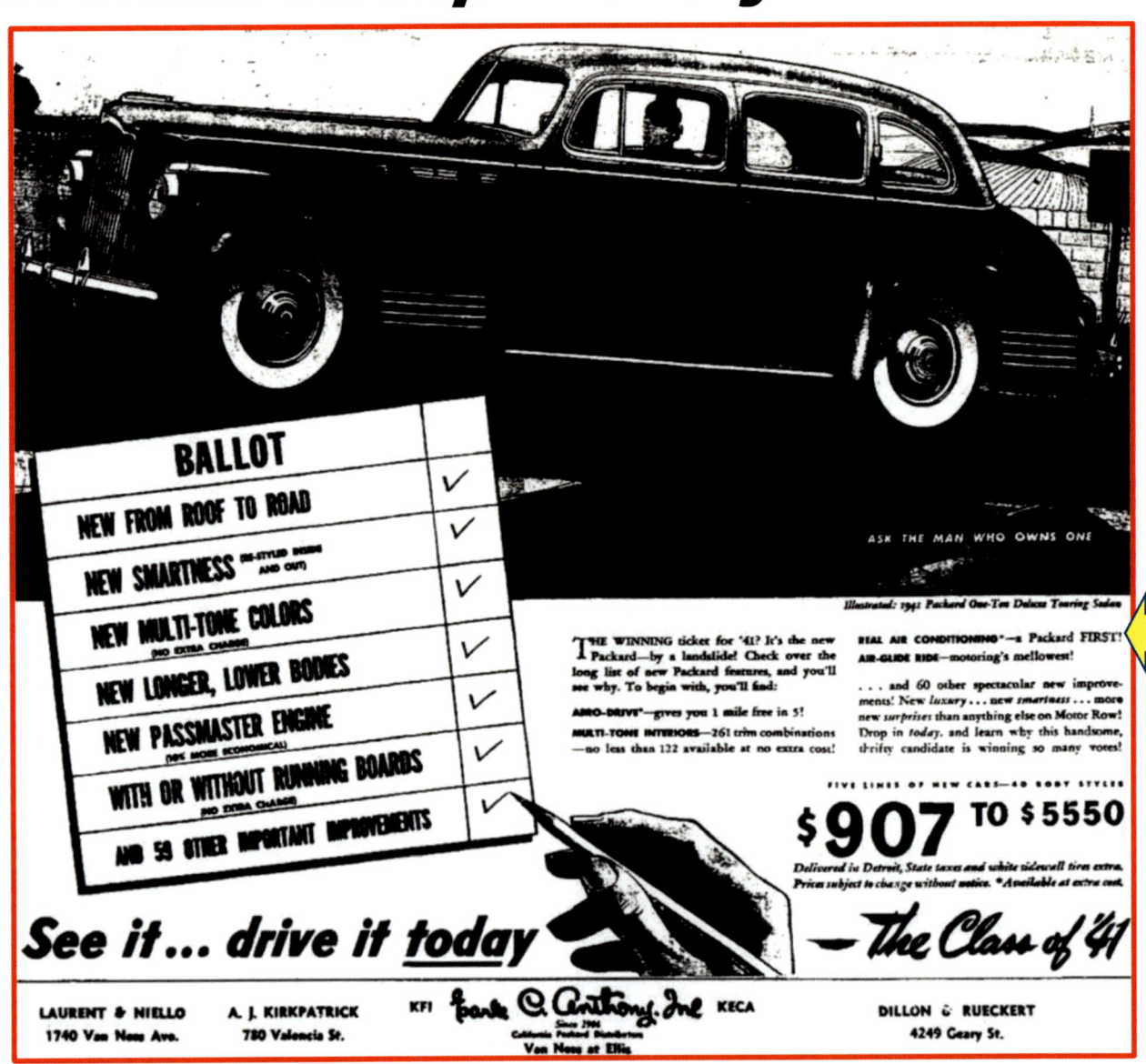

Figure 134. 1941 Packard One Ten Deluxe Touring Sedan ad with political theme, by Packard dealer Earle C. Anthony
San Francisco Chronicle, November 10, 1940   San Francisco Library

Factory Air: Cool Cars in Cooler Comfort 67

A seldom encountered 1941 Packard spiral-bound dealer album displayed a section devoted to its air-conditioning system. (Figs. 135–137) The *Genuine Air-Conditioning* heading in Figure 136 stated:

> Here is another Packard "first" to add to the long list of Packard contributions to motoring comfort and enjoyment—genuine air-conditioning.
>
> It filters, dehumidifies, and refrigerates the air you breathe while driving! Only after many months of designing, experimenting, and perfecting did Packard introduce the system, diagrammed here—but still far in advance of any other manufacturer. Packard Air-conditioning is now acclaimed (as) the most welcome comfort improvement in years by the literally hundreds of owners who have experienced its blessing.

Figure 135. 1941 Packard dealer album on display   Jack Swaney

Figure 136. 1941 Packard dealer album, air-conditioning section, upper page   Jack Swaney

The caption under the image of the louvered blower grille states, "An electric blower, behind the rear seat, directs cooled air over the heads of passengers—circulates it throughout the car."

The compressor's caption states, "This small but mighty compressor mounted on the engine has triple the capacity of the average household refrigerator unit."

An opened trunk caption states, "The evaporator assembly—where air is cooled—takes up no more room in the luggage compartment than a large suitcase…only 10" deep."

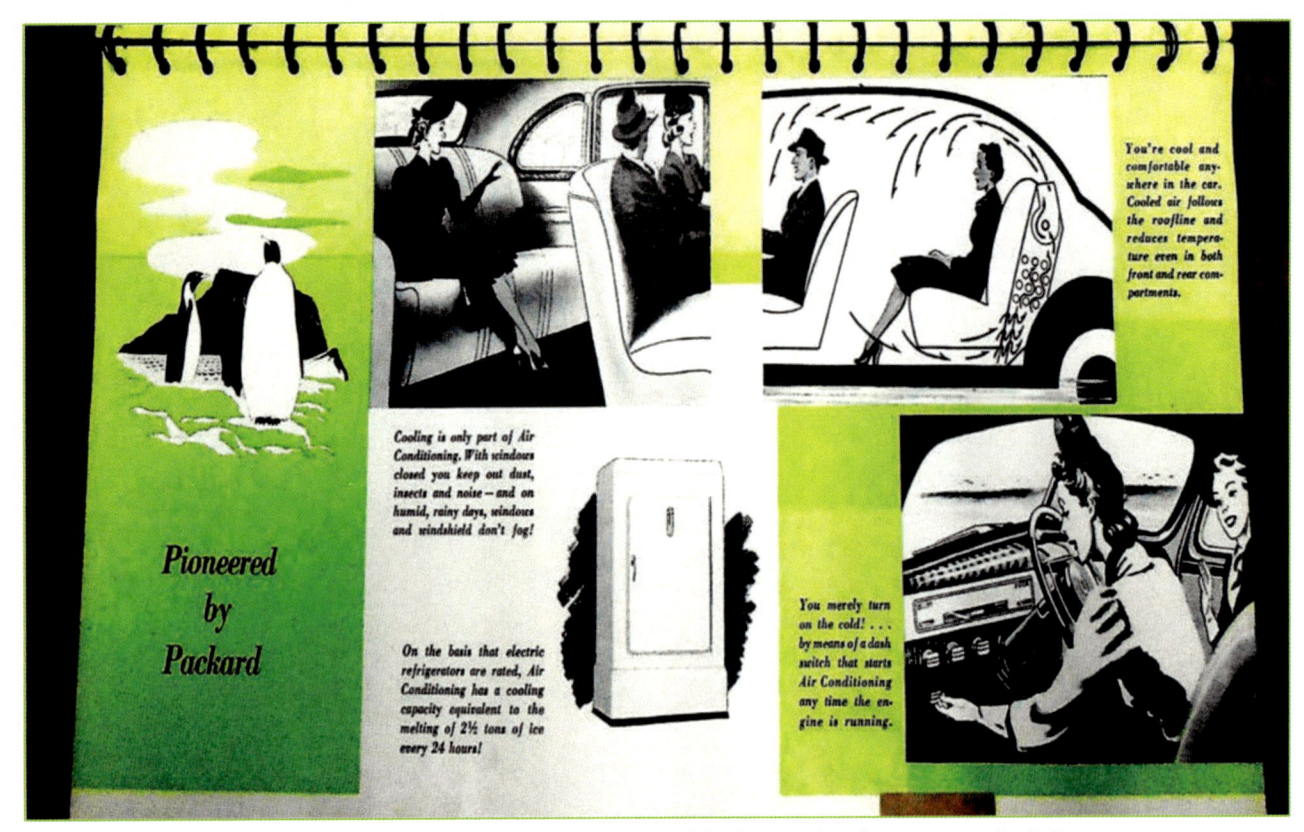

68   Chapter 2: 1941 Packard Factory Air-Conditioning

Figure 137. 1941 Packard dealer album, air-conditioning section, lower page   Jack Swaney

Figure 137. The caption underneath the penguin illustration states, "Pioneered by Packard."

The caption of the passenger interior states, "Cooling is only part of Air-conditioning. With windows closed, you keep out dirt, insects, and noise—and on humid, rainy days, windows and windshield don't fog!"

The caption of interior airflow states, "You're cool and comfortable anywhere in the car. Cooled air follows the roofline and reduces temperature even in both front and rear compartments."

The caption by the refrigerator states, "On the basis that electric refrigerators are rated, air-conditioning has a cooling capacity equivalent to the melting of two and one-half-tons of ice every twenty-four hours!"

The caption of the driver reaching for the control switch states, "You merely turn on the cold!…by means of a dash switch that starts air-conditioning any time the engine is running."

Factory Air: Cool Cars in Cooler Comfort 69

CELLARETTE: ULTRA-LUXURIOUS REFRESHMENT CENTER FOR 1941

EXCLUSIVELY AVAILABLE IN PACKARD'S ULTRA-LUXURIOUS, TOP-OF-THE-LINE PACKARD CUSTOM SUPER-8 ONE EIGHTY TOURING SEDAN. (Fig. 138)

Figure 138. 1941 Packard Cellarette air-conditioning accessory advertised in the Chicago Auto Show flyer, October 26 to November 3, 1940   Chicago Auto Show, Mitch Frumkin, Historian/Archivist of CAS

The New York National Automobile Show ran from October 12, 1940, to October 20, 1940, scheduled before the 1941 Chicago Auto Show dates of October 26, 1940, through November 3, 1940.

Mr. Guscha, a *PackardInfo.com Forum* researcher, presented an obscure article, "Mirrors of Motordom," by the Detroit Editor, A. H. Allen, for *Steel Magazine.* On October 21, 1940, his article commented that the 1941 Packard display at the New York National Automobile Show offered "one car equipped with complete 'Cellarette'– a small bar with ice-making machinery, glasses, and other necessary accouterments."

Could this specially equipped 1941 Packard One Eighty be the same car in both closely scheduled events? Could this same car be in the San Francisco auto show, presented in mid-November 1940?

Unfortunately, no logistical records or production data exist to satisfy the inquiry.

Packard introduced the Cellarette-optioned sedan in one model only, a 19th Series, 1941 Packard Custom Super-8 One Eighty Four-Door Touring Sedan, Model 1907, Body 1442, on a 138-inch wheelbase. The five-passenger One Eighty sedan weighed 4,350 pounds and sold for $2,632.

Unverified, Packard experts believe the Cellarette option's price equaled the air-conditioning system's $325 price, yet others have speculated a near-$1,000 price.

Unique in concept,
Unique in design,
Unique in its capabilities,
Unique in its presentation,

the 1941 Packard Cellarette hosted passengers with an exclusive, ultra-chic private car refreshment center.

The One Eighty Series differentiated itself from the similar One Sixty Series by its custom-made models, made available only by special order from independent coachbuilders in six of its eleven body styles.

The *Packard Data Book for 1941* described the Packard Cellarette's availability for "the 180 Four-Door Touring Sedan," construction, storage capacity, freezer accommodations, and ice-making capabilities in a "Supplement to page 80–Sept. 1940." (Fig. 139)

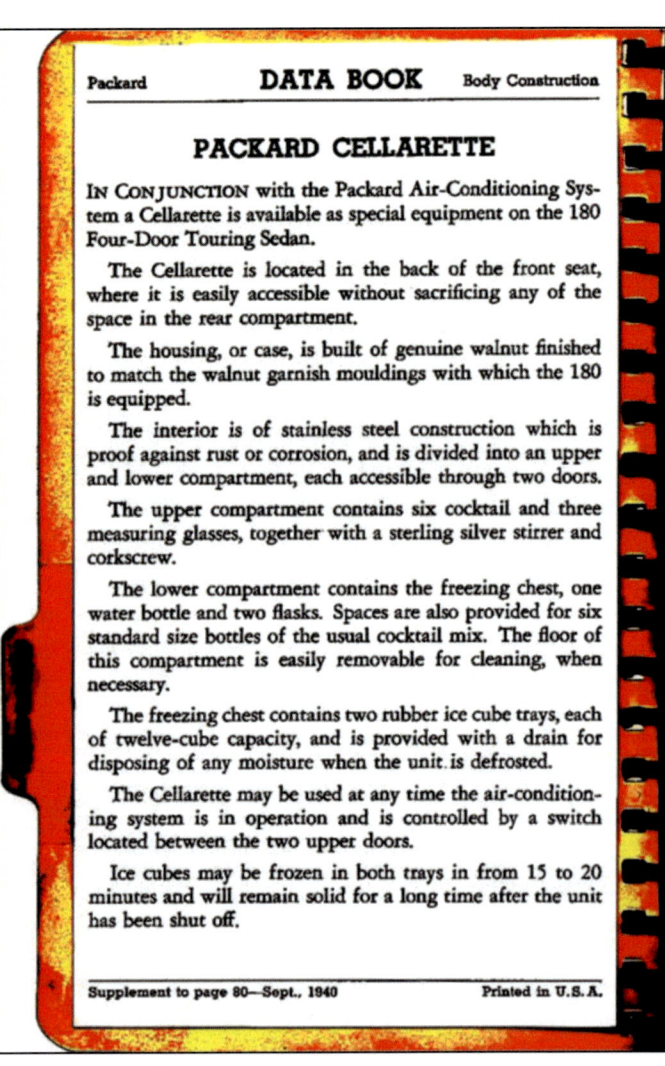

Figure 139. Packard Data Book for 1941 Cellarette Supplement to page 80   Andrew Hook via Steven Kelley

The *Packard Data Book for 1941* described interior elegance enhanced with pewter-inlaid walnut garnish moldings, exclusive to the Packard One Eighty models.

Complementing the luxurious interior theme, Packard custom-crafted a Cellarette, a sleek, genuine walnut cabinet inset into the Body 1442 Four-Door Touring Sedan front seatback. (Fig. 140)

Figure 140. 1941 Packard One Eighty, Body 1442, Cellarette cabinet   Dwight Heinmuller

The lower right drop-down cabinet door of the Cellarette contained one water bottle and two liquor flasks, in addition to six standard-size cocktail mixer bottles. (Fig. 141)

Figure 141. 1941 Packard One Eighty, Body 1442, Cellarette bottom right cabinet door opened for a view of the liquor storage area   Joe Block

Figure 142. 1941 Packard One Eighty, Body 1442, Cellarette cabinet doors opened with view of contents. Yellow arrow and inset display "CONDITIONER" or "CELLARETTE" switch.   Joe Block

Figure 143. 1941 Packard One Eighty, Body 1442, Cellarette cabinet doors opened with an ice cube tray ready for mixing drinks   Dwight Heinmuller

The Packard Cellarette Supplement page described the Cellarette's stainless steel interior and the removable floor guarded against rust or corrosion damage. (Figs. 139, 142)

The passenger moved a switch upward on the upper area of the console for the (Air) "Conditioner" or downward for the "Cellarette" freezer selection. (Fig. 142, yellow arrow, and inset)

Only one system worked at a time.

A freezer housed two rubber ice cube trays behind the lower-left Cellarette door, forming fifteen ice cubes rather than the twelve ice cubes in the tray described on the Supplement page.

Underneath the freezer shelves, Freon-12 tubing from the air-conditioning system froze the ice cube trays' water.

According to the *Packard Data Book for 1941* "Supplement to page 80–Sept. 1940:" (Fig. 139)

Ice cubes may be frozen in both trays in 15 to 20 minutes and will remain solid for a long time after the unit has been shut down.

Factory Air: Cool Cars in Cooler Comfort 73

Awoman posed as she mixed cocktails with glassware and cocktail mix from the Cellarette. An ice cube tray and liquor flask lay by, ready to complete the drink. Note that the bottle she held for the photo remained capped. (Fig. 144)

Figure 144. 1941 Packard One Eighty, Body 1442, Cellarette demonstration by a woman mixing cocktails. Note the capped bottle.   Dwight Heinmuller

74   Chapter 2: 1941 Packard Factory Air-Conditioning

Demonstrating the 1941 Packard Cellarette bar, a woman poured a cocktail mixer into a glass.

The image showed a fifteen-cube tray in the open freezer of the left-side lower drop-down door.

Yellow arrows show the glassware storage shelving. (Figs. 145 and yellow arrows, 146)

The lower drop-down doors displayed a glass, shot glass, liquor flask, combination corkscrew-bottle opener, and an ice cube tray.

Figure 145. 1941 Packard One Eighty, Body 1442, Cellarette demonstration by a woman mixing a cocktail   Joe Block

Figure 146. 1941 Packard One Eighty, Body 1442, Cellarette demonstration by a woman mixing a cocktail   Joe Block

The *San Francisco Examiner* featured an article about the Cellarette on Sunday, November 17, 1940, shortly before the San Francisco Auto Show. From an unknown origin, an image appeared of a woman preparing cocktails from the opened Cellarette. (Fig. 147)

**THE SAN FRANCISCO EXAMINER
SUNDAY, NOVEMBER 17, 1940**

The Packard motor car Company presents these twin miracles coincident with the introduction of the new Packard cars of 1941. With the miracle of swift miles far behind it, Packard reannounces mechanical refrigeration for its speeding cars and introduces another feature for the dusty throated, according to W. D. Venn, manager here for Earle C. Anthony.

At extra cost, Packard now offers a companion piece to air conditioning, Venn said.

It is the new, refrigerated *cellarette* manufactured for installation in the rear compartment of Packard sedans.

Housed in a handsome, compact cabinet mounted on the back of the front seat, the cellarette provides a freezing compartment, with replenishing trays of ice cubes manufactured from the refrigerating unit that runs the air conditioning system.

In addition to trays of sparkling ice, the cellarette has space for glasses, beverages, ice water and the necessary accessories to serve six people.

An air conditioned Packard, equipped with the ice bar, will be on display at the fall automobile show.

*A woman prepares cocktails from a Cellarette refreshment bar in a 1941 Packard
Source unknown*

*Figure 147. 1941 Packard One Eighty, Body 1442, offered optional Cellarette
San Francisco Examiner, November 17, 1940   Image source unknown   ABS*

The lavish luxury of an open bar for passengers led Packard to offer the Cellarette bar. Contrary to pre-WWII customs, no modern-day automaker now offers such a decadent option, citing the guise of political correctness. Currently, only private livery artisans install refreshment bars in stretch-type limousines and coaches.

Concurrent with the following *Packard News Service* bulletin, the *Santa Ana Register,* October 9, 1940, published an article titled, "New Packard Has Ice Unit." The first paragraph stated:

The Packard Motor Car Company presents twin miracles coincident with the introduction of the new Packard cars of 1941, according to W. R. Townsend, a local Packard dealer. With the miracle of swift miles far behind it, Packard re-announces mechanical refrigeration for its speeding cars and introduces another feature for the dusty-throated, he said. At extra cost, Packard now offers a companion piece to air-conditioning. It is the sensational, new, refrigerated Cellarette manufactured for installation in the rear compartment of Packard sedans.

The last paragraphs of this article repeated the wording of the *Packard News Service* publicity department bulletin.

The *Packard News Service* bulletin described the Cellarette in an unusual wording style:

All up and down the far frontiers, the bones of hardy plainsmen rest uneasy in their graves. To all that dusty-throated legion who braved the deserts of the West, comes news of a mechanical innovation that will green their molding dust with envy. Kit Carson heeds and remembers the burning dust of the Mojave. Captain Kearny rolls a wistful eye and recalls the blazing plains of New Mexico and the thick, cool flood of the Colorado before the plunge into the desert beyond Yuma.

Out of the molten maws of Detroit's industrial machines comes a new miracle of surcease to burning throats and temperate climate for fiery deserts...

And so, will another mechanical miracle write its record over the hot miles of the western deserts this year. Over the long reaches of the Butterfield stage route, over the torturing distances of the Santa Fe Trail, across the parching waste of the Utah desert, air-conditioned Packards will make their swift, comfortable way.

Against the smacking impact of hurtling tires, over the soft tinkle of ice-cubed glasses, the traveler may hear the vast stirring of a thousand ghosts—these will be the legions of the dusty-throated turning wistfully in their unmarked graves to view the passage of another miracle.

The search for an elusive Cellarette-equipped 1941 Packard culminated with someone from the author's family.

Packard collector David E. (Dave) McGahey, of Pampa, Texas, owned a 1941 Packard One Eighty, Body 1442, equipped with a Cellarette. The author's cousin, Annette Simons Story, and her husband, Jim, visited Jim's stepfather, Dave McGahey, in the Texas panhandle town of Pampa in 1977. A few miles away, he stored the vehicle in his McLean, Texas, work barn and drove them for a brief ride in the country. Ms. Story recalled:

1. The Packard exterior color was dark blue or black.
2. The interior was roomy enough for her six-foot, six-inch husband, Jim.
3. From separate storage, Mr. McGahey displayed two clear liquor flasks from the Cellarette cabinet.
4. Ms. Story confirmed the Cellarette cabinet's four drop-down doors and the inset cabinet position into the rear of the front seat, like the factory photo. (Fig. 148)
5. The lower left cabinet door opened to display the freezer.
6. Although functional, Mr. McGahey did not demonstrate the air conditioner or the freezer during the drive.
7. Mr. McGahey transported the badly deteriorated car from California in the 1970s, then restored it himself.

Unfortunately, no photos or records exist, as Mr. and Mrs. McGahey and Ms. Story's husband, Jim, are deceased.

Figure 148. 1941 Packard One Eighty, Body 1442, Cellarette cabinet   Dwight Heinmuller Ms. Story verified the Cellarette from a 1977 ride in Mr. Dave McGahey's Cellarette-equipped Packard.

Factory Air: Cool Cars in Cooler Comfort   77

When equipped with optional air-conditioning, the 1941 Packard One Eighty, Model 1907, Body 1442, five-passenger Touring Sedan exclusively offered the upscale Cellarette option.

However, the standard equipped sedan provided a storage bin "Robe Compartment," a Packard term for lap robe storage. The upholstered pull-down storage bin set into the rear of the front seat. Displayed here are images from Mr. Royce Baier and his wife, Sheila. (Figs. 149–154) A right profile view of the upholstered front seat design compares with the walnut trimmed Cellarette-equipped profile. (Figs. 148, 151) Closed, the upholstered door seals flush. (Fig. 152) The opened door rests close to the floor and reveals the storage space within, (Figs. 153, 154) confirming the inset area designed to accommodate the optional Cellarette cabinet.

Figure 149. 1941 Packard One Eighty, Body 1442 Touring Sedan   Owner M/M Royce Baier   Charlie Kietzman photo

Figure 150. 1941 Packard One Eighty, Body 1442 Touring Sedan   Owner M/M Royce Baier   Charlie Kietzman photo

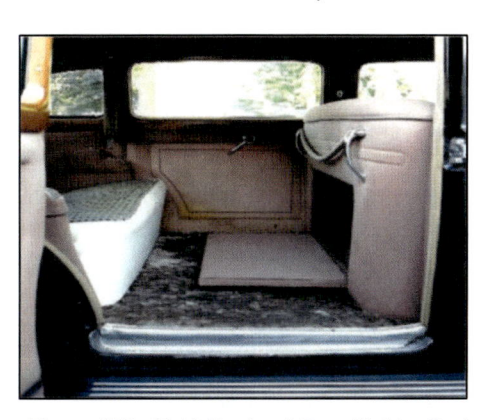

Figure 151. 1941 Packard One Eighty, Body 1442, "Robe Compartment" storage bin closed   Owner M/M Royce Baier   Charlie Kietzman photo

Figure 152. 1941 Packard One Eighty, Body 1442, "Robe Compartment" storage bin closed   Owner M/M Royce Baier   Charlie Kietzman photo

Figure 153. 1941 Packard One Eighty, Body 1442, "Robe Compartment" storage bin open   Owner M/M Royce Baier   Charlie Kietzman photo

Figure 154. 1941 Packard One Eighty, Body 1442, "Robe Compartment" storage bin open   Owner M/M Royce Baier   Charlie Kietzman photo

78 Chapter 2: 1941 Packard Factory Air-Conditioning

At the end of the 1940 model year, Packard offered an art deco-styled "Air-Conditioned" emblem as an accessory "ornament." (Fig 155)

A 1941 Packard One Eighty, Model 1907, Body 1442 Touring Sedan owned by Mr. Sal Saiya and wife, JoAnn, sported the 1942-style reverse-paint scheme, Crescent Blue over Silver French Gray. For 1941, Packard acknowledged the innovative and expensive status symbol of their mechanically refrigerated cars by affixing the art deco-style emblem as standard equipment onto the side panel (Figs. 156–158) and on the upper fender of the 1941 Packard Clipper. (Figs. 159, 160)

Figure 155. 1940 Packard factory photo of "Air-Conditioned" art deco-style lettering accessory, May 29, 1940, prelude to 1941 standard equipment   DPL NAHC   Romie Minor, Carla Reczek

Figure 156. 1941 Packard One Eighty, Body 1442, equipped with air-conditioning and emblem (yellow arrow)   Owner M/M Sal Saiya   Dwight Heinmuller photo

Figure 157. 1941 Packard One Eighty, Body 1442 "AIR-CONDITIONED" emblem placement on side panel   Owner M/M Sal Saiya   Dwight Heinmuller photo

Figure 158. 1941 Packard One Eighty, Body 1442 close-up of art deco-styled "AIR-CONDITIONED" emblem on side panel   Owner M/M Sal Saiya   Dwight Heinmuller photo

Figure 159. 1941 Packard Clipper "Air-Conditioned" emblem placement on upper fender   Owner Terry Weiss   ABS photo

Figure 160. 1941 Packard Clipper close-up of art deco-style "AIR-CONDITIONED" emblem on upper fender   Owner Terry Weiss   ABS photo

Factory Air: Cool Cars in Cooler Comfort 79

Photographed at another venue, the 1941 Packard One Eighty, Body 1442, presented its new grille styling and Packard's first inset, faired-in headlamps. (Fig. 161)

Under the bonnet, Packard instructed their contractor, the Bishop & Babcock Mfg. Co., to install the Servel two-cylinder air-conditioning compressor on the passenger side of the engine. (Fig. 162) The following images reflect the driver side compressor view (Fig. 163), the under-dash position of the air-conditioning control knob assembly (Fig. 164 and yellow arrow), close-up views of the assembly lettering showing "COOLING," "OFF," "HIGH," "MED," and "LOW" fan speeds (Fig. 165), with similar printed markings on the 1942 Packard control assembly. (Fig. 166)

Figure 161. 1941 Packard One Eighty, Body 1442, equipped with air-conditioning and cormorant Owner M/M Sal Saiya   Jack Swaney photo

Figure 162. 1941 Packard One Eighty, Body 1442, passenger side view of Servel air-conditioning compressor   Owner M/M Sal Saiya   PackardInfo.com Dave Czirr photo

Figure 163. 1941 Packard One Eighty, Body 1442, driver side view of Servel air-conditioning compressor Owner M/M Sal Saiya   PackardInfo.com Dave Czirr photo

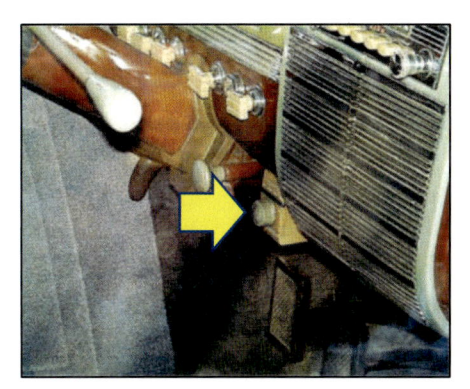

Figure 164. 1941 Packard One Eighty, Body 1442, air-conditioning control located below dash (yellow arrow)   Owner M/M Sal Saiya Jack Swaney photo

Figure 165. 1941 Packard One Eighty, Body 1442, air-conditioning control lettering, "COOLING," "OFF," "HIGH," "LOW," Owner M/M Sal Saiya Jack Swaney photo

Figure 166. 1942 Packard One Eighty Limousine air-conditioning control lettering, "COOLING," "OFF," "HIGH," ("MED"), "LOW" Former owner and photo Paul Fluckiger

80 Chapter 2: 1941 Packard Factory Air-Conditioning

Rear views of the 1941 Packard One Eighty, Body 1442 include the air-conditioning "cool air outlet," previously termed "blower grille" in 1940, on the package tray visible through the backlight. (Fig. 167) The trunk-mounted evaporator (Fig. 168) offers close-up, enlarged views of the nameplates for "Packard Air Conditioner" on the upper evaporator panel and the "Bishop & Babcock Mfg. Co." patent information on the lower panel. (Figs. 169, 170) The evaporator featured brown primer paint.

Figure 167. 1941 Packard One Eighty, Body 1442, rear view of air-conditioning cool air outlet
Owner M/M Sal Saiya   Jack Swaney photo

Figure 168. 1941 Packard One Eighty, Body 1442, air-conditioning evaporator with nameplates
Owner M/M Sal Saiya   Jack Swaney photo

Figure 169. 1941 Packard One Eighty, Body 1442, Packard Air-Conditioner nameplate
Owner M/M Sal Saiya   PackardInfo.com   Dave Czirr photo

Figure 170. 1941 Packard One Eighty, Body 1442, Bishop & Babcock Mfg. Co. nameplate with patent information
M/M Sal Saiya   PackardInfo.com   Dave Czirr photo

Factory Air: Cool Cars in Cooler Comfort 81

Packard sent an air-conditioning and Cellarette fold-over type "mailer" to potential customers early in 1941. The mailer casually marketed the mechanical refrigeration information using penguin characters as a suggestive cooling theme for illustrative purposes. (Figs. 171–177)

The author substituted a color image from *The Saturday Evening Post* for the black-and-white 1941 Packard One Ten Deluxe Touring Sedan image in the mailer. The new styling features for the 1941 model included "sparkling speedline stripes" on the lower fenders and the first faired-in inset headlamps on the front fenders.

Figure 171. 1941 Packard One Ten Deluxe Touring Sedan in mailer, p. 1a
Saturday Evening Post, November 9, 1940   ABS

Figure 172. 1941 Packard air-conditioning mailer, "Cool Comfort," p. 1b   Dwight Heinmuller

82 Chapter 2: 1941 Packard Factory Air-Conditioning

REAL AIR CONDITIONING ...A Packard First

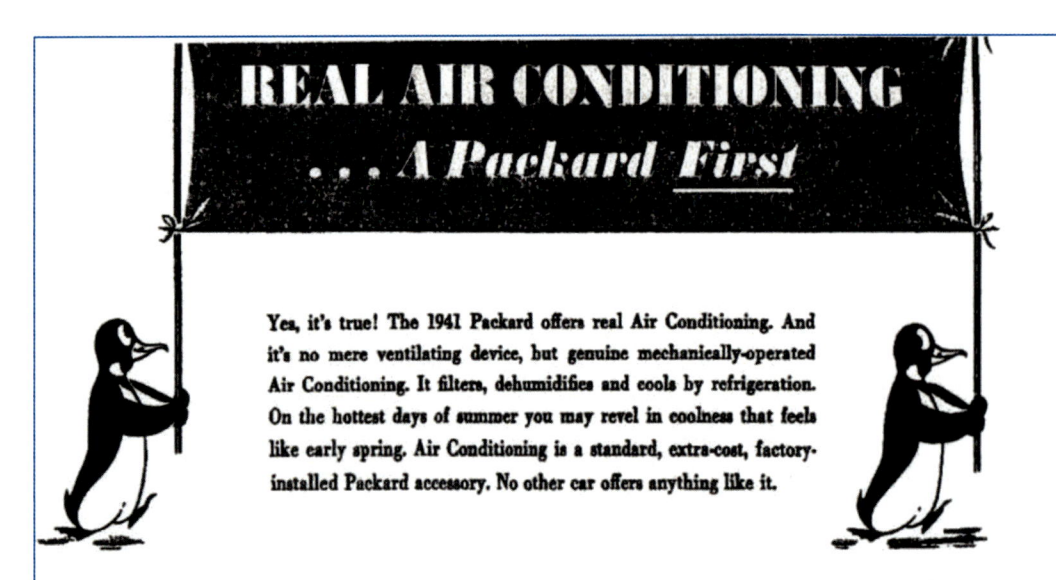

Yes, it's true! The 1941 Packard offers real Air Conditioning. And it's no mere ventilating device, but genuine mechanically-operated Air Conditioning. It filters, dehumidifies and cools by refrigeration. On the hottest days of summer you may revel in coolness that feels like early spring. Air Conditioning is a standard, extra-cost, factory-installed Packard accessory. No other car offers anything like it.

"IT COULD MAKE A MOUNTAIN OF ICE EVERY 24 HOURS"

The cooling effectiveness of Packard Air Conditioning is demonstrated by the fact that, rated like household refrigerators, it has a cooling-capacity rating of two and one-half tons of ice every 24 hours at a car speed of 60 miles per hour. Cooling starts instantly when the car engine is started, and you feel it throughout the car whether traveling at high speed or low. Temperature inside the car may be substantially reduced below outside temperature. Cooled air is circulated evenly throughout the car without drafts or discomfort.

"JUST AS SIMPLE AS TURNING ON THE LIGHTS!"

There's nothing complicated about Packard Air Conditioning. It works on the same proved principle as your electric household refrigerator. Turning it on or off is as simple as turning on the lights — by merely regulating a dash control. This same control may be adjusted to regulate the volume of cooled air which enters the car, just as you regulate a car heater. No defrosting or special servicing is required.

"IT'S COOL...IT'S CLEAN ...AND IT'S QUIET"

Cool because air is refrigerated... and dehumidified... clean because air is filtered—you will find the additional comfort of quietness in an Air Conditioned Packard because windows may be kept closed. You may converse easily, in normal tones, between front and back seats.

In half a dozen ways your first ride in an Air Conditioned Packard will be a revelation!

Figure 173. 1941 Packard air-conditioning mailer, p. 2   Dwight Heinmuller

Packard's penguin characters held a banner up for *Real Air-Conditioning A Packard First*. (Fig. 173)

It has the cooling capacity rating of two and one-half tons of ice every 24 hours at a car speed of 60 miles per hour.

Author note: The *Packard Data Book for 1941,* page 78, stated a two-ton capacity at sixty mph.

Initially equipped with a right-side paddle-type switch control in the 1940 Packard, the 1941 Packard changed to a knob-type switch control installed on the steering wheel's left or right side.

Packard's information stated, "No defrosting or special servicing is required." "IT'S COOL...IT'S CLEAN...AND IT'S QUIET," summed up the passenger's comfort level while riding in a 1941 Packard.

Factory Air: Cool Cars in Cooler Comfort   83

The penguin character at the blackboard summed up the benefits of cool, clean, and quiet air by stating, "Refrigerates, Dehumidifies, Filters, and Circulates" the air, much like WWII refrigerators of that era. (Fig. 174)

The coils absorbed heat as air passed over them, thereby cooling the air. The Packard air-conditioner, however, circulated the air with a built-in fan, not a feature of vintage-era refrigerators.

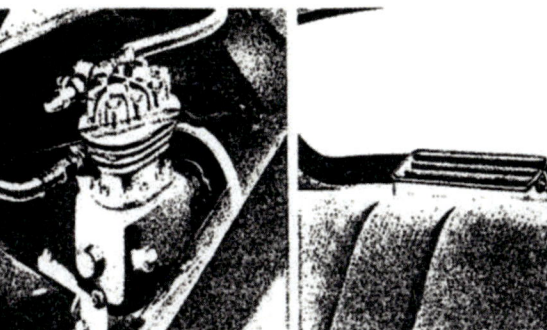

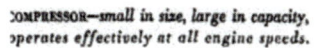

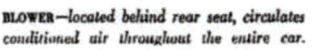

"Proved by Packard Standards" offered customer's the assurance that the air conditioner system's high-quality materials passed rigorous testing in the Packard Proving Grounds.

Figure 174. 1941 Packard air-conditioning mailer, p. 3   Dwight Heinmuller

Packard illustrated and itemized its four air-conditioning components, comprising (1) an engine-mounted compressor, (2) a condenser, (3) a frame-mounted refrigerant receiver or reservoir, and the (4) trunk-mounted evaporator. (Fig. 175)

It described the refrigerant flowing through each component, followed by exchanging the warm cabin air into refrigerated air.

Packard used a catchy slogan, *Make Ice Cubes as You Drive with a Packard Cellarette*. (Fig. 176)

The mailer featured images of a woman who opened the built-in walnut refreshment cabinet. She mixed drinks with glassware, liquor, and drink mixers for six, including ice from freezer trays.

It stated, "As an extra luxury (at extra cost) in an Air-conditioned Packard Super-8, you may have a 'Cellarette'– that makes two trays of ice cubes!"

Figure 175. 1941 Packard air-conditioning mailer, p. 4a   Dwight Heinmuller

Figure 176. 1941 Packard air-conditioning mailer, p. 4b   Dwight Heinmuller

Factory Air: Cool Cars in Cooler Comfort   85

"Comfort is *complete* in an Air-Conditioned Packard."

"An oasis in the desert" described the cool, filtered, clean air. Driving with the windows closed offered a quiet cabin and subdued traffic noises. (Fig 177)

IT'S COOL
When the thermometer soars like a duffer's golf score—*when it's really hot*—to step into an Air Conditioned Packard is like finding an oasis in a desert. Air is really cool—you feel its pleasant effects instantly!

IT'S CLEAN
When it's so dry that you're deluged by dust, just roll up the windows in your Air Conditioned Packard and know you'll arrive as spick-and-span as when you started. Air is clean because it's filtered.

IT'S QUIET
With windows closed, you enjoy the added comfort of quietness. Traffic noises are subdued. No need to talk above a normal tone to your fellow passengers. Comfort is *complete* in an Air-Conditioned Packard.

Figure 177. 1941 Packard air-conditioning mailer, p. 1c   Dwight Heinmuller

New for 1941, a Packard factory photo shows the addition of a small T-shaped handle for louver adjustment attached to the side of the cool air outlet, formerly termed "blower grille." Previously, 1940 passengers moved each louver individually to adjust the cool airflow. (Fig. 178 and yellow arrow)

James Hollingsworth, author, and longtime Packard club member, exhibited his 1941 Packard One Twenty Club Coupe at an Ark-La-Tex Packard club meet. His two-toned blue Coupe displayed the air-conditioning control knob, one of the three controls installed on the left of the steering column. (Figs. 179, 180, and yellow arrow)

A close-up evaporator view displays the "Packard Air Conditioner" and "Bishop & Babcock" nameplates. (Fig. 181)

Figure 178. 1941 Packard factory photo of a new control handle for the air-conditioning cool air outlet louvers (yellow arrow)   DPL NAHC Romie Minor, Carla Reczek

Figure 179. 1941 Packard One Twenty Club Coupe, equipped with air-conditioning Owner James Hollingsworth David W. Temple photo

Figure 180. 1941 Packard One Twenty Club Coupe interior with air-conditioning control (yellow arrow)   Owner James Hollingsworth David W. Temple photo

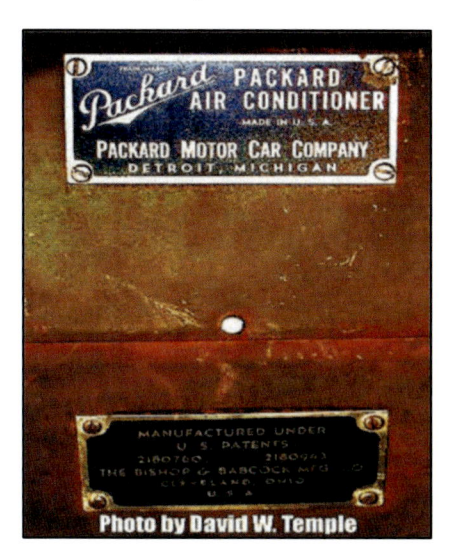

Figure 181. 1941 Packard One Twenty air conditioner nameplates on evaporator Owner James Hollingsworth David W. Temple photo

Chapter 2: 1941 Packard Factory Air-Conditioning

With an eye to the future of streamlining, Packard debuted the 1941 Packard Clipper on March 4, 1941, and announced the introduction with an insert into the 1940 Annual Report mailed to stockholders mid-March 1941, before its sale date in April 1941.

Packard management felt that its famous "Packard Look" needed freshening. The design and tooling, begun in 1939, cost a reported $350 million. Developed behind the scenes, Packard denied reports of a new body style and kept the entire project a secret.

Notable Clipper styling cues included a narrower, longer die-cast grille flanked at the bottom by horizontal grille bars, identified as "chromed louvers," faired-in headlamps, "FadeAway" front fenders, significantly lowered height, widened body, and a limited amount of chrome trim. (Figs. 182–186) The 1941 Clipper sold 16,600 units out of the 72,855 Packard units sold in 1941.

Figure 182. 1941 Packard Clipper introduction in the 1940 Packard Annual Report   Dwight Heinmuller

Figure 183. 1941 Packard Clipper shown in an insert to the 1940 Packard Annual Report, Mid-March 1941   Dwight Heinmuller

Factory Air: Cool Cars in Cooler Comfort 87

The new eight-cylinder 1941 Packard Clipper, model 1951, body 1401, featured the same 127-inch wheelbase as the 1941 Packard 120 and weighed 3725 lb., compared to 3535 lb. The single-model body offered a low center of gravity, the results of a new frame and suspension.

A 1941 "Packard Clipper Four-Door Sedan $1375" ad promoted it as the "first streamlined car planned for 'lookers' and riders!" (Fig. 186) The Clipper body width, viewed as a harbinger to the future, exceeded its height by nearly a foot, resulting in its low, sleek, roomy style.

Packard President Alvan Macauley asked Howard "Dutch" Darrin to submit a new Packard Clipper proposal. He based sketches upon custom body Packard Darrin models and completed the new Clipper design in ten days.

Figure 184. 1941 Packard Clipper grille view   Automotive Industries   Wesley Boyer

Figure 185. 1941 Packard Clipper rear view   Automotive Industries   Wesley Boyer

Figure 186. 1941 Packard Clipper ad   ABS

The *Packard Clipper Data Book, April 1941,* described the car best by stating: (Fig. 187)

*Front View—*As the new Packard Clipper approaches head-on, one is immediately impressed with the low, wide lines of the whole car.

A normally tall man can easily see over it.

Built close to the road, it portrays perfectly its fleetness, its steadiness, and its safety in every curving line and sweeping contour.

The interior dimensions provided "the widest front seat width of any car built today," "the front shoulder room…widest of competitive cars," and the rear "shoulder room is greater than that of any other car." (Fig. 188)

Figure 187. 1941 Packard Clipper Data Book
Regress Press   Jeff Steidle

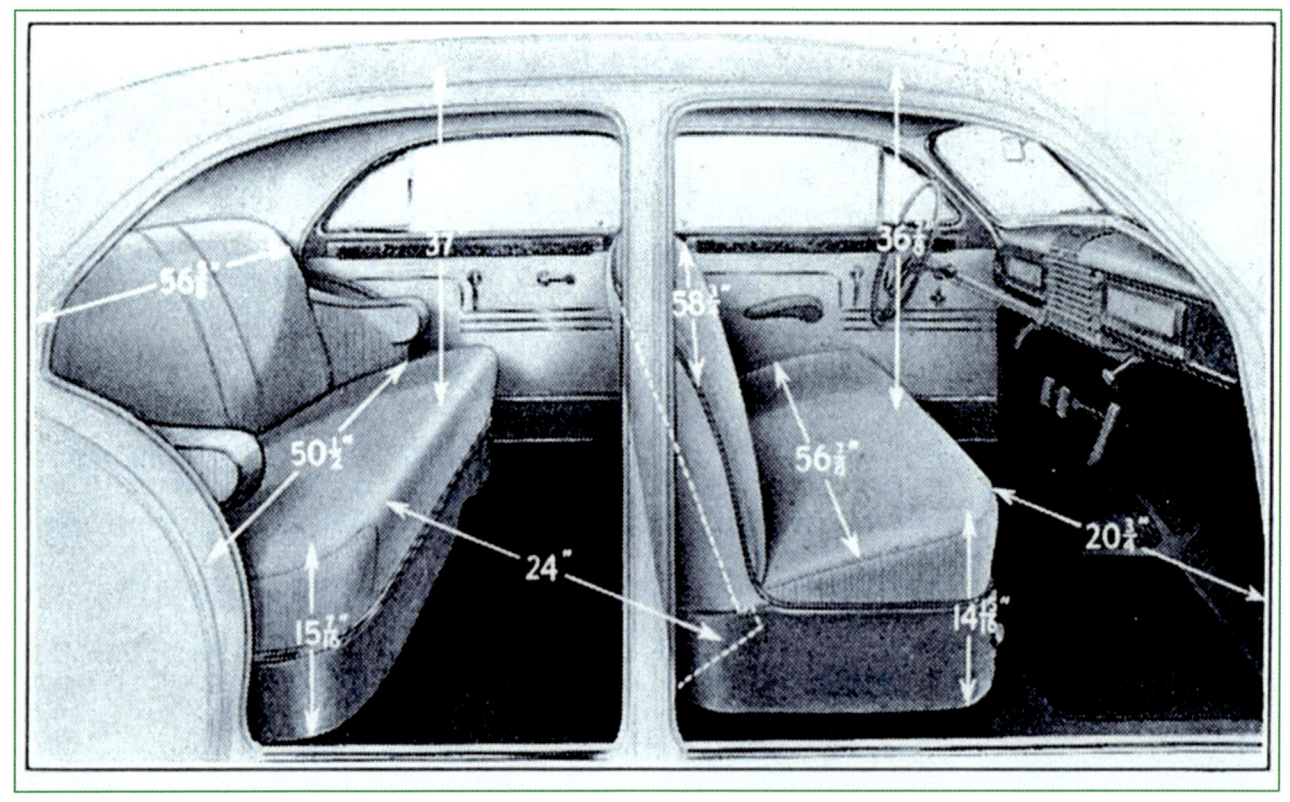

Figure 188. 1941 Packard Clipper interior dimensions   1941 Packard Clipper Data Book   Regress Press   Jeff Steidle

Spotlight on Packard

Presenting the 1941 Clipper Custom, Model 2011, equipped with factory air-conditioning

Owned by Terry Weiss

Photos by Allen B. Simons

At first glimpse, the grille of the 1941 Packard Clipper invites a more thorough inspection of the beautiful four-door sedan contours introduced in April 1941. (Fig. 189)

Gone is "That Packard Look" of yesterday, the grille-work created on a massive scale, meant to announce itself before the driver exited his car. Instead, this new Packard retains a vestige of the famous oxbow grille, but now it quietly speaks, "I am the new Clipper; there has never been a Packard like me."

Moving to the side, an observer agrees that there has never been a Packard like this before. It takes his breath away. From the elongated hood adorned with the eye-catching, chromed "AIR-CONDITIONED" emblem (Fig. 190 and yellow arrow), and the innovative "FadeAway" fenders that blend into the front door over concealed running boards, he senses a new proportion in this design. It is long, it is low, and it is wide.

The rear seat view reveals a louvered air-conditioning cool air outlet. (Fig. 190 and yellow arrow)

Figure 189. 1941 Packard Clipper Custom's elongated, narrow grille flanked by horizontal chrome bars   Owner Terry Weiss   ABS photo

Figure 190. 1941 Packard Clipper Custom "AIR-CONDITIONED" emblem (left yellow arrow) and air-conditioning cool air outlet (right yellow arrow)   Owner Terry Weiss   ABS photo

Owner Terry Weiss of Texas purchased this 1941 Packard Clipper in 2013 from the late James Hollingsworth, author of *Packard 1940: A Pivotal Year,* an avid Packard collector, restorer, and Packard club member.

Inside, the finely pinstriped tan upholstery lends a quiet dignity that continues to the art deco-style chrome trim on the door panels.

The Clipper advertised additional knee room from the inset design of the front seatback. Note the armrest placement inside the car, not attached to the door. The interior armrest placement signifies that this is a Clipper Custom, Model 2011. (Fig. 191)

Packard contracted with Bishop & Babcock Mfg. of Cleveland, Ohio, for the 1940–1942 model years. Packard Vice President of Engineering, W. H. Graves, reported that Packard produced approximately 2,000 air-conditioned cars during that time.

The adjustable, louvered cool air outlet displays the forward angle of the cool air discharge from the package tray, allowing conditioned air to flow along the headliner to cool the front passengers and circulate over the rear passengers. (Fig. 192)

It is difficult to imagine how welcome the air-conditioning system felt in Texas' blistering summer heat. Mr. Weiss recounted how a former owner of his Packard Clipper found *a pile of ice* (emphasis by author) on his back seat after arriving from a drive. As the air was extremely humid, ice forming on the cooling coils blew out from the vent discharge, then settled onto the rear seat cushion.

Figure 191. 1941 Packard Clipper Custom interior trim, inset front seat back, rear armrest location, and door panel trim   Owner Terry Weiss   ABS photo

Figure 192. 1941 Packard Clipper Custom rear seating and louvered cool air outlet   Owner Terry Weiss   ABS photo

Factory Air: Cool Cars in Cooler Comfort 91

The art deco-style symmetrical theme enhances the interior. The walnut-grained dash surrounds a chromed centered area, flanked by left-side instrument gauges and a right-side electric clock covering the glove box. Mr. Weiss installed an unobtrusively sized and colored air-conditioner evaporator under the dash. (Fig. 193)

The dash contrasts with the other 1941 Packard models as the stylish, chromed center area displayed lettered switches for standard and accessory controls.

The switches on the left column specify "*HEAD LITE*" (Packard spelling), "*ELECTROMATIC*" (optional clutch operation), "*INSTRUMENTS,*" and "*MAP LITE*" (Packard spelling). The air-conditioning push-pull "*COOLING*" switch heads the top right switch column. (Fig. 194, 195 and yellow arrows)

A close-up view reveals the debossed, indented lettering for "*COOLING*" (Fig. 195 and yellow arrow), "*HEATER,*" "*DEFROSTER,*" and "*CIGAR LITE* "(Packard spelling). One or two outward pulls of the "*COOLING*" switch operated the air-conditioning two-speed blower fan.

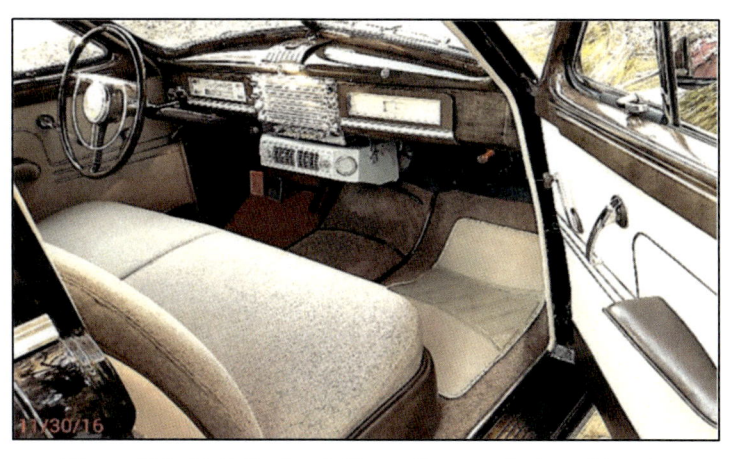

Figure 193. 1941 Packard Clipper Custom dash and interior Owner Terry Weiss   ABS photo

Figure 194. 1941 Packard Clipper Custom center dash with lettered, chrome switches. "COOLING" push-pull switch, upper right side (yellow arrow)   Owner Terry Weiss   ABS photo

Figure 195. 1941 Packard Clipper Custom dash air-conditioning push-pull control switch, marked "COOLING" (yellow arrow) Owner Terry Weiss   ABS photo

Figure 196. 1946 Packard Accessory book publishing error displayed a "Cooling" switch (yellow box), although not offered post-WWII.   Howard Hanson

Not displayed in 1941 Packard Clipper literature, the 1946 Packard Accessory brochure inadvertently continued a dashboard image of the "*COOLING*" switch in the chromed center area, although Packard discontinued the accessory availability in the post-WWII era.

The Clipper offered convenient releases that opened the hood by a lever under the driver's or passenger's dash area. (Fig. 197)

Bishop & Babcock of Cleveland, Ohio, installed the air-conditioning systems in the 1940–1942 Packards, the 1941 Cadillac, and the 1941–1942 Chrysler cars.

For the 1941 model, Packard installed the Clipper's compressor on the driver's side of the engine, perhaps because of spacing and fender bracing issues. The Packard factory photo shows a close-up of the compressor and placement. (Fig. 198)

Mr. Weiss installed a modern Sanden compressor for increased cooling efficiency. (Figs. 199, 200) He found that by running the Sanden compressor with the Clipper's original five-bladed fan and the trunk-mounted Packard evaporator, there was not enough cool air for East Texas' extreme summer heat. He maintained the evaporator for esthetic purposes and installed an under-dash unit.

The Sanden compressor placement mimics that of the original compressor. (Figs. 199, 200)

Packard's characteristic green engine color highlights the famous straight-eight 282-CID, 125-hp Clipper engine. (Fig. 200)

Figure 197. 1941 Packard Clipper Custom hood opens from either side, held with a prop rod   Owner Terry Weiss   ABS photo

Figure 198. 1941 Packard Clipper Servel air-conditioning compressor driver-side placement   DPL NAHC

Figure 199. 1941 Packard Clipper Custom view of Sanden compressor   Owner Terry Weiss   ABS photo

Figure 200. 1941 Packard Clipper Custom view of Sanden compressor, original 5-blade fan, and Packard's green engine paint   Owner Terry Weiss   ABS photo

Factory Air: Cool Cars in Cooler Comfort 93

Figure 201. 1941 Packard Clipper Custom with owner Terry Weiss.
"AIR-CONDITIONED" emblem (left yellow arrow) and air-conditioning cool air outlet shown (right yellow arrow)   ABS photo

Owner Terry Weiss presents his mirror-like black 1941 Packard Clipper. Yellow arrows indicate the "AIR-CONDITIONED" emblem on the hood and the air-conditioning cool air outlet behind the rear seat. (Fig. 201 and yellow arrows)

The Clipper's sleek "FadeAway" front fender lines flow rearward and taper over the trunk through the extended rear fender. (Fig. 202) The lines reflect similar design cues of Packard's custom-made Darrin bodies by styling genius Howard "Dutch" Darrin.

The 1941 Clipper evolves "That Packard Look" grille into a tall narrow shape that extends downward to horizontal chrome bars, identified as wide "chrome louvers," emphasizing the lower, wider body. (Fig. 203)

Figure 202. 1941 Packard Clipper Custom rear three-quarter view accentuates the Darrin-esque front FadeAway fender treatment and smoothly tapered trunk line   Owner Terry Weiss   ABS photo

Figure 203. 1941 Packard Clipper Custom distinctive grille work Owner Terry Weiss   ABS photo

94   Chapter 2: 1941 Packard Factory Air-Conditioning

Packard Clipper's vertically placed spare tire lies close to the evaporator. The best images show the lower-left evaporator's flexible hoses to and from the enclosed evaporator cooling coils. (Figs. 204, 205)

The air-conditioning expansion valve placement inside the evaporator contrasts with the exterior-mounted 1940 Packard expansion valve placement. (Figs. 206, 207)

Varying photo exposures differently reflect the original brown primer evaporator paint color.

Figure 204. 1941 Packard Clipper Custom air-conditioning evaporator and hoses leading to and from the cooling coils   Owner Terry Weiss   ABS photo

Figure 205. 1941 Packard Clipper Custom air-conditioning evaporator concealed by spare tire   Owner Terry Weiss   ABS photo

Figure 206. 1940 Packard exposed expansion valve positioned next to evaporator   Owner and photo West Peterson

Figure 207. 1940 Packard expansion valve close-up   Owner and photo West Peterson

Factory Air: Cool Cars in Cooler Comfort   95

The exterior *pièce de résistance* is the "AIR-CONDITIONED" emblem affixed beneath the decorative spear on either side of this 1941 Packard Clipper's bonnet. (Figs. 208–211 and yellow arrows)

Not only is this emblem informative, but it also draws attention to an ultra-exclusive, expensive accessory that sets this 1941 Packard Clipper apart from all other automobiles. Packard's mechanical climate control's luxury and prestige attributes set the bar for the highest level of passenger comfort.

Figure 208. 1941 Packard Clipper Custom "Air-Conditioned" emblem on bonnet   Owner Terry Weiss   ABS photo

Figure 209. 1941 Packard Clipper Custom driver side view of the "AIR-CONDITIONED" emblem at the top of the fender below the bonnet (yellow arrow)   Owner Terry Weiss   ABS photo

*Figure 210. 1941 Packard Clipper Custom placement of the "AIR-CONDITIONED" emblem at the top of the fender below the bonnet (yellow arrow) and view of the air-conditioning cool air outlet behind rear seat (yellow arrow)
Owner Terry Weiss   ABS photo*

*Figure 211. 1941 Packard Clipper Custom "AIR-CONDITIONED" emblem. (yellow arrow)
The setting rays of Texas sunlight accentuated the Clipper's streamlined beauty.   Owner Terry Weiss   ABS photo*

96   Chapter 2: 1941 Packard Factory Air-Conditioning

Packard photographed the 1941 Packard Clippers after the final assembly line. (Fig. 212)

Figure 212. 1941 Packard Clippers shown after final assembly line   DPL NAHC

Packard named its new 1941 model "Clipper," while Pan American World Airways had flown passengers in their Clipper flying boats over the Pacific Ocean since 1936.

In 1939, Pan Am inaugurated the first trans-Atlantic air travel from the United States to Lisbon, Portugal, with a stop in Foynes, Ireland, the airport now known as Shannon, Ireland.

A 1941 Packard Clipper posed in front of a Pan American Airways' Boeing 314 Clipper flying boat as it prepared for a Lisbon departure from its presently standing Marine Air Terminal, New York, now known as La Guardia airport. (Fig. 213)

Passengers often used the phrase, "I'm taking the CLIPPER." However, Packard's

Figure 213. 1941 Packard Clipper next to a PAN AMERICAN WORLD AIRWAYS BOEING 314 CLIPPER flying boat   ABS

use of the Clipper name spurred a legal battle with Pan Am a decade later in 1955, contending that Packard's Clipper name and other names used violated their Clipper trademark. Pan Am lost the suit in 1958, based upon the opinion that their Clipper trademark applied to an airline and an airplane, while Packard's Clipper name applied to an automobile company and an automobile.

Packard, in April 1941, introduced both the innovative 1941 Packard Clipper and published its long-delayed, eleven-page, *1941 Packard Air Conditioning Operation, Service and Owner's Instruction.* (Figs. 214–224)

1941
PACKARD AIR CONDITIONING OPERATION, SERVICE AND OWNER'S INSTRUCTION

ABS/DH 10-19-16

4-1-41

Packard Air Conditioning, which is supplied on 1941 Packard cars as factory-installed special equipment at extra cost, is a mechanical refrigeration system which provides cool, filtered, dehumidified air for passenger comfort.

With the compressor belt installed and the engine running, Air Conditioning is always available. It can be turned on by merely regulating a switch on the instrument panel. Cooling starts instantly when the car engine is started.

The Instrument Panel Switch is conveniently located on the instrument panel, to the left of the steering column. When the switch is turned on the cooling fan is in operation. There are three positions on the switch, so that the amount of cool air delivered into the car can be controlled.

If Air Conditioning is not desired, simply turn the switch to the "off" position.

The Adjustable Louvres, located in the air stream from the cooling fan discharge, direct the flow of air into the car. They may be adjusted to any desired position by simply turning a small knob.

Ventilation—For maximum cooling, the cowl ventilator and all windows should be closed. Ordinarily, ventilation is adequate under these conditions. However, if additional ventilation is required, it may be obtained by partially opening the cowl ventilator or window vents.

Refrigeration Cycle—Fundamentally, the refrigerant, which is circulated through the system by the compressor, picks up heat at the cooling coil, carries it to the condenser, and there discharges it to the outside air.

The refrigerant, Freon (F-12), a non-toxic, non-inflammable and practically odorless gas, is stored in the Receiver in a liquid state under relatively high pressure.

From the Receiver, the liquid passes through the Expansion Valve and into the Cooling Coil, where it expands into a gas at relatively low pressure. This expansion or evaporation from a liquid to a gas, absorbs heat from the metal coil and the air drawn over it by the Blower Fan, thus cooling the air.

The Compressor, mounted on the engine block, draws the refrigerant gas from the Evaporator at relatively low pressure and discharges it at high pressure into the Condenser. In the Condenser, the gaseous refrigerant is cooled sufficiently to condense into a liquid. From the Condenser the liquid refrigerant flows into the Receiver, and the cycle starts again.

Air Circulation—When the Cooling Fan is turned on, by means of the 3-position Switch located on the instrument panel, air in the car is drawn under the rear seat, through the Air Filter and Cooling Coil. It is then discharged into the car, through the Adjustable Louvres, so that it follows the contour of the roof and circulates throughout the car.

Air Filtering is accomplished by passing all the air discharged into the car by the Blower Fan through an oil coated fibre board Air Filter, which removes dust and other impurities from the air.

Air Cooling takes place when the air drawn from the car passes over the Cooling Coil. The air gives up its heat to the coil, where the temperature is normally from 40 to 50 degrees, and is discharged into the car at a temperature a few degrees higher than the coil temperature.

Winter Operation—During cold weather, when Air Conditioning is not required, remove the Compressor drive belt. Nothing else need be done.

When Air Conditioning is again desired, simply install the belt and the unit is ready for operation.

Air Filters—To insure proper air filtering and maximum cooling efficiency, the Air Filters, located at the inlet to the Cooling Coil, should be changed at least once a year, preferably in the Spring.

BODY

32

Litho. in U. S. A.

Figure 214. 1941 Packard Air-Conditioning Operation, Service and Owner's Instruction, p. 32   Dwight Heinmuller

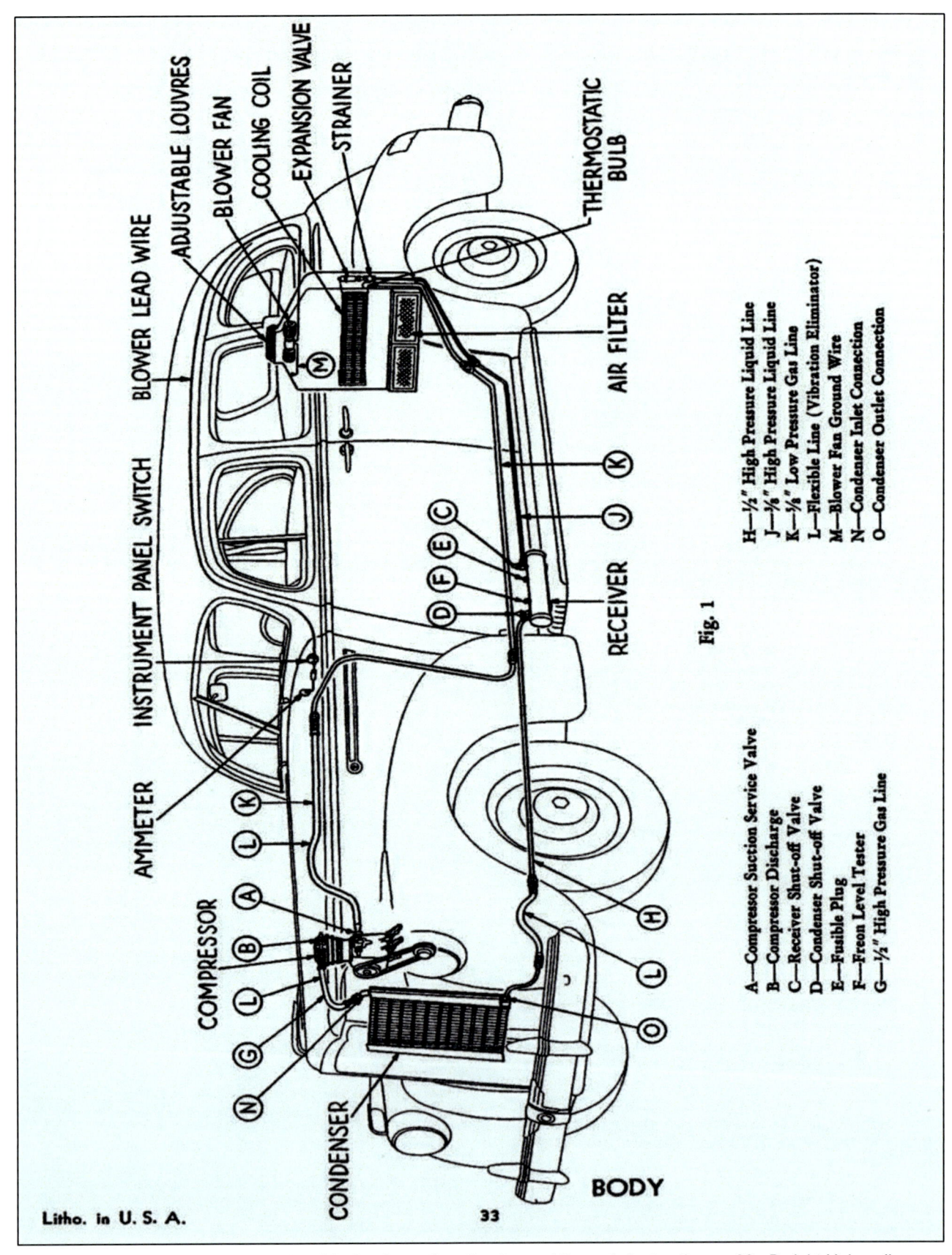

98   Chapter 2: 1941 Packard Factory Air-Conditioning

Figure 215. 1941 Packard Air-Conditioning Operation, Service and Owner's Instruction, p. 33   Dwight Heinmuller

OPERATION AND SERVICE

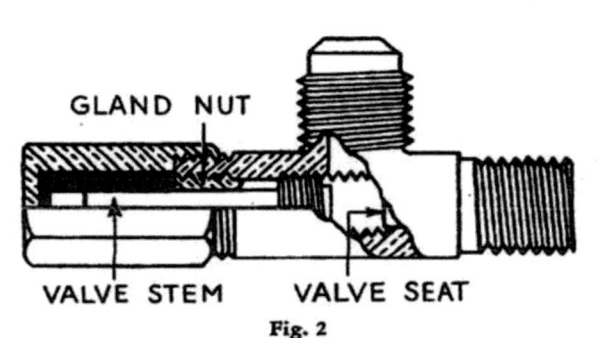

The Compressor Suction Service Valve "A," Fig. 1, is located at the Compressor inlet connection. It is of the double-seating type which seats or closes when it is screwed in all the way and also when it is screwed out all the way. (See Fig. 2.) When the valve is in the "in" position, the suction line is shut off from the Compressor. When the valve is in the "out" position or "back-seated" it is in the operating position, the Service Port Plug may be removed to permit attachment of the low-pressure or compound gauge or the charging line.

The Compressor Discharge Service Valve "B," Fig. 1, is located at the Compressor outlet connection. This valve is identical to the Compressor Suction Service Valve, except that it is for $1/2''$ tubing instead of $5/8''$ tubing. The Discharge Service Valve controls the outlet for refrigerant gas from the Compressor to the Condenser, and provides a connection for the attachment of the high pressure gauge.

Note: Always back seat valve before removing plug "A" for attaching gauges to Compressor Service Valves.

The Receiver Shut-off Valve "C," Fig. 1, is of the single-seating type as shown in Fig. 3. It is located at the outlet of the Receiver and is used when it is desired to pump all the refrigerant into the Receiver to permit removal of some part of the refrigeration apparatus.

The Condenser Shut-off Valve "D," Fig. 1, is of the same type as shown in Fig. 3. It is located on the inlet of the Receiver and is used to retain the gas in the Receiver, after pumping back, so that the Condenser may be removed without loss of charge.

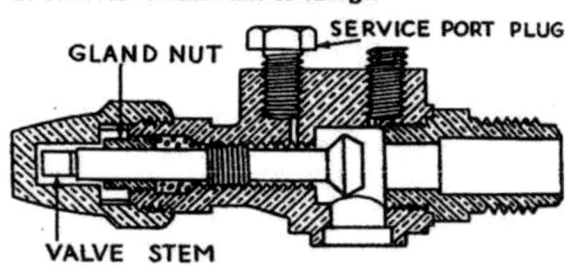

The Expansion Valve is mountd in the $3/8''$ high pressure liquid line at the inlet of the Cooling Coil, and its Thermostatic Bulb is clamped to the $5/8''$ low pressure gas line at the outlet of the Cooling Coil. The purpose of the Expansion Valve is to meter the amount of liquid refrigerant passing into the Cooling Coil. The flow of refrigerant through the valve is controlled thermostatically by the Feeler Bulb. If too much refrigerant passes into the coil it will not evaporate completely and some liquid will flow into the low pressure gas line. This liquid will cool the Thermostatic Bulb, which causes the valve to close partially, thus reducing the amount of liquid entering the Cooling Coil. If not enough liquid is entering the coil, the bulb will warm up, opening the Expansion Valve wider to let more liquid enter the Cooling Coil.

Do not attempt to adjust the Expansion Valve.

The Expansion Valve is of the Thermostatic type, $3/8''$ S.A.E. Inlet, $1/2''$ S.A.E. Outlet, $5/32''$ Orifice, 55 pound Freon-12, Gas charged power element, with $24''$ capillary.

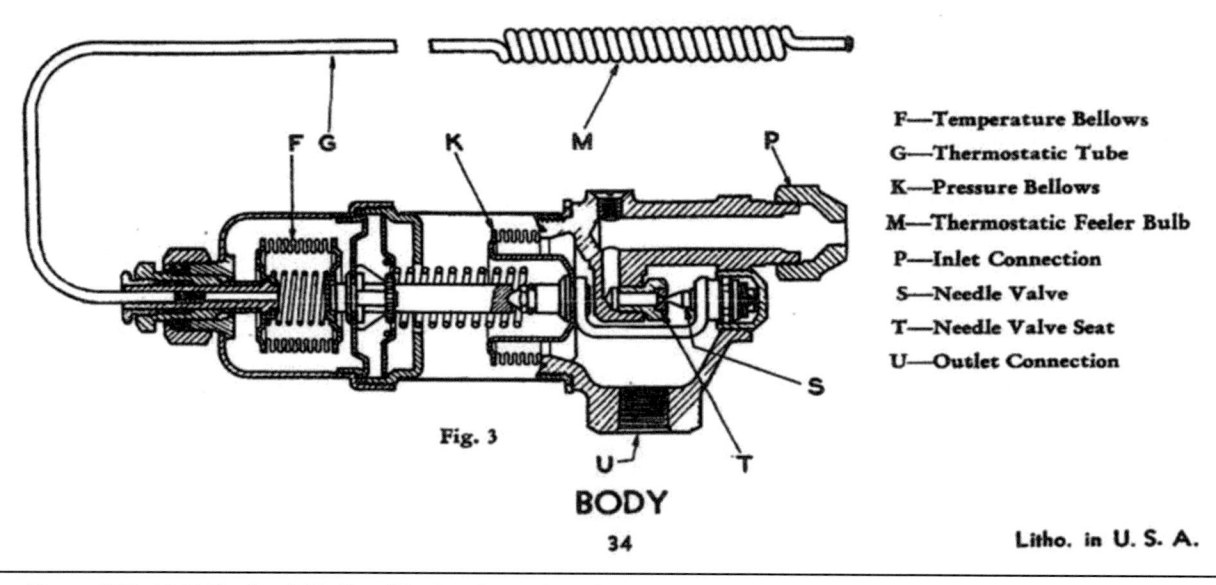

F—Temperature Bellows
G—Thermostatic Tube
K—Pressure Bellows
M—Thermostatic Feeler Bulb
P—Inlet Connection
S—Needle Valve
T—Needle Valve Seat
U—Outlet Connection

Figure 216. 1941 Packard Air-Conditioning Operation, Service and Owner's Instruction, p. 34   Dwight Heinmuller

FUSIBLE PLUG

The Liquid Receiver is equipped with a fusible plug set to discharge at 190° F. This is a safety device to prevent excessive pressure in the event the system is overcharged with Freon gas. When an excessive pressure is reached the Fusible Plug will melt, allowing all of the Freon to escape. It is then necessary to replace this plug with a new plug having the same temperature setting (190° F.) before recharging. In *replacing the plug make sure it is screwed in tightly.*

Note: If the fusible plug melts, be sure to check the condenser to see if it is stopped up with bugs and dirt. If it is, clean it before recharging the system.

AIR FILTERS

The Air Filters used in the Packard Air Conditioner are a replacement type of filter, constructed of oil coated, corrugated fibre board.

The filters should be replaced with new clean filters at least once a year, in the Spring. This insures that the air supplied by the Air Conditioner is properly cleaned at all times. Also, it insures maximum operating efficiency of the Air Conditioner.

To replace the filters, remove the rear seat, slide the old filters out of position and replace with new filters.

TO CLEAN AND DEODORIZE THE COOLING COIL

In any air conditioning system, smoke and body odors will be picked up by the circulating air and carried back to the Cooling Coil, where they will be deposited. As a result, unless the Cooling Coil is cleaned occasionally, the odors will accumulate and be carried back into the air conditioned space.

To prevent the possibility of objectionable odors thus created being carried into the passenger compartment, it is recommended that the Air Conditioner Cooling Coil be cleaned occasionally with a solution of Diversol or other bacteria insecticide which will not attack the tin plated interior parts of the cooling coil assembly. Use a solution strength of 4 ounces of cleanser to a gallon of water.

To clean and deodorize the coil, remove the cover plate on the coil housing in the trunk compartment and spray the coil, using about a gallon of the cleaning solution. After passing over the coil, the solution will drain from the return duct through the drain tube located in the bottom of the duct.

TO TEST FOR LEAKS
HALIDE TORCH METHOD

Connect the Halide Leak Detector S.T. 10105 to a Presto-Lite Tank or to the acetylene tank on the welding equipment.

Light the torch on the Halide Gas Leak Detector. Pass the end of the Leak Detector searching tube around the joint or connection to be tested. (After service work, all joints and connections in the system should be checked.) If there is a leak in the joint the color of the flame in the torch will turn to a brilliant green. This is a positive indication of a leak.

If a leak is detected in a flare connection, draw up the flare nut tightly. If the leak still exists the flare on the tubing is probably defective. The gas will then have to be pumped into the Receiver, as described in the section "Pumping Down Entire System," and the tubing disconnected and another, new flare made in the tubing.

If a leak is detected in a soldered joint, relieve the pressure in that part of the system down to zero on the gauge, either by pumping the refrigerant from that part of the system to another part, or, in the case of the Receiver, purging the gas out of the system to the air or into a small Refrigerant cylinder (Service Drum). Do not attempt to solder the joint while there is pressure in that part of the line.

If the system has lost its charge of refrigerant, there must be a leak, and it must be found rather than simply to add more refrigerant which will in turn be lost unless the leak is found and repaired. Do not give up until the leak is found.

Note: If the system has completely lost its charge, it will be necessary to add some refrigerant before the leak can be found.

To determine whether there is any refrigerant in the system, open the Liquid Tester "F," Fig. 1, on the Receiver, using Liquid Receiver Key S.T. 10074. If gas escapes from the tester, there is refrigerant in the system. If no gas escapes, the refrigerant has been lost and it is quite probable that air has entered the system. In this case, proceed as described in the first four steps in the section "To Completely Recharge the System from a Freon Drum."

SOAP AND WATER SOLUTION METHOD

Make a solution of soap (yellow laundry soap) and water. Prepare it at least an hour or so before using so that all bubbles have disappeared and the solution is of a thick "ropy" consistency about the same as heavy oil. Spread this solution on all joints or connections with a soft brush. Examine closely under a strong light. Leaks will show up by the presence of bubbles under or bursting through the film of the solution. Use a small mirror in examining the rear sides of joints otherwise not directly visible.

ATTACHING GAUGES TO READ SUCTION AND DISCHARGE PRESSURES

1. Remove cap from Compressor Suction Service Valve "A," Fig. 1.
2. Back seat valve by turning valve stem counter-clockwise as far as possible.
3. Remove plug from service port (Fig. 2).
4. Attach hose from Compound Gauge of Charging and Testing Gauge Unit S.T. 5198 to service port.
5. Purge gas line by opening valve on Compound Gauge side of gauge manifold and then opening Compressor Suction Service Valve ½ turn. This will allow the refrigerant to blow the air out of the low pressure line through the center charging line. After a few seconds close valve in gauge block.
6. Repeat the above operations with the Compressor Discharge Service Valve "B," Fig. 1, attaching the High Pressure Gauge line to service port in this valve and purging the line as before by opening the valve on the High Pressure Gauge side of the gauge block.

BODY

Figure 217. 1941 Packard Air-Conditioning Operation, Service and Owner's Instruction, p. 35   Dwight Heinmuller

7. Start engine and adjust Suction and Discharge Service Valves until gauges show pressure with minimum fluctuation of the needles.
8. Check all connections for leaks.

PUMPING DOWN LOW PRESSURE SIDE

1. Attach gauges to Compressor as described under section "Attaching Gauges To Read Suction and Discharge Pressures."
2. Remove cap from ⅜" Receiver Shut-off Valve "C," Fig. 1, and close valve by turning valve stem clockwise as far as it will go.
3. Start engine and run at slow speed. The Compressor will now pump all the gas from the ⅜" Liquid Line, the Expansion Valve, the Cooling Coil, and the ½" Low Pressure Gas Line (Suction Line) and store it in the Receiver.
Run the engine until the Compound Gauge reads 2" vacuum. Then stop the engine. It is desirable not to open the refrigerant lines while the Compound Gauge registers a vacuum. If necessary, "crack" the Receiver Shut-off Valve "C" until the Compound Gauge reads one pound pressure, when the lines may be opened.
4. It is now possible to remove any piece of equipment on the low side of the system without losing the Freon charge.

Watch both gauges while pumping down the system. If the High Pressure Gauge goes above 250 lbs., stop the engine, as either the Compressor has been run at too high a speed or the installation has received too much Freon gas when initially charged. If the pressure is due to excessive Compressor speed the pressure can be reduced by allowing the Compressor to remain idle for a few minutes until the pressure drops to normal, when the engine can be restarted.

However, if there has been too much Freon charged initially the pressure will not drop when the engine is stopped and it will be necessary to vent the excess gas by loosening the flare nuts on the inlet and outlet of the Expansion Valve (See Fig. 4) and allowing the gas to escape slowly to the atmosphere until the lines are empty.

IMPORTANT: Do not run Compressor while low pressure side is pumped down.

PUMPING DOWN ENTIRE SYSTEM

1. Follow directions given in first three steps under section "Pumping Down Low Pressure Side."
2. When all the Freon has been pumped from the low pressure side into the Receiver close the Condenser Shut-off Valve "D," Fig. 1, located at the front of the Receiver.
3. It is now possible to remove any piece of equipment from the system without losing the Freon charge.

Note: It is not possible to pump all the gas in the Condenser into the Receiver and therefore a small amount of gas will be lost when a connection is broken on the high pressure side. In some cases it may be necessary to add Freon to the system to make up for this loss.

IMPORTANT: Do not run Compressor while system is pumped down.

AIR AND MOISTURE IN THE SYSTEM
DEHYDRATORS

All outside air has some moisture in it in vapor form. Just how much moisture it has, varies considerably from the dry air of Arizona to the humid air of any location on a hot, sultry or rainy day. If moist air is allowed to get into the refrigeration system both the air and water are harmful. If the air itself were perfectly dry it would do no more harm than to increase the discharge pressure (thus decreasing the capacity of the Compressor), cause excessive heating of the Compressor and in general take up room in the system that could be advantageously used for the Freon which performs the cooling.

The introduction of moisture is more serious as it may freeze up in the Expansion Valve and Cooling Coil and stop or retard the cooling action and also corrode many of the finely finished metal surfaces such as the Discharge and Suction Valves and other parts of the Compressor and the seat and needle of the Expansion Valve.

Air and moisture can be accidentally sucked into a system if a leak is present on the evaporator or other portions of the low pressure part of the system when for any reason the low pressure part of the system is on a vacuum; or they may be introduced by allowing lines or apparatus to stand open for long periods when the system is being opened for repairs either to the system or to some part of the car that is inaccessible except by removal of a part of the conditioning system.

Air has a tendency to collect in the Condenser as it does not condense or turn into a liquid and pass on to the Receiver. Air can be purged by removing the plug from the Service port of the Compressor Discharge Service Valve, closing the Compressor Suction Service Valve, and "cracking" the Discharge Service Valve. This can best be done after the engine has been stopped a few minutes as the air tends to collect in the upper part of the Condenser. It may be necessary to run and stop the engine a few times to remove all the air from the system, and thus reduce the excessive discharge pressure.

Moisture—There is only one effective correction to moisture in the system—remove it! Special non-freezing liquids may be put in the system that combine with the moisture and, diluting it, prevent freeze-ups at the Expansion Valve needle and seat. However (disregarding freeze-ups), the moisture is still in the system to cause corrosion and some of these liquids are corrosive in themselves. Do not use this method.

The Dehydrator, S.T. 5186, is a cylinder with inlet and outlet connections in which is a material known as a desiccant. The approved desiccant is "Drierite," but activated alumina or silica gel may also be used. (DO NOT USE CALCIUM-CHLORIDE.) These are in granular form that allow Freon to pass but absorb the moisture in the Freon.

If it is suspected or known that there is moisture in the system proceed according to the instructions under the section "Pumping Down Low Pressure Side," and then as follows:

1. Loosen nut and remove ⅜" line from inlet connection to Strainer, Fig. 1.

BODY
36

Litho. in U.S.A.

Figure 218. 1941 Packard Air-Conditioning Operation, Service and Owner's Instruction, p. 36   Dwight Heinmuller

2. Connect the Dehydrator to the ⅜" line and, by means of another short piece of ⅜" tubing, to the inlet connection to the Strainer, leaving ⅜" nut on the Strainer connection loose.
3. "Crack" the Receiver Shut-Off Valve and purge the air from the ⅜" tubing and dehydrator.
4. Tighten the ⅜" nut at the Strainer inlet connection.
5. Open the Receiver Shut-off Valve and start engine.
6. Allow the Dehydrator to remain in the line while the cooling system is in operation for at least an hour, then remove by repeating operations for pumping the refrigerant out of this line.
7. Remove Dehydrator and replace original ⅜" connection to inlet on Strainer, pull up tightly, open Receiver Shut-off Valve and test for leaks.
8. System should now be free of moisture and ready for use.

Note: Always be sure that both ends of the Dehydrator are sealed when not in use. If this is not done the desiccant will absorb moisture from the air and may become saturated. If it were used in this condition it would do more harm than good. The safest procedure is to always put a fresh charge of desiccant in the Dehydrator each time it is to be used.

TO REPLACE EXPANSION VALVE

1. Pump all the Freon from the low pressure side of the system as described in section "Pumping Down Low Pressure Side."
2. Remove the Thermostatic Bulb from its clamp on the ⅝" line leaving the Cooling Coil.
3. Remove Expansion Valve by loosening inlet and outlet flare nuts. Some gas may escape, but the loss should be negligible if the Receiver Shut-off Valve is closed tightly.

NOTE: WHEN THE EXPANSION VALVE IS REMOVED OR AT ANY TIME WHEN THE LINES ARE OPEN CARE MUST BE TAKEN NOT TO START THE ENGINE UNLESS THE COMPRESSOR HAS BEEN UNBELTED.

4. Install new valve and pull flare nuts tight.
5. Clamp Thermostatic Bulb to ⅝" line as before.
6. "Crack" Receiver Shut-off Valve "C" until Compound Gauge reads 30 to 40 pounds, then close.
7. Test around Expansion Valve nuts for leaks.
8. Open Receiver Shut-off Valve "C" by turning valve stem counterclockwise. Tighten gland nut around valve stem. Replace cap tightly.
9. Start engine and watch gauges. The Compound Gauge should read 20 to 40 pounds, and the High Pressure gauge 140 to 190 pounds, after the engine runs at slow speed for a few minutes (with the blower running).
10. "Back seat" the Compressor Discharge and Suction Service Valves "B" and "A" by turning the valve stems counterclockwise as far as they will go. Remove gauges, replace plugs, tighten gland nuts and replace caps tightly.

TO REMOVE AND REPLACE CONDENSER

1. With the Blower running, start the engine and run at slow speed until the Condenser is warm.
2. Stop engine, remove caps from the Compressor Discharge Service Valve "B," Fig. 1, and the Condenser Shut-off Valve "D," Fig. 1, and turning stems clockwise as far as they will go.
3. Loosen flare nut on the Condenser Inlet Connection "N," Fig. 1, and allow the gas in the Condenser to escape slowly.
4. Unscrew this flare nut after all the gas has escaped and also unscrew the flare nut on the Condenser outlet connection "O," Fig. 1.
5. Remove Condenser, cover inlet and outlet to keep damp air out of Condenser.
6. When it is desired to replace the Condenser, put it back in place.
7. Connect flare nut to Condenser Outlet Connection "O" and tighten.
8. Connect flare nut to Condenser Inlet Connection "N" and tighten.
9. Loosen service port plug "A" in Compressor Discharge Service Valve (See Fig. 2).
10. "Crack" the Condenser Shut-off Valve "D", allowing gas from the ½" line to the Receiver to pass up through the Condenser and out at the loose plug on the Compressor Discharge Service Valve, thus purging the air from the Condenser. It is necessary to purge only a few seconds.
11. Tighten the plug in the Compressor Discharge Service Valve.
12. Open Compressor Discharge Service Valve "B" and Condenser Shut-off Valve "D" by turning counterclockwise. Replace and tighten valve caps.
13. Check all joints for leaks.

TO REPLACE COMPRESSOR

1. Run engine at slow speed for a few minutes until Compressor is warm.
2. Stop engine, remove caps from Compressor Discharge and Suction Service Valves "B" and "A," Fig. 1, and close both valves by turning stems clockwise as far as they will go.
3. Loosen Plug (See Fig. 2) in Service Port of Suction Service Valve "A" and allow gas in the Compressor to slowly escape. Also remove plug in service port of Discharge Service Valve "B" and allow gas in Compressor head to escape.
4. Remove cap screws securing the Compressor Service Valves to the Compressor and carefully lift the valves away from the Compressor.
5. Loosen bolts securing the Compressor to the Compressor base, and remove the drive belt.

BODY
37

Litho. in U. S. A.

Figure 219. 1941 Packard Air-Conditioning Operation, Service and Owner's Instruction, p. 37   Dwight Heinmuller

6. Remove bolts and Compressor from base.
7. Take off nut holding Compressor pulley to shaft and remove pulley. See that Woodruff key is in place on replacement Compressor and transfer pulley to replacement Compressor. Slide it in place and tighten shaft nut.
8. Put replacement Compressor on base, put in bolts loosely, put on belt and pull up Compressor until belt is tight—not too tight, just so it does not slip. Also see that belt and pulleys are lined up. Tighten bolts.
9. Put in new copper flange gaskets between the Service Valves and Compressor. Put valve flanges down against the copper gasket, run in cap screws finger tight. Tighten cap screws evenly so that flange fits flat against the gasket.
10. Replace plug in the service port of the Suction Service Valve and crack this valve by opening ¼ turn to blow gas from the suction line up through the Compressor and out the service port of the Discharge Valve. Turn Compressor pulley over by hand to assist in purging the air from the Compressor.
11. Tighten plug in Service port of Suction Service Valve, open valve stem to back seat, tighten gland nut, put on and tighten valve cap.
12. Put in and tighten plug in service port of Discharge Service Valve, back seat valve stem, tighten gland nut, replace and tighten cap.
13. Test for leaks around Service Valve flanges, service ports, and caps. If there are no leaks, the Compressor is ready for use.

CHECKING OIL LEVEL IN COMPRESSOR

If there has been a loss of liquid Freon from the system some of the oil (which mixes readily with Freon and is present in all parts of the system) may also have been lost. The level in the Compressor may be checked as follows:

1. With the blower running, start the engine and run at slow speed for a few minutes until the Compressor crankcase is warm. Then stop engine.
2. Remove cap and close (clockwise) the Compressor Suction Service Valve as far as it will go. Loosen the plug in the Service port and allow the gas in the Compressor to slowly escape.
3. When there is no further escape of gas, remove the Oil Filler Plug, a hex head plug on the side of the Compressor crankcase.
4. Insert a clean rod to the bottom of the Compressor crankcase, and measure the height of the oil level. It should be up to, or within ⅜" below the centerline of the Compressor shaft.
5. If the oil level is low, add oil as necessary. USE SPECIAL PACKARD COMPRESSOR OIL ONLY.
6. Replace Oil Filler Plug, tighten plug in Service Port on Suction Service Valve, open valve to back seat, tighten gland nut and replace, and tighten cap. Test for leaks. No purging of the Compressor is necessary for this operation.

CHECKING FREON LEVEL

It is very simple to determine if there is a sufficient amount of Freon in the system for normal operation. With the Compressor running, open (slightly) the Liquid Tester "F," Fig. 1, which is a small test cock on the Receiver. If there is liquid refrigerant up to the level of the Tester, liquid Freon will come out in a milky white flow. If gas only blows out, this indicates that the system is short of refrigerant and some should be added, as described below.

If neither liquid nor gas escapes from the Tester, the system must be completely recharged, either from a Freon drum or from a charged Receiver. Both of these methods are described below. A complete charge is 6¼ pounds of Freon.

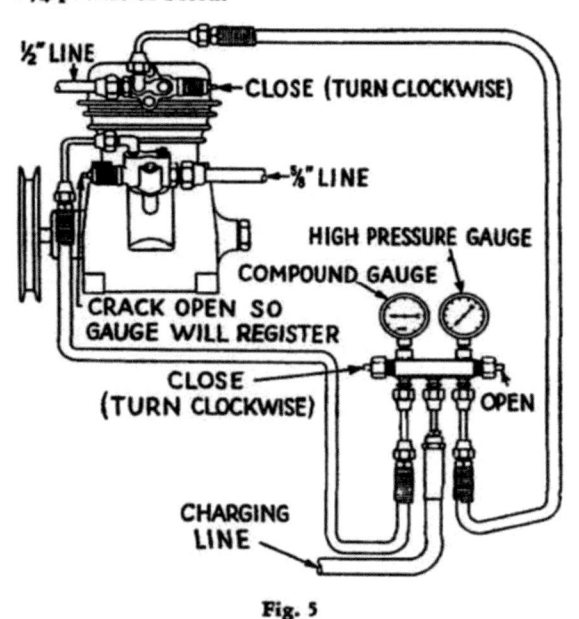

Fig. 5

TO PUMP AIR FROM SYSTEM

When the entire charge of Freon has been lost, it is quite probable that air has been introduced into the system at the point where the leak occurred. It is necessary to remove this air before adding a new charge of Freon. By using the Compressor as a suction pump, the air can be removed as follows:

1. Install gauges and adjust valves as shown in Figure 5.
2. Be sure both valves in Receiver Tank are open.
3. Start engine and run at slow speed until Compound Gauge reads 20-28 inches vacuum.
   Note: If oil is discharged through the charging line along with the air during this operation, stop the engine for a few minutes, then start up again and proceed until the proper vacuum is reached.
4. When the system has been pumped down to 20-28 inches of vacuum, stop the engine. If the vacuum holds for several minutes, this is an indication that

BODY

38

Litho. in U. S. A.

Figure 220. 1941 Packard Air-Conditioning Operation, Service and Owner's Instruction, p. 38   Dwight Heinmuller

any leak in the system is comparatively small and it will be safe to recharge the system before finding the leak.

> **Note:** If the vacuum does not hold when the engine is stopped, there must be a bad leak in the system. Be sure to find this leak before recharging the system. To find the leak, attach the Freon drum to the end of the charging line and open the valve in the drum, allowing Freon to enter the lines until both gauges register 60-70 pounds. Now check the entire system for leaks as described under the section "To Test for Leaks."
>
> When the leak has been found and repaired, again pump air from system and proceed as described in "To Completely Recharge System from a Freon Drum."

If all the Freon charge has not been lost it is not necessary to pump air from the system. Proceed as follows.

TO ADD FREON TO SYSTEM

1. Install gauges and adjust valves as shown in Fig. 6.
   Note: Be sure to purge the air from both gauge lines, also from the charging line.
2. Open valve on Freon drum.
3. Start the engine and run at slow speed. The Compressor is now drawing gaseous Freon out of the drum.
   Note: Stand the Freon drum upright so that only gaseous Freon is drawn into the Compressor. Do not invert the drum or lay it on its side as this will allow liquid Freon to enter the Compressor, perhaps damaging the Compressor valves and causing oil pumping and slugging. If the drum gets cold, set it in pail of warm water to hasten the vaporizing of the liquid Freon.
4. Keep trying the Liquid Tester on the Receiver. When a milky white spray comes from it, shut the valve on the Freon drum and stop the engine.
5. Find out where the Freon leaked out and repair the leak.
6. Start engine and check liquid level in receiver tank. Charge in more refrigerant if necessary.
7. Remove gauges from Compressor.

TO COMPLETELY RECHARGE THE SYSTEM FROM A FREON DRUM

When the refrigerant charge has been completely lost, all air must be removed from the system as described in section, "To Pump Air From System." When air has been completely exhausted, proceed as described in "To Add Refrigerant."

TO COMPLETELY RECHARGE THE SYSTEM BY THE CHARGED RECEIVER METHOD

1. Remove the old Receiver from car.
2. Install the new Receiver, which is charged with Freon.
   Note: Receiver should contain 6¼ lbs. of Freon.
3. Loosen flare nut on ½" line at Compressor Discharge Service Valve "B," Fig. 1.
4. Purge ½" line from Receiver to Compressor by cracking Condenser Shut-off Valve "D," Fig. 1, and allowing gas to pass through the line, forcing the air out ahead of it through the loose connection at the Compressor. When all the air is purged and Freon begins to escape (use Halide torch to detect the Freon), tighten the ½" nut at Compressor.
5. Loosen flare nut on ⅝" line at Compressor Suction Service Valve "A," Fig. 1.
6. Purge the ⅝" and ½" lines from Receiver to Compressor in same manner as ½" line was purged, by cracking Receiver Shut-off Valve "C," Fig. 1, and allowing air to escape through loose connection on ⅝" line at Compressor. When the air is purged, tighten the ⅝" nut.
7. Check entire system for leaks.

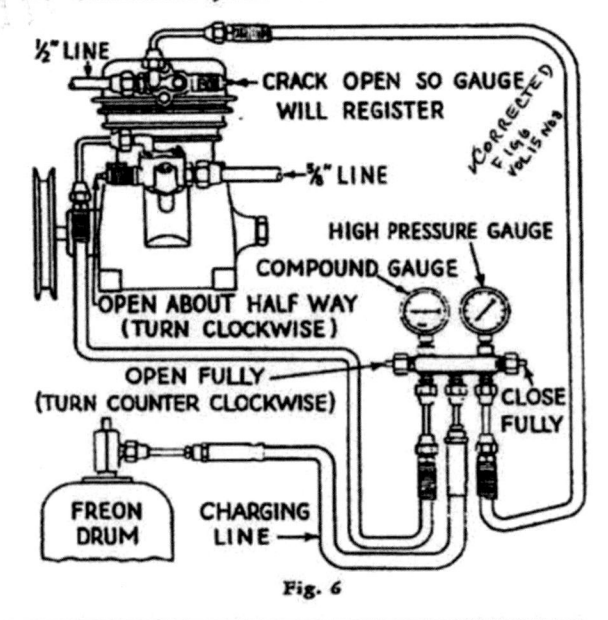

Fig. 6

NOTE: Paste in 1941 Air Conditioning Service Instructions Booklet, mailed with April 15th Service Letter, replacing corresponding page (8).

Figure 221. 1941 Packard Air-Conditioning Operation, Service and Owner's Instruction, p. 39

SERVICE CHART

CAUSE	CHECK	CORRECTION
Air from Evaporator Not Cold		
1. Compressor not running, or running slowly. Belt broken or loose and slipping.	1. Obvious upon inspection.	1. Loosen bolts from Compressor to base and shift Compressor to tighten belt. Be sure to line pulleys up properly.
2. Loss of Refrigerant.	2. Open Liquid Tester on Receiver. If gas only comes out, some Freon has been lost.	2. Test entire system for leaks thoroughly. Find and repair leaking joint, then add refrigerant as described under "Adding Refrigerant to the System." *If Fusible Plug in Receiver has melted, check for dirt or bugs in Condenser.*
3. Expansion Valve Strainer stopped up with foreign matter.	3. Low suction pressure. Suction tube out of Cooling Coil warm.	3. Stoppage will probably be found in Strainer at inlet of Expansion Valve. Remove and wash in clean naphtha. Follow instruction under "Pumping Down Low Pressure Side" to remove Strainer.
4. Condenser stopped with dirt, bugs or lint.	4. High discharge pressure. Discharge line from Compressor extra hot.	4. Clean Condenser thoroughly with hose.
5. Moisture in Expansion Valve.	5. Same as 3 above, except symptoms may not appear every time unit is operated. Moisture sometimes passes valve and does not appear for several hours.	5. Same as 3 above and also install Dehydrator, see section "Air and Moisture in the System."
6. Loose or improperly insulated Expansion Valve thermostatic bulb.	6. Obvious upon inspection.	6. Tighten Thermostatic Bulb clamp and insulate Bulb from outside air.
Not Enough Air from Blower		
1. Air Filters stopped up.	1. Low suction pressure. Suction line cold.	1. Remove Air Filters and replace with new ones.
2. Blower Fan running under speed—loose or corroded connections, rheostat switch broken, battery charge low.	2. Insufficient air circulation.	2. Trace circuits for bad connections. Check Instrument Board Switch and replace if necessary. Check battery and recharge if low.
3. Cooling Coil stopped with dirt or lint.	3. Low suction pressure, suction line cold.	3. Cleanse Cooling Coil as described in section "To Clean and Deodorize the Cooling Coil."

BODY

40

Litho. in U. S. A.

Figure 222. 1941 Packard Air-Conditioning Operation, Service and Owner's Instruction, p. 40   Dwight Heinmuller

Interior of Car Not Cool, But Normal Amount of Cold Air from Blower

CAUSE	CHECK	CORRECTION
1. Windows or cowl ventilator open. Doors opened too much of time.	1. Obvious. **Note:** Ventilation requirements for crowded car, especially if occupants are smoking, may impose abnormal load, particularly in hot, moist weather.	1. Use care in regulating ventilation for smoke and in leaving doors open.
2. Car engine idling or running very slowly large portion of time.	2. High suction pressure, high discharge pressure.	2. Set engine idling speed somewhat higher. Pull out hand throttle when standing with engine running. Run Blower Fan full speed.
3. Extra moist weather—perhaps raining.	3. A large part of the cooling capacity is used to remove excess moisture from the air. The inside of the car will be more comfortable than an unconditioned car. However the temperature as read on an ordinary thermometer may not read any lower than a shaded thermometer outside the car.	3. Use very minimum of ventilation and operate Blower Fan at maximum speed.

Compressor Unit Noisy

CAUSE	CHECK	CORRECTION
1. Loose drive pulley or Compressor pulley.	1. Inspect if nut on Compressor shaft is tight, also key from Compressor shaft to pulley may be loose in keyway.	1. Tighten shaft nut and replace key if loose.
2. Squeaky drive belt, loose or greasy.	2. Belt should not have more than ½" slack. Inspect for oil on belt.	2. Tighten belt if loose. If greasy, wipe clean with naphtha and coat with powdered talc.
3. Expansion Valve Strainer stopped up with foreign matter.	3. Warm suction line. Vacuum reading low on Compound Gauge. (See Fig. 5.) This causes oil pumping by Compressor—results in knocking sound in Compressor.	3. Remove and wash Strainer in clean naphtha. Follow instructions under "Pumping Down Low Pressure Side" to remove Strainer.
4. Moisture in Expansion Valve.	4. Same as 3 above except noise does not appear every time unit is operated. Moisture sometimes passes valve and does not appear for several hours.	4. Install Dehydrator as described in section "Air and Moisture in the System."
5. Liquid Freon instead of gas being returned to Compressor through suction line. This causes oil pumping and inadequate lubrication of the Compressor. Usually caused by defective Expansion Valve or by thermostatic bulb loose on suction line.	5. Wet and cold suction line—even the Suction Service Valve and the Compressor cylinder housing and crankcase may be cold, although the engine and Compressor are running—causes a knocking sound similar to loose bearings.	5. If thermostatic bulb is loose from suction line, tighten it and also be sure it is properly insulated. If this does not correct the trouble, replace the Expansion Valve.

Litho. in U. S. A.

BODY

Figure 223. 1941 Packard Air-Conditioning Operation, Service and Owner's Instruction, p. 41   Dwight Heinmuller

CAUSE	CHECK	CORRECTION
6. Compressor loose on base.	6. Obvious upon inspection.	6. Tighten four Compressor hold-down bolts.
7. Air in system. Stoppage in valves or lines. Overcharge of Freon.	7. High discharge pressure. Causes pounding, laboring sound.	7. Correction depends on cause of the high discharge pressure. For air in system, purge air or discharge system and entirely recharge. Stoppage: remove stoppage. Overcharge of refrigerant: purge Freon to tester level.
8. Too much oil in Compressor crankcase.	8. Evidenced by a dull, thumping sound.	8. Check at Compressor oil filler plug as under "Checking Oil Level in Compressor."
9. Not enough oil in Compressor crankcase.	9. Usually not evidenced until bearings and pistons are worn and knocking or seal leaking. Denoted by very hot crankcase. Seal may be squeaking due to insufficient oil in Compressor or oil passages to seal stopped with foreign matter.	9. This condition frequently the result of leakage of oil from the system and if noticed and caught soon enough, may be corrected by adding oil to the Compressor, but if damage has already resulted, it will be necessary to change the Compressor.
10. Broken parts in Compressor.	10. Evidenced by a knocking sound similar to loose bearings.	10. Replace entire Compressor.
11. Compressor valve noise telephoned to dash.	11. Tubing fastened to dash may be improperly insulated from dash so that noise carried from Compressor is amplified by dash.	11. Insulate tubing from dash with soft rubber.

Noisy Operation Other Than Compressor

1. Loose tubing, straps, brackets or sheet metal parts.	1. See correction.	1. Trace sound for location and cause and repair as may be required.
2. Noisy Blower Fan motor.	2. See correction.	2. Remove Blower motor and Fan and replace motor.
3. Low Freon charge.	3. Hissing sound at Expansion Valve. May be no more than normal, but if excessive, may indicate low Freon charge.	3. Check Liquid Tester on Receiver. If necessary, add Freon until proper level is reached.
4. Freon passing through Cooling Coil.	4. Gurgling sound in Cooling Coil. Not audible except under very quiet conditions.	4. Entirely normal. No correction required.

Figure 224. 1941 Packard Air-Conditioning Operation, Service and Owner's Instruction, p. 42   Dwight Heinmuller

The *1941 Packard Air-Conditioning Operation, Service and Owner's Instruction* also covered service for the preceding year of the 1940 Packard Weather Conditioner air-conditioning system. It remains a mystery why Packard delayed the service manual's publication of this first-year, mechanically innovative and complex accessory until mid-year of the second year's production, especially when technicians required training to support this new technology.

Perhaps the manufacturer of the air-conditioning components, Bishop & Babcock, delayed publication by not completing and forwarding the manual until 1941. Unfortunately, no records reflect an answer to this query.

Packard management's preliminary service instructions continued from introducing the 1940 Weather Conditioner through to 1941.

Contrary to Packard standards, it recommended other than Packard authorized technicians to service the system. Based on the "Packard Service Letter, March 1, 1940," it stated, "If service is required, a good refrigeration man should be called in." (Fig. 225 and yellow arrow)

Figure 225. "1940 Packard Service Letter, March 1, 1940" (yellow arrow)   PackardInfo.com   Kevin Waltman

For reasons unknown, the typically published air-conditioning service manual that accompanied a new, major accessory in 1940 appeared an entire model year after introduction. The April 1941 publication of Packard's first air-conditioning service manual covered the basic information but offered few illustrations for the technician who encountered needed service work on Packard's mechanical refrigeration.

Later, in the so-called postwar era, the 1953 autos re-introduced factory air-conditioning. The 1953 service manuals presented the technicians with step-by-step instructions for easier comprehension, accompanied by numerous illustrations and photographs taken while working on the car.

The brevity of the 1941 manual left many questions and issues unanswered and not covered, such as:

1. The specifications, tolerances, and manufacturer's information of the Servel-based compressor.

2. Photos showing the car's componentry, such as the compressor, condenser, bracketing, hoses, receiver, evaporator with/without the inspection panel to show the cooling coils, expansion valve, a technician holding and reading the gauges during Freon measurement, the procedure to remove/replace the air filters, removal, and installation of the compressor belt, outside temperature/inside cabin temperature charts under varying outside conditions, and the CFM, Cubic Feet per Minute airflow from the cooling register at differing blower speed settings.

3. Photos of the air-conditioning control panel for the 1940 system, the knob's location for the 1941 system, and louver adjustments for the cooling register.

Author note: The lack of photos is surprising, especially considering the manual's delayed publication.

Packard's brief eleven-page service manual proved nearly identical to similarly written manuals for the one-year-only offerings of the 1941 *Cadillac Air-Conditioning Manual* and the 1941 *Chrysler Air Refrigeration System*, as the Bishop & Babcock Mfg. Co. installed each company's air-conditioning components. (Figs. 226, 227)

Refer to Chapters 4 and 5, the 1941 Cadillac and 1941 Chrysler factory air conditioning chapters.

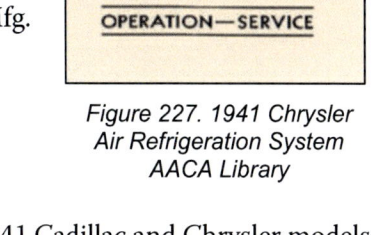

Packard enthusiastically embraced technological advancements in their innovative years by increasing passenger comfort. Competitively, the 1941 Cadillac and Chrysler models confirmed they fell behind Packard's air-conditioning leadership in 1940 by contracting for factory air-conditioning and installation with Packard's supplier, the Bishop & Babcock Mfg. Co., of Cleveland, Ohio.

Once again, 1941 Packard passengers enjoyed air-conditioned comfort and style by driving their

Figure 226. 1941 Cadillac Air-Conditioning Manual, CLCMRC (Cadillac & LaSalle Club Museum and Research Center) Paul Ayres

Figure 227. 1941 Chrysler Air Refrigeration System AACA Library

FACTORY AIR -equipped,

COOL CARS IN

COOLER COMFORT.

Chapter 3

1942 Packard Factory Air-Conditioning

The 1942 Packard line, presented on August 25, 1941, expanded upon the innovative, mid-year introduction of the streamlined 1941 Clipper sedan. For 1942, *Styling was Key*.

Packard metamorphosed from the original Clipper sedan model of 1941 into four distinct six and eight-cylinder Clipper lines. (Figs. 228, 233–237)

Figure 228. 1942 Packard Clippers brochure cover   AACA Library

The 1942 *Packard Clippers* brochure featured covers that played on the Pan American World Airways' Clipper name. Packard displayed Pan Am's unique aircraft triple tail image on the front cover and the aircraft image of its famous trans-Atlantic and Pacific Boeing 314 Clipper flying boat on the rear cover. (Figs. 228, 229)

Pan Am's slogan, "The Line of the Flying Clipper Ships," applied not only to the airline, as Packard led customers to believe, but the name also symbolized Packard's own modern and sleek Clipper automobiles. (Fig. 230)

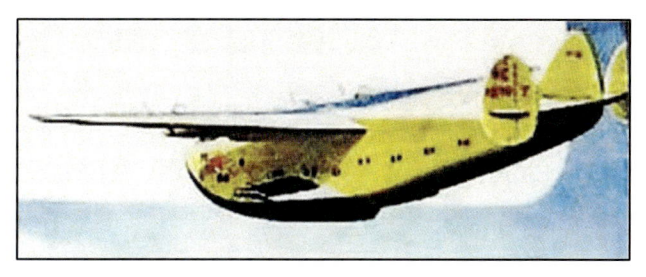

Figure 229. 1942 Packard Clippers brochure rear cover featured Pan Am Boeing 314 flying boat   AACA Library

Figure 230. 1940 Pan American Airways System and slogan
AACA Library

Additionally, the occasionally used 1941 Packard phrase, "Skipper the Clipper," served as the 1942 Packard advertising campaign's signature slogan. (Fig. 231)

Figure 231. 1942 Packard promotional slogan,
"Skipper the Clipper, Brand New for '42"
Benson Ford Research Center, The Henry Ford

Packard's white-capped persona of a Pan Am skipper offered a caveat for the future winds of war as the caption read, "Looking Ahead? You may want to keep your next car longer than usual..." (Fig. 232)

Figure 232. 1942 Packard Clippers brochure   Benson Ford Research Center, The Henry Ford

Author note: It was not until 1955 that Pan Am sued Packard for trademark infringement using its Clipper name and numerous Pan Am designations for various Packard models, promotions, and advertising campaigns. The lawsuit continued until 1958 when Packard agreed to cease using the names; Pan Am dismissed the lawsuit but incurred millions in its legal defense.

112 Chapter 3: 1942 Packard Factory Air-Conditioning

Style wise, the 1942 Packard continued the sleek, 1941 Clipper "FadeAway" fender silhouettes. (Figs. 234-237) The 1942 Clippers removed the "sparkling speedline stripes" featured in 1941 models, replacing them with rear wheel shields (fender skirts) on most models. (Figs. 235–237)

Figure 233. 1942 Packard Clippers brochure announced Clipper styling in all price ranges Benson Ford Research Center, The Henry Ford

Figure 234. 1942 Junior Clipper Six and Eight models 1942 Packard Clippers brochure Benson Ford Research Center, The Henry Ford

Figure 235. 1942 Junior Clipper Custom Six and Eight models 1942 Packard Clippers brochure Benson Ford Research Center, The Henry Ford

Factory Air: Cool Cars in Cooler Comfort   113

The brochure offered an "added value" choice of an eight-cylinder Clipper that cost only $55 more than the standard six-cylinder Clipper models. (Fig. 238)

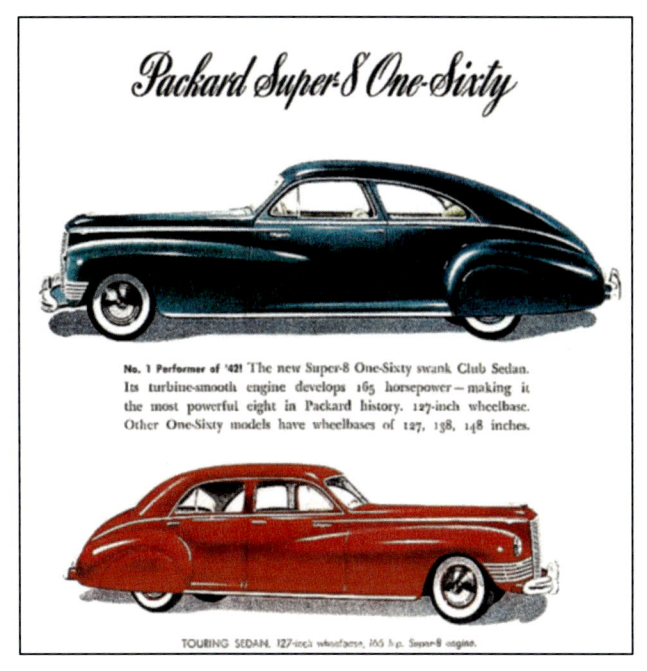

Figure 236. 1942 Senior Packard Clipper Super-8 One Sixty 1942 Packard Clippers brochure
Benson Ford Research Center, The Henry Ford

Figure 238. 1942 Packard Clipper Special and Custom models advertised the choice of an 8-cylinder engine for $55 more than a 6-cylinder model   1942 Packard Clippers brochure   Benson Ford Research Center, The Henry Ford

Figure 237. 1942 Senior Clipper Super-8 Custom One Eighty   1942 Packard Clippers brochure Benson Ford Research Center, The Henry Ford

Figure 239. 1942 Packard One Eighty Touring Sedan featured air-conditioning
Photo by Hyman Ltd. Classic Cars

Several models of the Senior Packard Super-8 One Sixty and Custom Super-8 One Eighty series retained their former styling with the slightly revised front end, as did the air-conditioned 1942 Packard Custom Super-8 One-Eighty Touring Sedan displayed here. (Fig. 239)

Chapter 3: 1942 Packard Factory Air-Conditioning

The 1942 Packard brochure continued offering its air-conditioning accessory. Its headline advertised, *Now in Its Third Successful Year! Packard Real Air-Conditioning.* (Fig. 240)

Figure 240. 1942 Packard offered air-conditioning for the third year   1942 Packard Clippers brochure Benson Ford Research Center, The Henry Ford

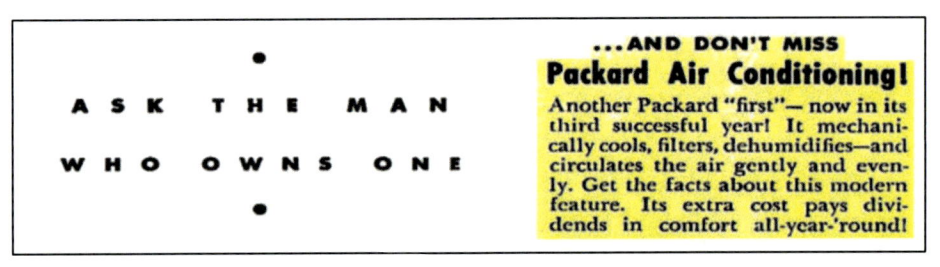

Figure 241. 1942 Packard brochure reminded customers about Packard air-conditioning and to "Ask the Man Who Owns One"   1942 Packard Clippers brochure Benson Ford Research Center, The Henry Ford

Packard followed the practice of producing colorful artwork in glossy brochures to attract potential customer's attention. (Figs. 240, 241)

For pricing information, however, the customer needed to contact a salesperson. The *Information on 1942 Packard Cars,* dated October 1, 1941, listed the Suggested Price of $325.00 for "Air-conditioning, without a heater (All models except convertible (Figs. 242–243)

Author note: The CPI determined that the $325 price for the 1942 air-conditioning accessory would add the equivalent of $5,414 to a new car in January 2021 dollars.

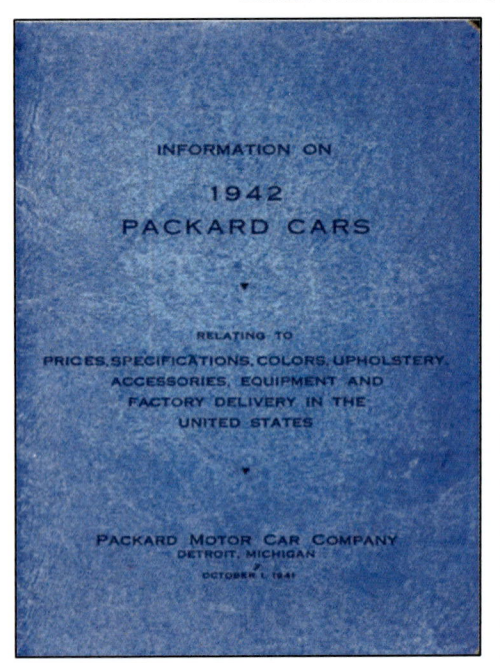

Figure 242. Information On 1942 Packard Cars, October 1, 1941 AACA Library

Figure 243. Information On 1942 Packard Cars, "Accessories and Equipment"   AACA Library

Always considered the salesperson's most comprehensive reference source, the (1942) *Packard Data Book* identified model choices, standard equipment, styling cues, and specifications. (Fig. 244)

The *Packard Data Book* stated: (Fig. 245)

Packard Air-Conditioning goes far beyond the mere ventilators or air filters offered on some cars. Packard Air-Conditioning means cooling by mechanical refrigeration, as well as filtering and ventilating.

Figure 245. 1942 Packard Data Book air-conditioning information, p. 40   Dwight Heinmuller

Figure 244. 1942 Packard Data Book cover AACA Library

Air-Conditioning: Packard supplied pertinent information about its air-conditioning accessory. (Figs. 245–247)

The operator needs no mechanical knowledge to operate the system based on similar household refrigeration principles. A dashboard control switch regulates the speed and volume of the circulated, cooled, and filtered air.

Packard introduced mechanical refrigeration as an optional accessory for the 1940 model year. According to the Packard Vice President of Engineering, W. H. Graves, the company sold nearly 2,000 air-conditioning units from 1940 through the truncated sales year of 1942.

Packard's phantom rendering of a 1942 Clipper Sedan illustrated the air-conditioning componentry, including the compressor, condenser, receiving tank, and cooling coils. (Fig. 246)

The Cooling Cycle: The Freon-12 refrigerant entered the compressor in gaseous form, compressed it to 180 pounds per square inch, and entered the condenser in front of the radiator. Airflow over the condenser cooled the gas to a liquid, moved it to a receiving tank, and then traveled through a smaller diameter tube to the expansion valve and cooling coils in the trunk-mounted evaporator.

Liquid Freon-12 expanded and changed to a gas as it absorbed heat from the cooling coils when the warm cabin recirculated air passed over them, releasing cold air. Now in the gaseous form again, the refrigerant returned to the compressor, where the cycle repeated.

Cool Air Circulation: Warm, recirculated air passed over the chilled cooling coils. A sirocco-type (squirrel-cage) electric fan blew the cooled air upward through a louvered cool air grille along the headliner towards the front seat passengers.

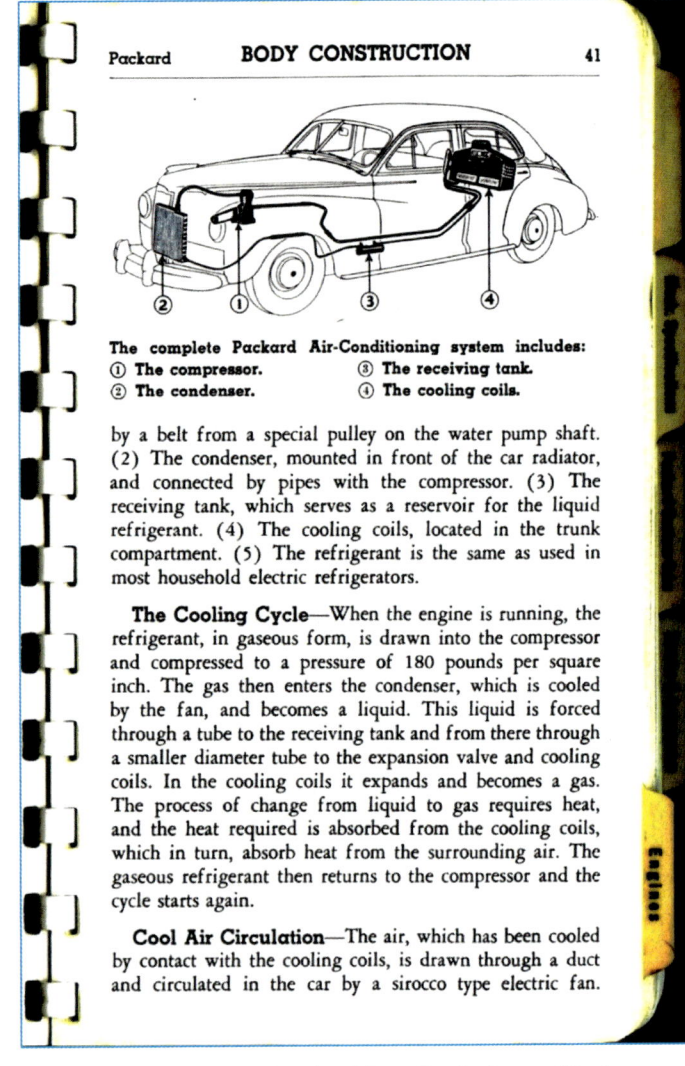

Figure 246. 1942 Packard Data Book air-conditioning information, p. 41   Dwight Heinmuller

Fan-forced cooled air circulated past the passengers, returning under the rear seat from the fan's suction action, through air filters into the trunk-mounted evaporator, where this warmer, recirculated air passed again over the cooling coils. (Fig. 247)

Dehumidification: As the recirculated air passed over the cooling coils, moisture condensed on the coils and drained to the ground.

Air Filtration: Recirculated air passed through air filters for removal of dust, dirt, and pollen.

Author note: Refer to the 1940 Packard chapter for air filter images and the return airflow pathway beneath the rear seat and through the cooling coils.

Cooling Effectiveness: Packard rated the cooling capacity as "two tons of melting ice in a twenty-four-hour period" at a car speed of sixty mph.

Insulation: Air-conditioned cars received specially applied insulation in the roof, sides, and floor.

Author note: Refer to the 1941 Packard chapter for comprehensive data book information about standard and specially applied insulation in air-conditioned cars.

Factory Installation: Air-conditioning components required factory installation because of special handling in production.

Author note: This obfuscated statement glossed over the factory location of the system's installation.

According to *Packard: A History of the Motor Car and the Company*, author L. Morgan Yost stated on page 365:

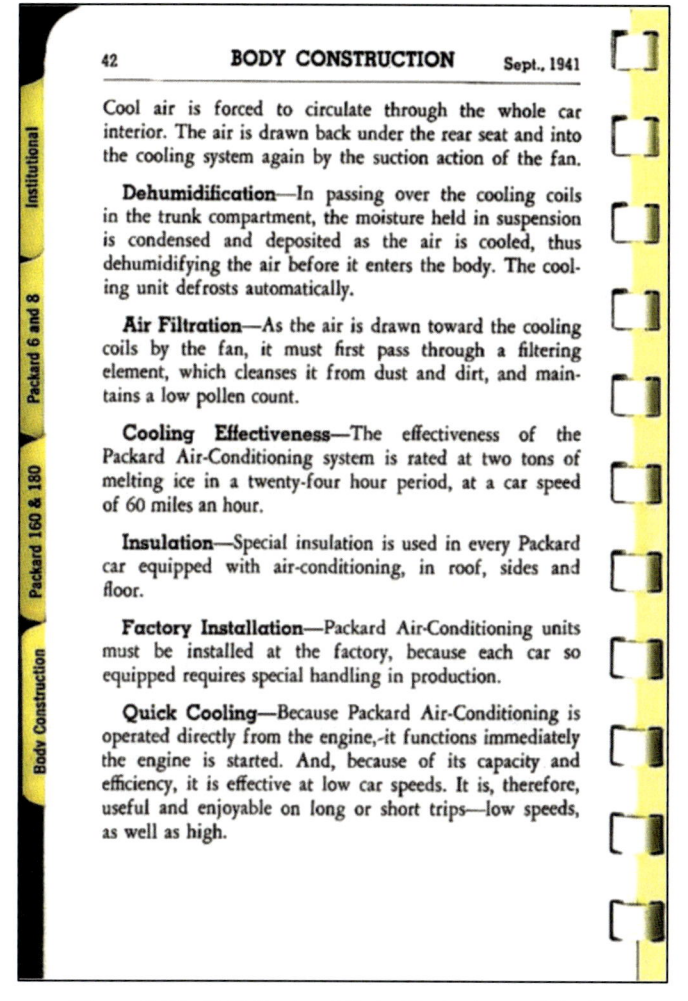

Figure 247. 1942 Packard Data Book air-conditioning information, p. 42   Dwight Heinmuller

Special insulation for cars so equipped was provided at the factory, but, although noted in Packard literature as "factory installed," the refrigerating unit itself was actually fitted into the car by Bishop & Babcock in Cleveland, and in theory, vehicles ordered with it were shipped there, the unit was installed and then the cars were forwarded to the dealers.

Quick Cooling: The air-conditioning system operated immediately through a direct engine connection. It supplied efficient, effective cooling at low and high speeds. "Because Packard Air-Conditioning is operated directly from the engine, it functions immediately [after] the engine is started."

Author note: Even though the control panel showed an "off" position, the compressor ran continuously, which potentially lead to a mechanical breakdown much sooner than the expected design lifespan. It also decreased the engine gas mileage.

A year after the reintroduction of automotive air-conditioning, the 1954 Cadillac, Buick, Oldsmobile, Pontiac, and Nash models achieved a technological milestone by introducing an electromagnetic clutch that disengaged the compressor when turned off, extending its compressor service life and saving gasoline.

118   Chapter 3: 1942 Packard Factory Air-Conditioning

Packard distributed *Packard Promotional Pointers,* dated November 17, 1941, to the dealership salespersons, titled, "What They Think About Styling." (Fig. 248)

It confirmed, *"*Every Packard salesperson will agree that the new Clippers for 1942 can be justly termed 'hot.'" Continuing, "They have aroused more enthusiastic comments than any of the other new 1942 models."

It based this reaction upon the number of 1940 and 1941 competitive cars traded in on the new Packard Clippers.

The report stated that its styling captivated the public, led by the streamlined FadeAway front fender design.

The New Yorker magazine automotive writer stated,

> To me the Packard FadeAway fender, which merges gently with the body near the middle of the door panel, is far more graceful than the flaring appendages that protrude from some of the extreme '42 cars.

In a subtle reference to Cadillac, the article continued,

> Flaring appendages that protrude isn't so bad and very politely describe what so many must think about those bolted-on-boxes, which vainly try to imitate the genuine fade-away effect. Real FadeAway fenders melt or disappear gently into the body, without any crack, seam, or caulking. (Fig. 249)

Figure 248. 1942 Packard Promotional Pointers Vol. VIII, No. 4, November 17, 1941 styling commentary   PackardInfo.com   Kevin Waltman

Figure 249. 1942 Cadillac 62 featured bolted-on front and rear pontoon fender extensions   80 YEARS OF CADILLAC LaSALLE   Walter McCall

A *Packard Senior Cars 1942* brochure included a small press kit titled *Never Was Quality So Important*. (Figs. 250, 251) It observed the threatening world war atmosphere while underscoring and reiterating Packard's quality construction.

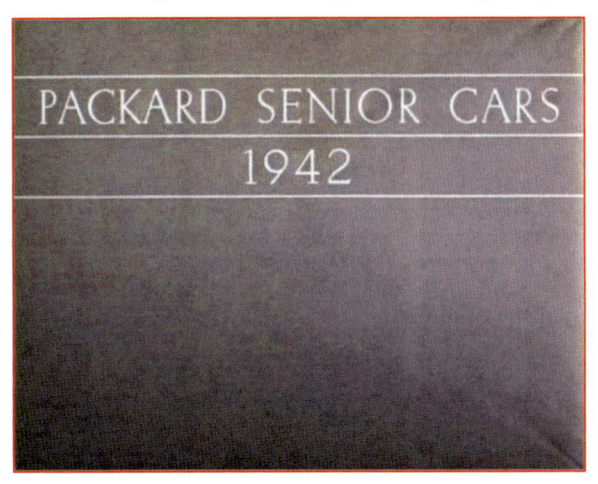

Figure 250. Packard Senior Cars 1942 Press Kit   Regress Press   Jeff Steidle

Figure 251. 1942 Packard brochure "Never Was Quality So Important," included in Senior Cars Press Kit   Regress Press   Jeff Steidle

Packard used 1942 through 1944 calendar images to emphasize the changing winds of war for the auto industry. (Fig. 252)

Importantly, it continued to offer "Genuine Air-Conditioning" for Packard customers.

It gained defense contracts to manufacture Supermarine PT boat engines and Rolls-Royce aircraft engines while maintaining its premier Packard automobile quality. (Fig. 253)

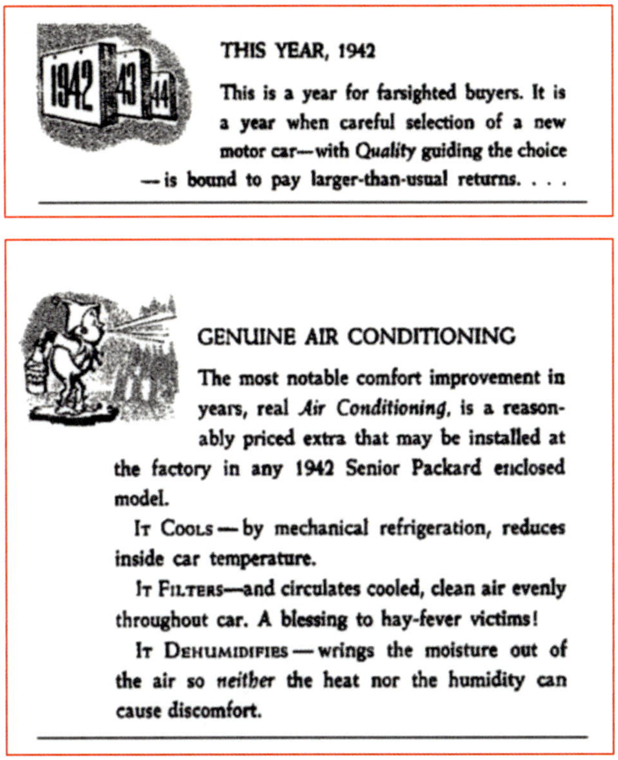

Figure 252. 1942 Packard brochure "Never Was Quality So Important"   Regress Press   Jeff Steidle

Figure 253. 1942 Packard brochure "Never Was Quality So Important,"   Regress Press   Jeff Steidle

120   Chapter 3: 1942 Packard Factory Air-Conditioning

The January 1942 *Popular Science* magazine, priced at $.15, announced the 1942 Packard air-conditioner with a Clipper phantom image denoting the cooling components' images of the compressor, the circulation pattern of the air-conditioned passenger cabin, and the trunk-mounted evaporator. (Figs. 254, 255)

Figure 254. Popular Science, January 1942 ABS

Air Conditioner Assures Year-Round Car Comfort

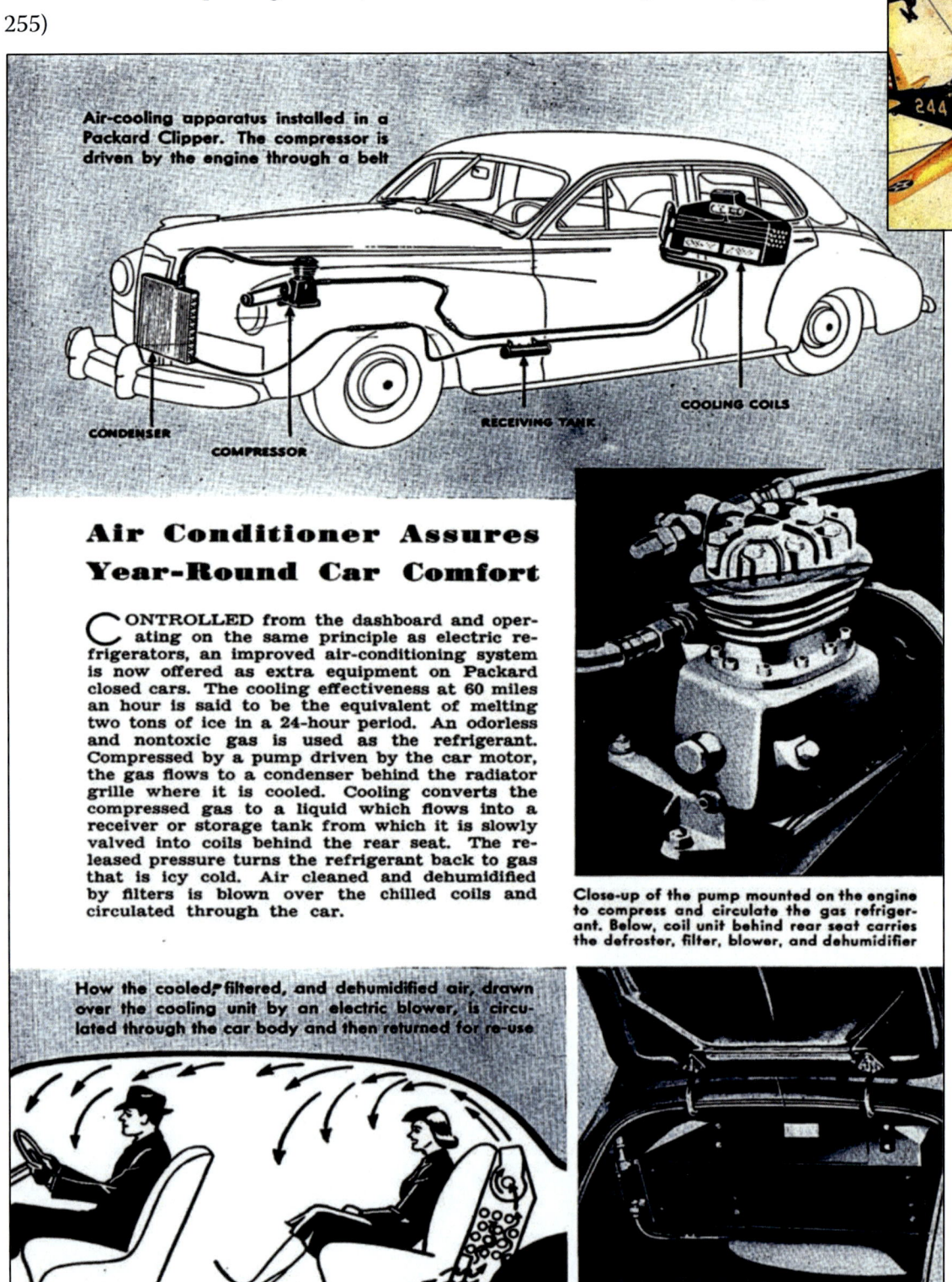

CONTROLLED from the dashboard and operating on the same principle as electric refrigerators, an improved air-conditioning system is now offered as extra equipment on Packard closed cars. The cooling effectiveness at 60 miles an hour is said to be the equivalent of melting two tons of ice in a 24-hour period. An odorless and nontoxic gas is used as the refrigerant. Compressed by a pump driven by the car motor, the gas flows to a condenser behind the radiator grille where it is cooled. Cooling converts the compressed gas to a liquid which flows into a receiver or storage tank from which it is slowly valved into coils behind the rear seat. The released pressure turns the refrigerant back to gas that is icy cold. Air cleaned and dehumidified by filters is blown over the chilled coils and circulated through the car.

Close-up of the pump mounted on the engine to compress and circulate the gas refrigerant. Below, coil unit behind rear seat carries the defroster, filter, blower, and dehumidifier

How the cooled, filtered, and dehumidified air, drawn over the cooling unit by an electric blower, is circulated through the car body and then returned for re-use

Figure 255. 1942 Packard air-conditioning article, "Air Conditioner Assures Year-Round Car Comfort" Popular Science, January 1942   ABS

Spotlight on Packard

Presenting the 1942 Packard Clipper Eight, Equipped with Factory Air-Conditioning

General Douglas MacArthur's WW II staff car
Story and photos by Jim Hollingsworth
Permission granted for article reprint by Bud Juneau, Editor
Northern California Packards, The Packard Club, P.A.C.

The Packard Club's Spring 1975 edition of *The Cormorant* (Fig. 256) featured an article written by Jim Hollingsworth, the well-known Packard authority. (1930–2012)

His article traced General Douglas MacArthur's 1942 Clipper Eight staff car used through his WWII service in Australia, the Philippines, and to the occupation of Japan after the war.

Particularly interesting is that his special order car included top-of-the-line accessories, including air-conditioning, Electromatic clutch, overdrive, radio, heater, defroster, and, as a unique touch, a painted Cormorant hood ornament.

A photograph of General Douglas MacArthur precedes the article. (Fig. 257)

Figure 256. 1942 Packard Clipper Eight former WWII staff car of General Douglas MacArthur, equipped with air-conditioning The Cormorant, Spring 1975   The Packard Club

Figure 257. General Douglas MacArthur "Back Then... MacArthur's '42 Packard" Bud Wells

Mr. Hollingsworth captured the essence of the history surrounding General MacArthur's 1942 Clipper Eight with the title "Destined to Survive." (Figs. 258–267, 270)

DESTINED TO SURVIVE

General MacArthur's 1942 Clipper is still alive and stock — after having many close calls with extinction.

by Jim Hollingsworth
Dallas, Texas

For a particular Packard Clipper built in the fall of 1941, fate was to dictate an unusual destiny. First, it was to have a famous owner, Gen. Douglas MacArthur; it would survive World War II; it would escape by three days an appointment to be demilitarized and lost to history; it would, by moments, survive a plan to make it a parts car; and finally by less than four weeks, once again destiny would dictate it would escape the possible fate of a crusher at some unknown salvage yard.

The car, a 1942 Packard 120 eight Clipper custom, motor number E 318750 D, serial number 1512-5747, was built in the fall of 1941 specifically for General MacArthur who had ordered a personal car from Packard Motor Car Company. Attached to a letter from Packard was the General's personal check, returned with an expression of gratitude for his confidence in their product and promising delivery, with their compliments, of a Packard for his personal use along with the next military order destined for Australia.

This Packard had every conceivable accessory offered for the civilian market. It was a 2011 series 120 Clipper with factory air conditioning, overdrive, electromatic clutch, radio, heater defroster, fender skirts, and strangely, a cormorant mounted on top of the standard 1942 Packard Clipper hood ornament. The car was olive drab, with the usual white star and other military markings. External military equipment included adjustable louvered shutters fitted over the headlights, two powerful driving lights, a blackout light, convoy lights, a siren, snap-fasteners for the canvas covers for the front and rear windshields, and flag mounts. Internal military hardware included a Thompson sub machine gun mount, a fire extinguisher, and a first aid kit.... all within easy reach of the driver. None of the exterior trim was ever chrome plated, but rather it was metal painted olive drab ... this included the cormorant. All interior chrome had been painted over with a light coat of "gold looking" paint which eliminated all bright work on the instrument panel. The military lights were controlled by three toggle switches mounted to the left of the instrument grouping. One of the three switches changed the operation of the horn ring from the standard horn to the siren.

The casting date of the engine block was 11-13-41, very late in the short production run for the 1942 model, which ended in early February of 1942. The Clipper identification chart shows motor numbers from E 300001 to E 319329. Therefore, this car with motor number E 318750 D was followed by only 579 other 120 Packard 8's.

This product would follow MacArthur throughout the war— from Australia to the Philippines, to the occupation of Japan. It was destined to survive World War II along with its famous owner.

2

Figure 258. 1942 Packard Clipper Eight article, "Destined to Survive," General Douglas MacArthur's former WWII staff car, equipped with air-conditioning   The Cormorant, Spring 1975   Jim Hollingsworth

Mr. Bud Wells wrote, "Back Then...MacArthur's '42 Packard," in November 2014. He quoted a letter from Packard's President, Mr. Max Gilman, to General MacArthur, dated February 25, 1942. The letter thanked him for his special order of a 1942 Clipper Eight, returned his check for $2,600, and informed him that Packard employees and families contributed the purchase funds for his impending delivery. (Fig. 259)

Author note: The CPI computes the $2,600 price for the 1942 Clipper Eight would cost the equivalent of $43,045 in 2021 dollars.

> We are deeply honored that your interest in the Packard Motor Car has prompted your personal order for one for military service. A shipment of our vehicles will be leaving our plant in March for the military; these will be the last cars to leave our facilities until the end of hostilities.
>
> Included in this shipment is one Packard sedan model 2011, motor number E318750D which is a gift to you from all the employees and their families of the Packard Motor Car Company. We hope this gift will express to you our heartfelt thanks, appreciation, and support for your fantastic efforts on behalf of our country during this period of great peril.
>
> Your two thousand six hundred dollars is herein returned with our compliments.

Figure 259. Packard President Gilman's letter of February 25, 1942 to General Douglas MacArthur with refunded check for his special order 1942 Clipper Eight and notice of a gift from Packard employees "Back Then...MacArthur's '42 Packard" Bud Wells

Figure 260. 1942 Clipper Eight WWII staff car once owned by General Douglas MacArthur, viewed in Dallas Texas The Cormorant, Spring 1975 Article and photo by Jim Hollingsworth

Mr. Hollingsworth displayed an image of the car in Dallas. (Fig. 260)

The paperwork included specifications for a different car, the 1942 Model 2003, Super-8 One Sixty. (Fig. 261, yellow arrow)

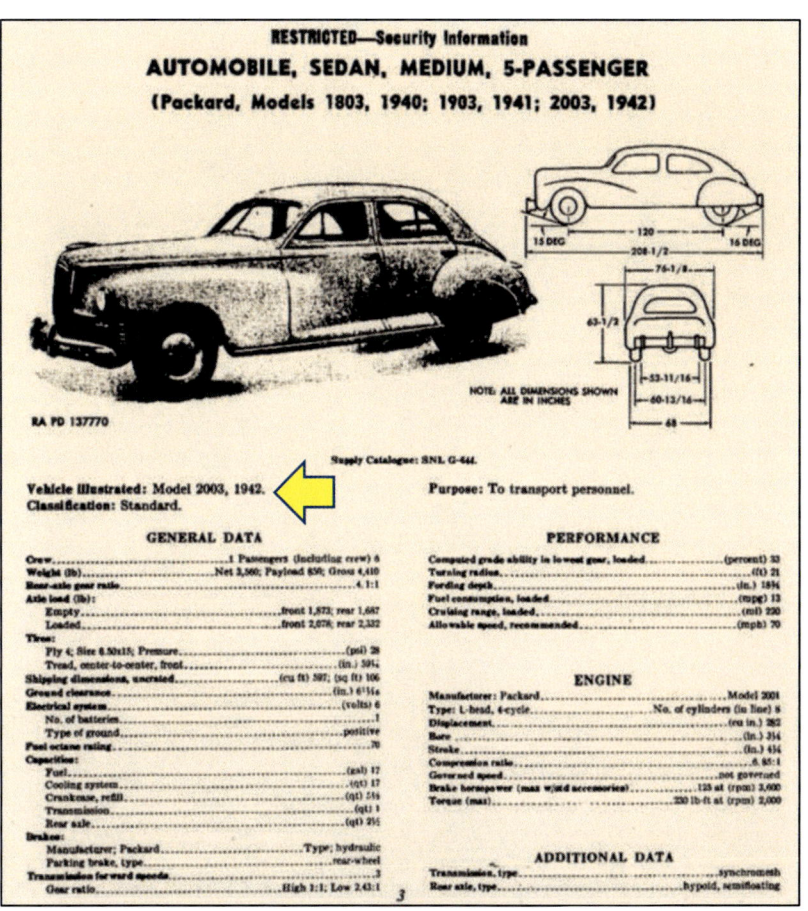

Figure 261. 1942 Clipper specification's sheet for a different model (yellow arrow) The Cormorant, Spring 1975 Article and data by Jim Hollingsworth

124   Chapter 3: 1942 Packard Factory Air-Conditioning

Canvas attached with snap fasteners covers the rear window for transporting. Packard equipped the front windshield similarly. (Fig. 262)

Figure 262. 1942 Clipper Eight, once owned by General Douglas MacArthur, equipped with air-conditioning. Shown with rear window canvas cover for transporting.
The Cormorant, Spring 1975   Article and photo by Jim Hollingsworth

The 1942 Clipper Eight, once owned by General Douglas MacArthur, featured WWII-required painted grille-work and headlights covered with adjustable louvered shutters.

The Clipper Eight featured air-conditioning and other optional accessories, including auxiliary driving lights, a siren, flag mounts, and snap fasteners for a canvas cover to protect the windshield and rear window during transport.

A painted Packard Cormorant hood ornament rested atop the hood. (Fig. 263)

Author Hollingsworth posed in uniform next to the Clipper Eight.

Figure 263. 1942 Clipper Eight former WWII staff car of General Douglas MacArthur. Author Hollingsworth posed beside painted grille that displayed optional installed accessories
The Cormorant, Spring 1975   Article and photo by Jim Hollingsworth

Mr. Hollingsworth related post-WWII owners, long-term storage, and a future owner's desire to save the air-conditioning system. (Fig. 264)

In September of 1948, following the formal occupation of Japan, the car was retired from service. General MacArthur would release the car to himself from military service and make it a gift to his favorite and loyal driver, who shall for purposes of this article remain anonymous (let's call him Smitty).

General MacArthur arranged through a navy friend, who was skipper of the aircraft carrier Princeton, for transportation of the car to San Diego. At that point it was loaded aboard a flat bed military truck and transported to Fort Sam Houston in San Antonio, Texas, where it was released to the custody of the former driver, Smitty.

Smitty drove the car to his Dallas, Texas, home with intentions of converting the car to his civilian use.

An appointment was made to remove all military hardware and repaint the car a bright new postwar color. Since the body shop was busy, Smitty agreed to bring the car back the following week. That night, after putting the Packard away in his small, dirt floored garage, Smitty died in his sleep.

Again, the famous Packard, destined to survive, had escaped a conversion to civilian status.

In 1968, Don MacLellan, the present owner of this car, was restoring a 1948 Packard limousine. The manager of the local parts store had mentioned to Don that a widow in the neighborhood had an old Packard for sale. After several attempts to interest Don in the car had failed, the parts store owner withheld a parts order from Don until he would at least "look" at the car and "talk" to the widow.

The garage, with no windows and a single door, had become almost inaccessible in the twenty years that had passed. A tree had grown up in front of the garage door which had to be cut down. The tree stump was too large to remove, so a notch was cut in the door to permit Don to gain entry into the dark garage with its dirty old Packard prisoner.

As the history of the car was still unknown at this point, Don purchased the car with the idea that at least the factory air conditioning could be cut out and used in the restoration of his 1948 limousine.

THE UNITED STATES GOVERNMENT
CERTIFICATE OF RELEASE OF A MOTOR VEHICLE

The undersigned DEPARTMENT or ESTABLISHMENT of the UNITED STATES GOVERNMENT hereby certifies that the motor vehicle described herein, the property of the UNITED STATES GOVERNMENT, has been transferred this 18 day of SEPTEMBER, 19 48, to:

GENERAL DOUGLAS MACARTHUR

GENERAL HEADQUARTERS ALLIED POWERS, TOKYO JAPAN

PACKARD 42 2001 CLIPPER PACKARD L-HEAD, 8-CYCLE

AUTOMOBILE SEDAN

IN LINE 8 GASOLINE N.A. N.A. PNEUMATIC 120"

N.A. N.A. N.A.

The undersigned DEPARTMENT or ESTABLISHMENT further certifies that this transfer was the first transfer of the herein described motor vehicle in ordinary trade and commerce subsequent to acquisition by the UNITED STATES GOVERNMENT.

RELEASED FROM THE PERSONAL SERVICE OF GENERAL DOUGLAS MACARTHUR BY ORDER OF SUPREME COMMANDER ALLIED POWERS PACIFIC GENERAL DOUGLAS MACARTHUR

UNITED STATES ARMY — TOKYO

PACIFIC DISPOSITION TASK FORCE — JAPAN

SUPREME COMMANDER (SCAP) — Douglas MacArthur

CLIPPER IDENTIFICATION CHART

YEAR		1942					
MOTOR NUMBER		E1501 E12906		E300001 E319329		E500001 E503366	CE500001 CE503366
MODEL and No. of CYLINDERS		110 Six		120 Eight		160 Super Eight	180 Super Eight
BODY STYLE	No. of Pass.	Spec. 2000	Cust. 2010	Spec. 2001	Cust. 2011	2003	2006
Sedan	6	1582	1502	1592	1512	1572	1522
Club Sedan	6	1585	1505	1595	1515	1575	1525
Business Coupe	3	1588		1598			

5

Figure 264. 1942 Clipper Eight continuation of owner history, post-WWII government release, and Clipper Identification Chart
The Cormorant, Spring 1975 Article and data by Jim Hollingsworth

126   Chapter 3: 1942 Packard Factory Air-Conditioning

The article continues with close-up views of the 1942 Clipper Eight: adjustable headlight shutters, optional siren, auxiliary driving lights, cowl-mounted Packard Motor Company delivery plate, and painted Cormorant hood ornament. (Fig. 265)

Detail shots showing the war lights and siren above, the cowl delivery plate below, and the curious-for-1942 hood ornament.

Figure 265. 1942 Clipper Eight, once owned by General Douglas MacArthur, with close-up views of head and fog lamps, five-star general plaque, siren, Packard Motor Car Company cowl-mounted delivery plate, and painted Cormorant hood ornament
The Cormorant, Spring 1975   Article and photos by Jim Hollingsworth

The author photographed the *pièce de résistance*, i.e., the historically significant air-conditioning componentry of General Douglas MacArthur's WWII-era 1942 Clipper Eight staff car. Photos reveal the louvered cool air grille, trunk-mounted evaporator, and the Servel air-conditioning compressor. (Fig. 266)

Three photos of the stock 1942 Packard Air Conditioning system: package shelf outlet (top left), trunk blower unit (top right), and compressor in engine compartment (below). Besides this rare option, the car was equipped with Overdrive and Electromatic Clutch, both of which work well in 1975.

Figure 266. 1942 Clipper Eight air-conditioning componentry views of the cool air grille with louver control handle, trunk-mounted evaporator, and the Servel compressor   The Cormorant, Spring 1975   Article and photos by Jim Hollingsworth

128 Chapter 3: 1942 Packard Factory Air-Conditioning

The central chromed instrument panel remained unchanged from the 1941 model, while Packard updated the driver side speedometer and instruments, and the passenger side clock. Two 1941 images supplement and enlarge the two-speed pull-out air-conditioning switch of the 1942 Clipper Eight. Debossed "Cooling" letters on the upper right side switch edge identify its function. (Figs. 267–269, and yellow arrows)

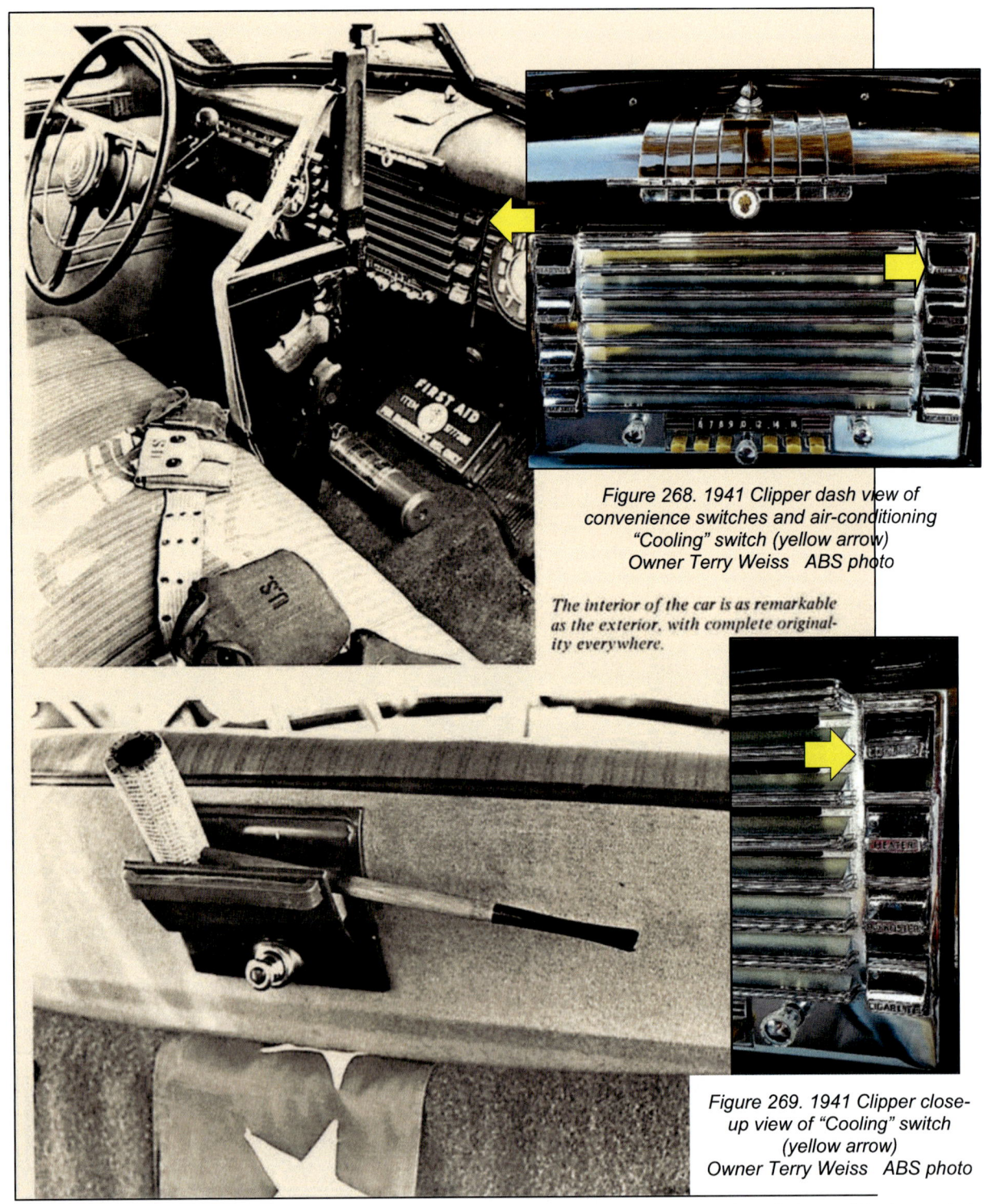

Figure 268. 1941 Clipper dash view of convenience switches and air-conditioning "Cooling" switch (yellow arrow)
Owner Terry Weiss   ABS photo

The interior of the car is as remarkable as the exterior, with complete originality everywhere.

Figure 269. 1941 Clipper close-up view of "Cooling" switch (yellow arrow)
Owner Terry Weiss   ABS photo

Figure 267. 1942 Clipper Eight Staff car, once owned by General Douglas MacArthur, shows location of air-conditioning control switch (yellow arrow)   The Cormorant, Spring 1975   Article and photos by Jim Hollingsworth

Factory Air: Cool Cars in Cooler Comfort   129

A photograph displays General Douglas MacArthur's 1942 Clipper Eight alongside a WWII vintage Jeep. Packard Life Member, Don MacClellan, owns the car as of the 1975 article. (Fig. 270)

Two rugged WWII vehicles, both original and valuable. Don MacClellan, a Life Member of our Club, has owned the MacArthur Packard since 1968.

Several months passed and the widow called to ask Don to please send her a letter relieving her of any responsibility for the car since he had not removed the Packard from her old garage. Forgetting his promise, after more weeks passed, the widow called again to ask Don if he wanted some old papers that belonged with the car. He said he did, but felt she really was reminding him to remove the car from her property.

The following Saturday, Don put his acetylene torch in the back of his pickup, ready to go pick up the Packard and remove the air conditioner. As he climbed into his truck, Don's postman waved and walked over with his mail. Thumbing quickly through the mail for several expected business letters, Don saw a large envelope from the widow. Upon opening the envelope, he found the military release, letters from Packard Motor Car Company, and all of the documents which identified this historical vehicle as something "special."

This time, by moments ... destiny had saved the Packard from becoming "just another parts car."

After removing the car, Don then discussed the history of the car with Smitty's widow and was shown family photo albums with pictures of MacArthur, the Packard, and Smitty, in various places throughout the Pacific. Although she would not part with the pictures at that time, Don felt a return visit might bring about an agreement to loan the pictures to him to be reproduced.

After a lengthy overseas business trip, Don attempted to call the widow but found that the telephone had been disconnected. A visit to the widow's house was in vain. She had died just after Don had removed the car from the garage and with only one distant relative, who was uninterested in her meager belongings and any involvement in her funeral expenses, the county removed her belongings from the rented house and arranged for the burial of the widow.

Had Don waited any longer to take possession of the Packard, it could well have gone the way of her remaining property ... destruction. Once more, the Packard was destined to survive.

This now famous Packard has remained untouched ... the dirt and grime of the past 33 years are still on its engine. After 74,000 miles, the car still drives nicely and all systems remain operational.

This is one PACKARD ... DESTINED TO SURVIVE.

Figure 270. 1942 Clipper Eight, General Douglas MacArthur's WWII staff car, equipped with air-conditioning. The air-conditioned car is photographed with vintage WWII Jeep.   Commentary by B. Wells
The Cormorant, Spring 1975   Article and photo by Jim Hollingsworth

Another 1942 Clipper Eight served as a staff car during World War II. The following image displayed General Eisenhower's meeting with General Omar Bradley at 12th Army Group HQ in August 1944. (Fig. 271)

His Packard, similar to General MacArthur's car, featured air-conditioning.

Author note: Packard also supplied traditional One Sixty model staff cars for the Army.

Figure 271. 1942 Clipper Eight staff car, equipped with air-conditioning. General Eisenhower (right) met with General Bradley in August 1944   olive-drab.com

Retaining the previous body style while incorporating an updated front end, an air-conditioned 1942 Packard One Eighty offered insights to Packard's most prestigious, luxurious model, courtesy of Hyman Ltd. Classic Cars. (Figs. 272–275, 277–289)

*Figure 272. 1942 Packard One Eighty Touring Sedan, equipped with air-conditioning
Photo by Hyman Ltd. Classic Cars*

Packard's 1942 One Eighty represented the traditional styling of the Senior cars. The 138-inch wheelbase, dark blue touring sedan, body type number 1542, featured a six-passenger seating capacity, air-conditioning, a gray-blue wool broadcloth upholstery, automatic window regulators (power windows, introduced in 1941), genuine walnut trim, rear compartment footrests, and a storage compartment behind the front seat.

Factory Air: Cool Cars in Cooler Comfort 131

An air-conditioned Packard required a special order and the addition of extra insulation in the body. After modification, it shipped the car to the Bishop & Babcock Mfg. Co. of Cleveland, Ohio, for installation of the air-conditioning componentry.

The following images reflect the air-conditioning controls, compressor placement, cool air outlet with louver control handle, and the trunk-mounted evaporator with nameplates. (Figs. 273–275, 277–281, 285)

Figure 273. 1942 Packard One Eighty brown colored, air-conditioning control knob (yellow arrow), on left side of the defroster control's beige color knob   Photo by Hyman Ltd. Classic Cars

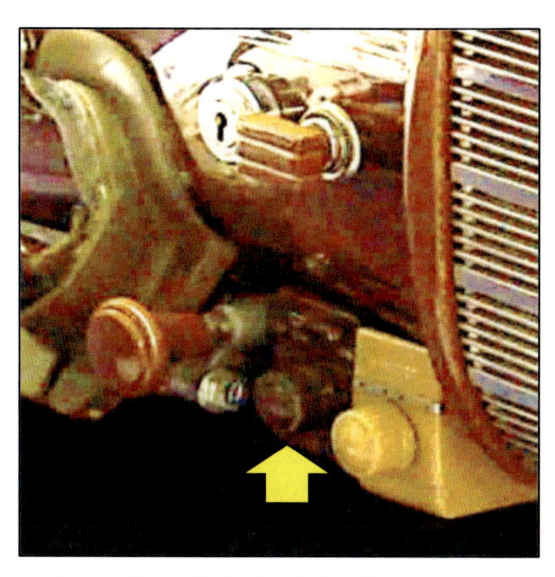

Figure 274. 1942 Packard One Eighty close-up of brown air-conditioning control knob (yellow arrow) on left side of defroster control's beige knob   Photo by Hyman Ltd. Classic Cars

The 1942 Packard One Eighty engine bay displays a replacement air-conditioning compressor that features the AC ports on the compressor's top. It compares to an original 1940 Packard One Eighty Servel compressor in design and placement. (Fig. 275, 276)

Figure 275. 1942 Packard One Eighty replacement air-conditioning compressor Photo by Hyman Ltd. Classic Cars

Figure 276. 1940 Packard One Eighty original Servel air-conditioning compressor Owner and photo West Peterson

132 Chapter 3: 1942 Packard Factory Air-Conditioning

The 1942 Packard One Eighty reveals *ultra-luxe* passenger comfort and style. Contributing features include air-conditioning, the adjustable louvered cool air grille behind the rear seat (Fig 278), luxurious wool broadcloth seating, and automatic window regulators, known now as power windows, located under each assist strap (Fig 277).

A center folding armrest encourages individual passenger seating with conveniently positioned reading lights. Built-in ashtrays with a lighter provide convenience for smokers. Pewter inlaid genuine walnut window garnish moldings enhance the cabin's ambiance.

Figure 277. 1942 Packard One Eighty luxury seating, amenities, and air-conditioning offer passenger comfort of the highest grade.   Photo by Hyman Ltd. Classic Cars

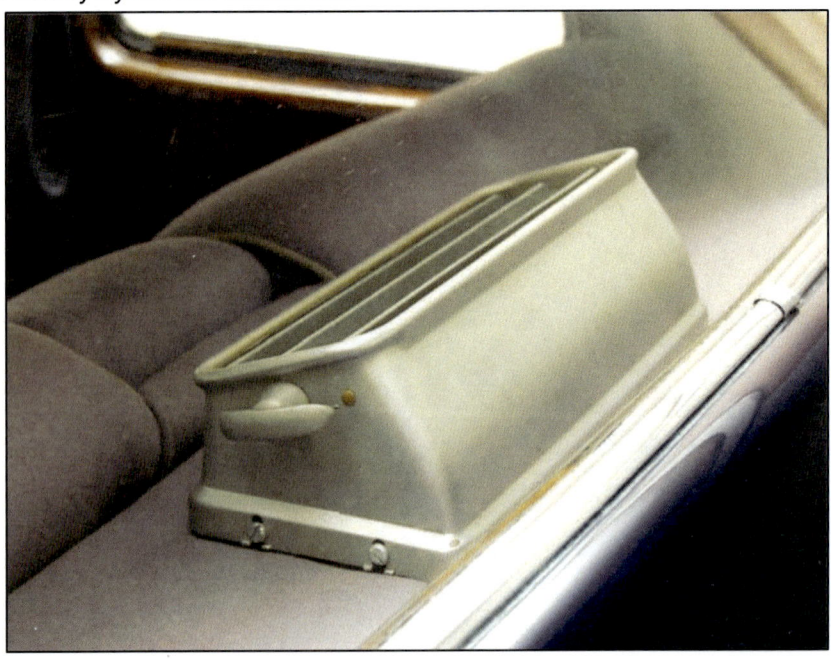

*Figure 278. 1942 Packard One Eighty close-up view of air-conditioning cool air outlet and new handle for louver adjustment
Photo by Hyman Ltd. Classic Cars*

Packard's air-conditioning system placed the cooling coil, called the evaporator, in the trunk's forward area. (Fig. 279) Tubing and hoses from the engine-mounted components connected to the enclosed expansion valve and the evaporator where a fan circulated the air through the cooling coils and blew the cooled air upward into the cabin as it followed the headliner. (Fig. 280)

Figure 279. 1942 Packard One Eighty trunk-mounted air-conditioning evaporator that housed the expansion valve, cooling coil, and blower   Photo by Hyman Ltd. Classic Cars

Figure 280. 1942 Packard One Eighty rear view shows air-conditioning cool air outlet in rear window (yellow arrow)   Photo by Hyman Ltd. Classic Cars

Packard contracted with The Bishop & Babcock Mfg. Co. of Cleveland, Ohio, to install the special ordered air-conditioning system in their cars.

Printed information on the evaporator nameplate reads, "Packard Air Conditioner by The Packard Motor Car Company, Detroit, Michigan." (Fig 281)

The lower nameplate identifies the manufacturer and its U. S. patents, Bishop & Babcock Mfg. Co.

*Figure 281. 1942 Packard One Eighty close-up views of nameplates identifying "Packard Air Conditioner," manufactured under U. S. Patents by "The Bishop & Babcock Mfg. Co., Cleveland, Ohio"
Photo by Hyman Ltd. Classic Cars*

134   Chapter 3: 1942 Packard Factory Air-Conditioning

The Senior Series 1942 Packard One Eighty represented the most prestigious automobile for wealthy passengers' transportation. This "Custom" model featured extra conveniences not available in lesser automobiles. Several of those special touches follow in images from Hyman Ltd. Classic Cars. (Figs. 282–289)

Figure 282. 1942 Packard One Eighty featured "automatic window regulators." (yellow arrows) Introduced in 1941, known now as power windows, the rear window switches on the upper dash duplicated those for the rear door window operations.   Photo by Hyman Ltd. Classic Cars

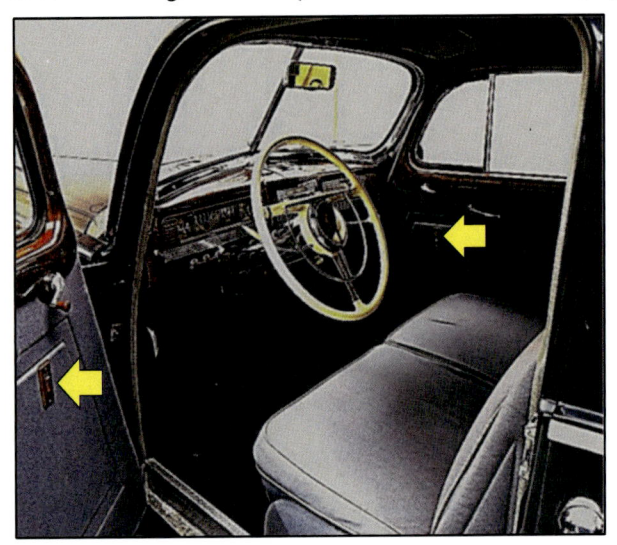

Figure 283. 1942 Packard One Eighty view of driver and passenger door power window switches (yellow arrows)   Photo by Hyman Ltd. Classic Cars

Figure 284. 1942 Packard One Eighty close-up view of driver door power window switch Photo by Hyman Ltd. Classic Cars

Figure 285. 1942 Packard One Eighty center opening rear door reveals inlaid walnut window garnish moldings, plush seating accommodations, and the air-conditioning cool air outlet behind the rear seat. (yellow arrow)   Photo by Hyman Ltd. Classic Cars

Figure 286. 1942 Packard One Eighty close-up view of rear passenger power window control located beneath assist strap (yellow arrow) Photo by Hyman Ltd. Classic Cars

Factory Air: Cool Cars in Cooler Comfort 135

Figure 287. 1942 Packard One Eighty continued the traditional dual side mounts, accented by art deco-inspired speedline stripes   Photo by Hyman Ltd. Classic Cars

Figure 288. 1942 Packard One Eighty close-up view of the cloisonné type "Bonnet Louver Nosepiece Medallion"   Photo by Hyman Ltd. Classic Cars

Figure 289. 1942 Packard One Eighty Cormorant hood ornament gracefully adorns this quintessential luxury automobile Photo by Hyman Ltd. Classic Cars

The Japanese bombing of Pearl Harbor, December 7, 1941, provoked President Franklin Roosevelt to declare war on Japan with Congressional approval on December 8, 1941.

Earlier, the United States maintained a so-called neutrality position after Poland's German invasion in September 1939. The automobile companies' watchful waiting position continued auto production during the 1941 model year; however, they contracted with munitions builders for massive arms production during that time. Increased war production involvement proved inevitable.

Automobile production dwindled as plants converted to war materiel production. About a month after the United States entered World War II, President Roosevelt issued Executive Order No. 9040, creating the War Production Board (WPB) to supersede the Office of Production Management (OPM). Automotive production fell under WPB auspices.

As of February 9, 1942, Packard built the last 1942 Clipper Eight at the Detroit East Grand Boulevard plant. (Fig. 290) Packard shut down to concentrate on the Supermarine 1,350-hp, 4,000 CID, marine V-12 PT boat engines (a Patrol Torpedo boat, torpedo-armed fast attack vessel used by the United States Navy and others in World War II, Fig. 291), and their redesign for Rolls-Royce Merlin aircraft engines installed in the P-40 and P-51 aircraft.

Rapidly, Detroit adopted the moniker the "Arsenal of Democracy" for its manufacturing contributions to the war effort.

Dwight Jon Zimmerman stated in the *Defense Media Network*, "In 1941, civilian automobile production totaled about 3.6 million vehicles. In 1942, that number dropped to less than 1.15 million. Postwar civilian production numbers did not reach 1941 levels until 1949."

Sales by model, according to *Packard: A History of the Motor Car and the Company*, for the truncated 1942 Packard model year ending February 9, 1942, yielded:

Figure 290. 1942 Clipper Eight marked as the last car produced on February 9, 1942. Packard President George Christopher stands at right. DPL NAHC

Figure 291. Packard built PT boat Supermarine engines during WWII. 1942 Packard brochure, "Never Was Quality So Important." ABS

Six	11,325
Eight	19,199
One Sixty	2,580
One Eighty	672
1942 PACKARD TOTAL SALES	33,776

The famous Packard ox-yoked-shaped grille of the 1940 Packard One Eighty featured the first sealed beam headlamps in bullet-shaped housings. (Fig. 292) Significantly redesigned, the 1941 model featured hood, headlamp, fender, and window modifications (Fig. 293), as described in a 1941 Packard brochure. (Fig. 294)

Packard introduced the 1941 Clipper with a stylized, narrow grill. (Fig. 295) In 1942, the Senior Packard's vertical grille and flanking bilateral, horizontal grille motif replaced the previous vertical cooling grilles shown in Figs. 292–294 (Fig. 296)

*Figure 292. 1940 Packard One Eighty grille
Owned and photo by West Peterson*

The front lines of beauty! ... distinctly new but still distinctively Packard! Headlamps faired into fenders and surmounted by streamlined parking lamps add to rakish, clean-cut appearance—enlivened by sparkling cooling grilles.

*Figure 293. 1941 Packard One Eighty grille
Owned by M/M Sal Saiya   Jack Swaney photo*

*Figure 294. 1941 Packard grille modifications
described in brochure   ABS*

*Figure 295. 1941 Packard Clipper grille
Owner Terry Weiss   ABS photo*

*Figure 296. 1942 Packard One Eighty grille
Photo by Hyman Ltd. Classic Cars*

138 Chapter 3: 1942 Packard Factory Air-Conditioning

The "Packard Motor Car Information Forum," on their website, *PackardInfo.com,* supplies a valuable knowledge base for Packard owners. The following articles refer to pertinent air-conditioning information published during the World War II years of 1942, 1943, and 1944, indexed from their website. (Figs. 297, 302–306)

PACKARD SERVICE LETTER   VOL. 15   NO. 22   NOVEMBER 15, 1941

AIR CONDITION REFRIGERANT

Because of defense restrictions, it is impossible to obtain an adequate supply of Freon for air conditioned equipped cars. Methyl Chloride will be used. This material was very generally used as a refrigerant for many years before Freon was developed. It is slightly toxic and contains a small percentage of odor identifier which will give adequate notice of leakage. Warning plates will be attached where the installation is made at the factory. One plate is attached to the Blower Motor Control Switch Knob and one to the Evaporator Case Assembly.

When Methyl Chloride is installed in the field, the warning plates should be obtained and attached.

Piece No. 379637—Air Conditioned Evaporator Case Assembly Warning Plate
Piece No. 379638—Blower Motor Control Switch Knob Warning Plate
Piece No. 379639—Screw (2)

When Methyl Chloride is used as a refrigerant, this fact and the information on the plates should be brought to the attention of the customer.

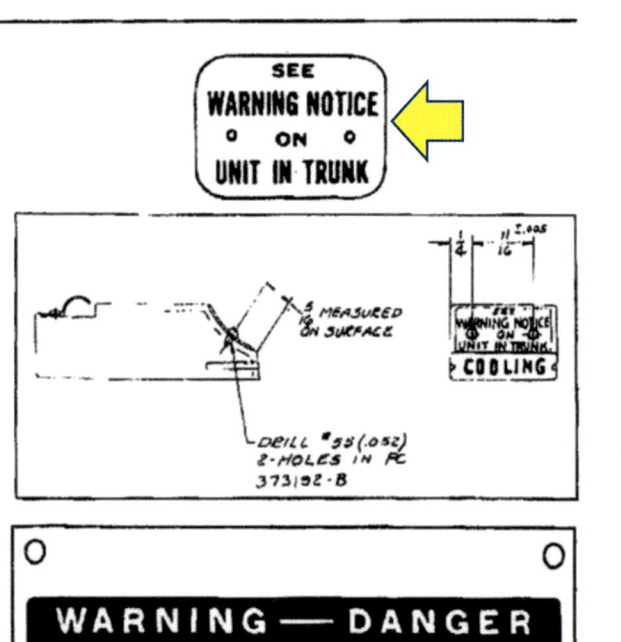

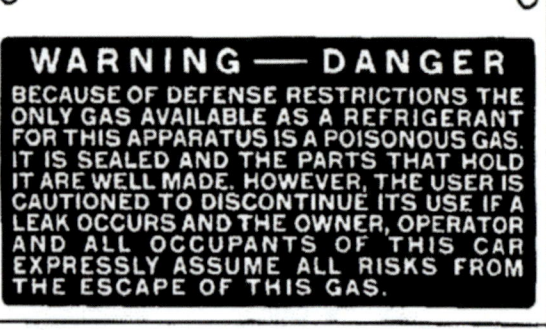

WARNING — DANGER
BECAUSE OF DEFENSE RESTRICTIONS THE ONLY GAS AVAILABLE AS A REFRIGERANT FOR THIS APPARATUS IS A POISONOUS GAS. IT IS SEALED AND THE PARTS THAT HOLD IT ARE WELL MADE. HOWEVER, THE USER IS CAUTIONED TO DISCONTINUE ITS USE IF A LEAK OCCURS AND THE OWNER, OPERATOR AND ALL OCCUPANTS OF THIS CAR EXPRESSLY ASSUME ALL RISKS FROM THE ESCAPE OF THIS GAS.

Figure 297. 1942 Packard Service Letter Vol. 15, No. 22, November 15, 1941. Warning Notices announced the substitution of Freon-12 refrigerant with Methyl Chloride, due to WWII defense restrictions   PackardInfo.com   Kevin Waltman

Historically significant Packard air-conditioning control assembly images represent Paul Fluckiger's contribution by Silverstone Motorcars, North Andover, Massachusetts.

Photographed from a formerly owned air-conditioned 1942 Packard One Eighty limousine, Model 2008, body number 1550, Mr. Fluckiger captured two images of the air-conditioning switch that retained the sticker referred to in Figure 297, (yellow arrow), "See Warning Notice on Unit in Trunk." The sticker notified the owner of the WWII-required Methyl Chloride substitution for the Freon-12 refrigerant in the air-conditioning system.

Located under the left dash, the 1942-era sticker on the air-conditioning switch appeared attached with adhesive, not screws, per the Service Letter. The sticker appeared transparent, allowing visualization of lettering on the switch. (Figs. 298–300)

Imaged from another angle, under the sticker, in white capital letters, "COOLING" described the control and appeared at the center top; for the fan speeds, "OFF" appeared below, bottom left, followed by "HIGH," then "LOW" lettering. (Figs. 299, 300)

Figure 298. 1942 Packard One Eighty Limousine air-conditioning switch under dash   Photo by Paul Fluckiger, Silverstone Motorcars, North Andover, MA

"DEFROST, OFF, HIGH, MED," and "LOW" lettering identified the defroster switch located under the right dash that matched the air-conditioning switch lettering style. (Fig. 301)

Figure 299. 1942 Packard One Eighty limousine warning sticker, "See Warning Notice on Unit in Trunk," attached above the block lettering on the air-conditioning switch   Photo by Paul Fluckiger

Figure 300. 1942 Packard One Eighty limousine air-conditioning switch warning sticker placed over air-conditioning operation lettering of, "COOLING, OFF, HIGH, LOW"   Photo by Paul Fluckiger

Figure 301. 1942 Packard One Eighty limousine defroster switch and lettering   Photo by Paul Fluckiger

SERVICING AIR CONDITIONING UNITS FOR SPRING OPERATION

There are certain instructions that should be followed when cars equipped with air conditioning units have been disconnected for the winter. We would suggest that you reread the instructions issued in April, 1941 and mailed with Service Letter Volume 15, No. 8 of April 15, 1941.

The automobile air conditioning system when placed in service for the summer months should be carefully checked to see that there has been no loss of refrigerant gas from the system due to slow leaks during the winter months when no cooling was required. This check is readily made by slightly cracking the test cock on the side of the liquid receiver under the car. A discharge of milk-like spray indicates a full charge. If there is no discharge except the invisible gas, the system is low in refrigerant and needs recharging. This test should be made only after the compressor has been running not less than 15 minutes.

In the event the above test indicates and insufficient charge of refrigerant, additional refrigerant should be added until the test cock shows the correct discharge of milk-like spray.

The oil level in the compressor should then be checked to determine if there has also been a loss of oil. As a rule, a loss of refrigerant due to a small leak does not necessarily mean that the oil has also been lost; however, it is recommended that the oil level be checked. This test is accomplished by allowing the compressor to run from 20 minutes to a half hour to make sure that the oil is properly distributed throughout the system. After this period, the compressor is then stopped and both service valves closed. The oil filler plug is then removed from the crankcase and a pencil or clean wire is inserted so that it touches the bottom of the crankcase. When removed, it should show an oil level of approximately $1\frac{1}{2}"$ in the crankcase. If the oil level is below $1\frac{1}{2}"$, additional oil should be added to give a normal level. *Only special compressor oil can be used.* Care should be taken in inserting the test rod as an undue agitation causes the oil in the crankcase, which is saturated with refrigerant, to foam.

In the event it has been necessary to add refrigerant to the system, a careful check should be made of all tubing connections, joints, etc. in the system so that the leak can be located and fixed.

Figure 302. 1942 Packard Service Letter, Vol. 16, No. 9, May 1, 1942. Recommended servicing of air-conditioning system for summer operation PackardInfo.com   Kevin Waltman

PACKARD SERVICE LETTER

VOL. 16, NO. 18 — SEPTEMBER 15, 1942

SERVICE MANAGER'S PERSONAL COPY

AIR CONDITIONER COMPRESSORS

We have recently examined several air conditioner compressors which were stuck so tightly that they would not turn.

These compressors stuck because of a copper deposit on the cylinder walls and pistons. The deposit, in turn, was caused by an acid condition in the refrigerant which attacked the copper in the system and carried it into the compressor.

The acid condition is apt to develop unless the system has been properly serviced, but it can be avoided if proper care is used.

We have also urged that equal care be used with the dehydrator if it is found necessary. Please refer to the Service Letter Supplement covering the air conditioner, particularly the portion describing the use of a dehydrator.

The complete Dehydrator Equipment is covered by ST-5186, and additional material may also be obtained from us. One pound cans are carried under ST-10088 and five pound cans under ST-10089.

The collection of the copper deposit in the compressor is serious not only because of its effect on the compressor itself but also because it indicates that corrosion has attacked other parts of the system and may make the air conditioner inoperative.

Corrosion can be avoided if the system is kept free from moisture and if the proper dehydrator and compressor oil are used.

Figure 303. 1942 Packard Service Letter Vol. 16, No. 18, September 15, 1942. Servicing of air conditioner compressors PackardInfo.com   Kevin Waltman

PACKARD SERVICE LETTER
VOL. 17, NO. 19 — **OCTOBER 1, 1943**

SERVICING AIR CONDITIONER UNITS REFRIGERANT LEAKAGE

In checking for refrigerant leaks in Clipper installations pay particular attention to the high pressure gas line running from the compressor to the condenser, forward of the radiator.

This line passes over the steel arch which forms the upper portion of the radiator core cradle and in some cases it will be found that the tube bears tightly against the cradle. When this is the case any "rock" in the motor must be taken by the portion of the line between the compressor and the cradle, and the load on the connection at the compressor will be greatly increased. The connection may develop a leak.

Arrow Indicates Contact Point

This condition can be corrected by heating the cradle at the point where the contact occurs (after removing the line) and driving it down with a hammer far enough to provide a clearance. The movement of the motor will then be taken up by the full length of the line.

AIR CONDITIONERS IN COLD WEATHER

If desired, the air conditioning system can be kept in operation during cold weather.

On the other hand, it is a simple matter to disconnect the compressor belt. When this is done the compressor does not operate during the period when it is not being used.

The removal of the compressor belt is not important as far as the slow driven, city operated car is concerned, but the benefit derived from removing the belt increases with the car speed.

The fan belt, particularly, is heavily loaded at the higher speeds, and its life will be increased if the compressor is disconnected on cars driven at higher than city speeds.

There is a further step which can be taken on all air conditioned cars. The system can be "pumped down" so that the refrigerant is stored in the receiver tank. This eliminates the possibility of leakage in the lines. When this is done the compressor belt should also be disconnected.

When the air conditioner is put into service in the spring it should be of course checked for leaks, and you should make sure that the compressor carries the proper amount of oil.

Figure 304. 1943 Packard Service Letter Vol. 17, No. 19, October 1, 1943. Servicing air-conditioner refrigerant leaks and cold weather preparations PackardInfo.com Kevin Waltman

AIR CONDITIONER FILTERS

We have used two types of filters in the air conditioning system, as shown in the attached print.

We do not know which type of filter went into any particular car but the filters are interchangeable if these directions are followed.

The filter shown in figure No. 1 contains a small wire gauze screen at the outlet end. When the threaded connection is tightened, a seal is obtained by forcing the cone on the end of the screen against the tapered seat.

The filter shown in figure No. 2 does not use the small screen. The seal is obtained by screwing the threaded connection against the flat washer which bears against the flat seat.

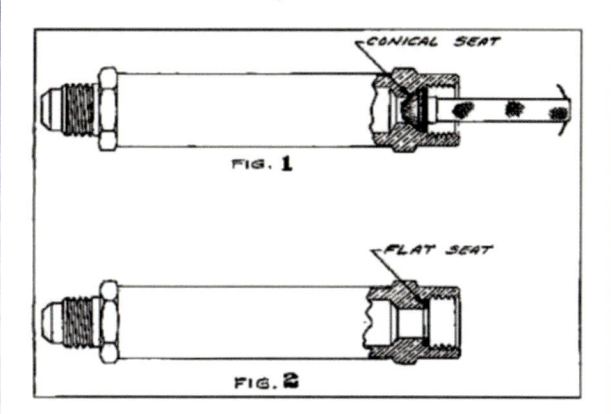

The small screen has been found unnecessary (the large screen inside the filter does the work), but it must be used with the filter having the tapered seat in order to form a tight joint.

When the filter with the flat seat is used the screen should be thrown away.

Figure 305. 1943 Packard Service Letter Vol. 17., No. 20, October 15, 1943. Servicing of air conditioner filters PackardInfo.com   Kevin Waltman

PACKARD Service Counselor

PARTS ★ ACCESSORIES ★ PRODUCT ★ PROFITS
INSTITUTIONAL   PROMOTIONAL

VOL. 18, NO. 6   JUNE, 1944

Shop Talk

METHYL CHLORIDE IN AIR CONDITIONING SYSTEMS

It is no longer possible to secure Freon for air conditioning systems. The available supply is being used for war purposes.

You will, therefore, use methyl chloride when the system is recharged. (See the Service Letter of November 15, 1941.) Methyl chloride has been in use for years for this purpose, and is familiar to every one who has worked on refrigerators or air conditioning systems.

If an air conditioning system containing Freon requires an additional charge, methyl chloride can be added. The remaining Freon need not be discarded. When the two gases are mixed there will be a slight change in the pressures shown on the gauges. The high pressure side will show the Freon pressure and the low pressure side will indicate methyl chloride, whose pressure is somewhat lower. The low pressure side, however, should still be within the old limit of 20 to 40 pounds.

Do not use the Halide torch for checking leaks when the system contains methyl chloride, since this method was developed to indicate Freon. Use a soap and water solution as described in the air conditioning instructions.

Except as noted above the two refrigerants are used in the same way and require the same precautions in order to operate satisfactorily. Leakage, moisture and dirt are the three enemies of any system regardless of the refrigerant used.

Figure 306. 1944 Packard Service Counselor Vol. 18, No. 6, June 1944. Use of Methyl Chloride in air-conditioning systems   PackardInfo.com   Kevin Waltman

As Packard entered a previously unknown business environment after February 9, 1942, it built war materiel, not automobiles. Its newly negotiated contracts required assembly line reconfiguration or a complete removal for the duration.

Packard's advertising department did, however, offer a War Bonds promotional headline that attracted attention, "Announcing the new Packard for '43." (Fig. 307)

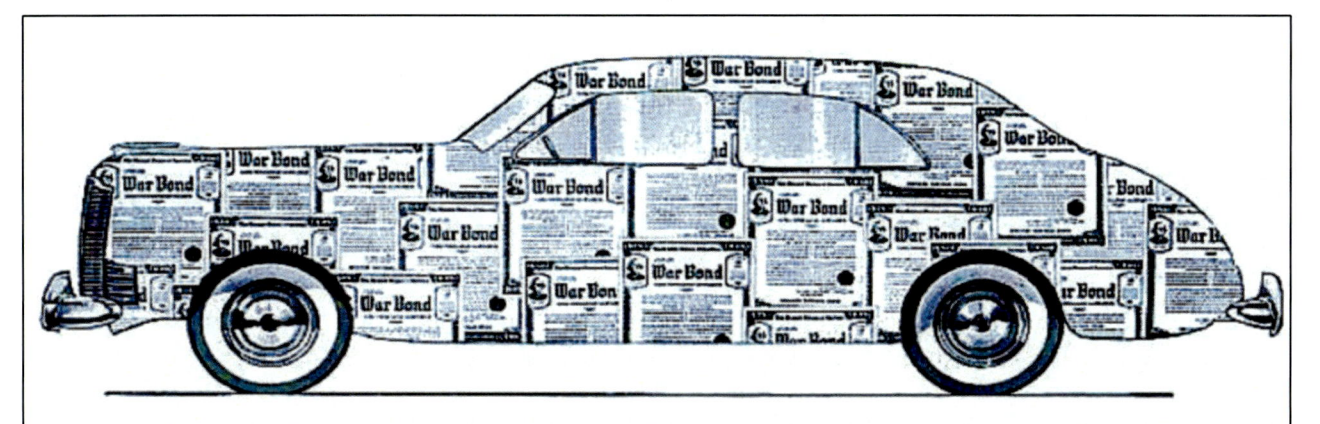

Figure 307. War Bond advertisement announced the new 1943 Packard   ABS

LOOKING FORWARD...PACKARD FACTORY AIR-CONDITIONING for 1953

VOLUME 1: THE INNOVATIVE YEARS, 1940 to 1942, represented a bold look into mechanical refrigeration comfort for Packard, Cadillac, and Chrysler passengers. Factory air-conditioning's initial momentum ended in February 1942, when President Roosevelt and the United States Congress halted all automobile sales in favor of war materiel manufacturing. The post-WWII consumer rushed to purchase all things manufactured, especially newly built, previously war-rationed automobiles. In the rapid conversion process, automakers did not offer such an expensive luxury accessory as air-conditioning. Not until 1949 did automotive production match prewar levels.

Based upon a group of Texas entrepreneurs in the late 1940s, nascent automotive air-conditioning companies produced thousands of units for southwestern car owners, so many that the Big Three Detroit automakers acquiesced and introduced factory-installed air-conditioning in 1953.

Follow the journey in VOLUME 2: 1953–THE MAGICAL YEAR.
Detroit automakers introduced factory air-conditioning in the following cars:

1. Packard contracted for Frigidaire units from General Motors.
2. General Motors launched Frigidaire factory air for Cadillac, Buick, and Oldsmobile.
3. Chrysler Corporation introduced Airtemp factory air for Chrysler, Imperial, De Soto, and Dodge.
4. Ford Motor contracted with NOVI, an independent company, for the Lincoln.

Six full-color chapters will examine the 1953 factory air-conditioned cars, accentuated by richly contrasted, legible images from factory brochures, service publications, owner's literature, magazine testing, and advertising. "Spotlight" sections display privately owned air-conditioned cars.

FACTORY AIR:

COOL CARS IN

COOLER COMFORT

AN ILLUSTRATED HISTORY OF AUTOMOTIVE FACTORY AIR-CONDITIONING
VOLUME 2: 1953–THE MAGICAL YEAR.

Continue with VOLUME 3: UP-FRONT FORERUNNERS and OLD-STYLERS,

featuring 1954–1956 factory air-conditioned cars.

Continue with VOLUME 4: UP-FRONT LATE BLOOMERS,

featuring 1957–1960 factory air-conditioned cars.

Chapter 4

1941 Cadillac Factory Air-Conditioning
and one produced in 1942
Exclusive color coverage of the only three air-conditioned 1941 Cadillacs in existence

The 1941 Cadillac continued its luxury status as "THE STANDARD OF THE WORLD." Its famous Flying Lady hood ornament decorates the stunning William L. "Bill" Mitchell-designed Cadillac Fleetwood Series Sixty Special, owned by Dr. Richard Zeiger. (Fig. 308)

Figure 308. 1941 Cadillac Fleetwood Series Sixty Special Flying Lady hood ornament
Owner Dr. Richard Zeiger   Ronald Verschoor photo

An adage in our society states, "To be the finest of anything is to be called the Cadillac of its kind."

From *80 Years of Cadillac LaSalle,* Walter M. P. McCall wrote in 1980:

> Ownership of a Cadillac, or even to be seen riding in one, is irrefutable evidence that one has indeed arrived. No rationalizing its solid engineering, remarkable resale value, or its velvety ride can deter the popular notion that this is, more than anything else, the ultimate symbol of material success.

Earlier, Henry Leland's reputation for part interchangeability and close manufacturing tolerances, based on his experience in small arms manufacturing, earned Leland's Cadillac the English Dewar Trophy for parts standardization in 1908.

After joining General Motors, Cadillac received another Dewar Trophy in 1912 for the battery-powered self-starter. (Appropriately, *The Self-Starter* is the Cadillac & La Salle Club's monthly magazine title.)

In 1914, Cadillac introduced the industry's first mass-produced V-8 engine while incurring other manufacturers' scorn, resulting in one of the most famous automotive ads ever written. In January 1915, Theodore F. MacManus wrote "The Penalty of Leadership." (Fig. 309) The prose offered a refreshing viewpoint—then and now.

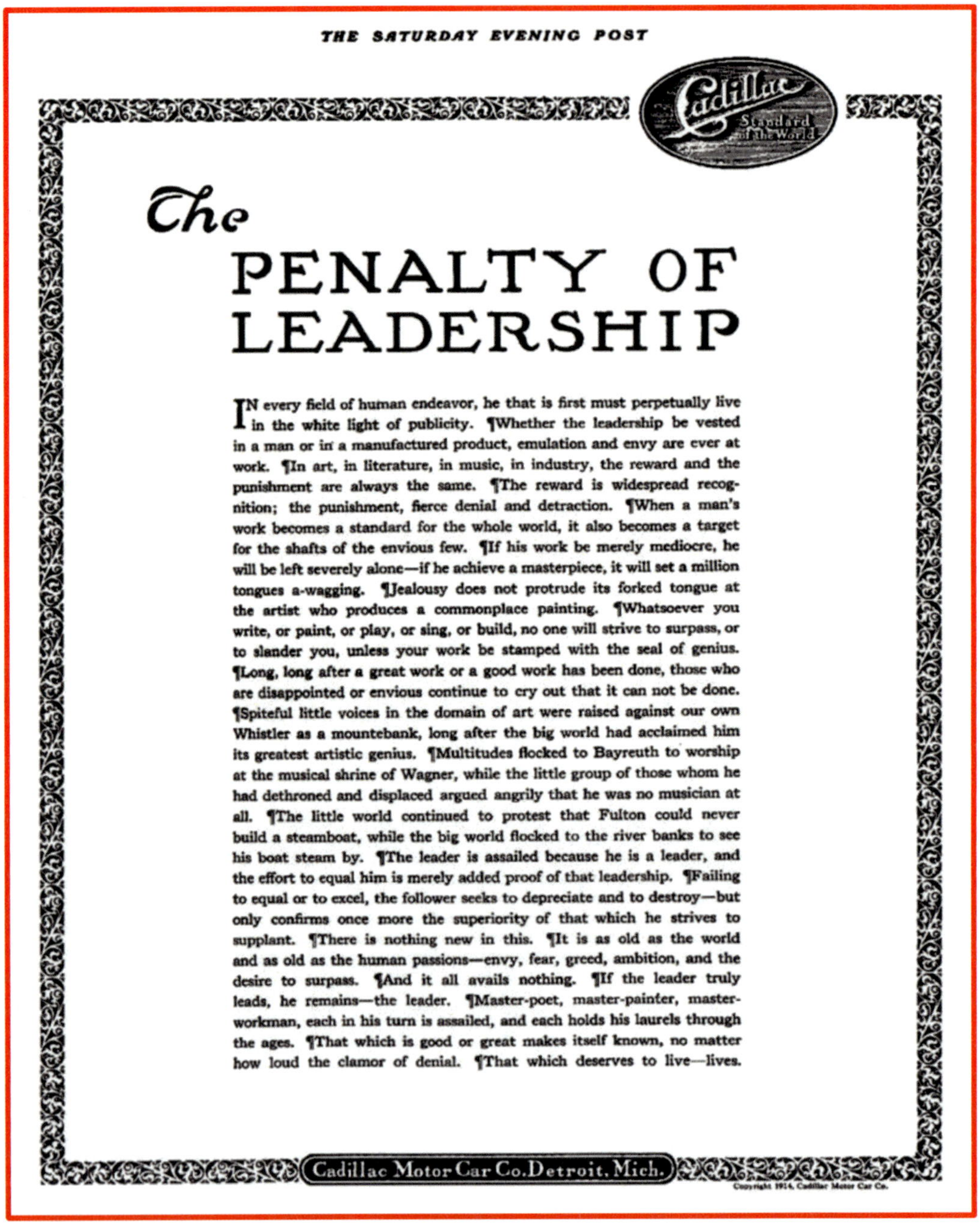
THE SATURDAY EVENING POST

Cadillac Standard of the World

The PENALTY OF LEADERSHIP

IN every field of human endeavor, he that is first must perpetually live in the white light of publicity. ¶Whether the leadership be vested in a man or in a manufactured product, emulation and envy are ever at work. ¶In art, in literature, in music, in industry, the reward and the punishment are always the same. ¶The reward is widespread recognition; the punishment, fierce denial and detraction. ¶When a man's work becomes a standard for the whole world, it also becomes a target for the shafts of the envious few. ¶If his work be merely mediocre, he will be left severely alone—if he achieve a masterpiece, it will set a million tongues a-wagging. ¶Jealousy does not protrude its forked tongue at the artist who produces a commonplace painting. ¶Whatsoever you write, or paint, or play, or sing, or build, no one will strive to surpass, or to slander you, unless your work be stamped with the seal of genius. ¶Long, long after a great work or a good work has been done, those who are disappointed or envious continue to cry out that it can not be done. ¶Spiteful little voices in the domain of art were raised against our own Whistler as a mountebank, long after the big world had acclaimed him its greatest artistic genius. ¶Multitudes flocked to Bayreuth to worship at the musical shrine of Wagner, while the little group of those whom he had dethroned and displaced argued angrily that he was no musician at all. ¶The little world continued to protest that Fulton could never build a steamboat, while the big world flocked to the river banks to see his boat steam by. ¶The leader is assailed because he is a leader, and the effort to equal him is merely added proof of that leadership. ¶Failing to equal or to excel, the follower seeks to depreciate and to destroy—but only confirms once more the superiority of that which he strives to supplant. ¶There is nothing new in this. ¶It is as old as the world and as old as the human passions—envy, fear, greed, ambition, and the desire to surpass. ¶And it all avails nothing. ¶If the leader truly leads, he remains—the leader. ¶Master-poet, master-painter, master-workman, each in his turn is assailed, and each holds his laurels through the ages. ¶That which is good or great makes itself known, no matter how loud the clamor of denial. ¶That which deserves to live—lives.

Cadillac Motor Car Co. Detroit, Mich.

Copyright 1914, Cadillac Motor Car Co.

Figure 309. "The Penalty of Leadership" Cadillac advertisement Saturday Evening Post, January 2, 1915 Theodore F. MacManus ABS

Although written over 100 years before, the last statement still rings true: "That which deserves to live—lives."

In the 1930s, those who drove the luxury car of choice chose Packard's V-12, 475-CID engine. Cadillac reintroduced the V-16, 452-CID engine for the 1938 model year to attain that prestigious position. Meanwhile, that year, William "Bill" Mitchell introduced the Cadillac Fleetwood Series Sixty Special, a fresh design from Cadillac that served as its halo car. The styling sensation model ran from 1938 to 1941. For the model year 1941, Cadillac introduced air-conditioning, one year later than Packard.

Before the air-conditioning discussion, a reflection of the 1941 Cadillac background information is in order. In 1940, Cadillac ended its V-16 and its 1924–1940 companion car, La Salle. In its place, Cadillac introduced the Series Sixty-One fastback-styling in coupe and sedan models for 1941 while promoting the $1345 price as an entry-level to Cadillac luxury car status. It sold 29,250 units at the lowest price Cadillac V-8 ever produced. (Fig. 310)

Figure 310. 1941 Cadillac Series Sixty-One, Five-Passenger Coupe, Model 6127, listed for $1,345   ABS

Cadillac sales thrived as it spread its retail sales prices from just above lower-priced cars to its two expensive limousines. In fact, according to the *Cadillac Milestones 1902–1942,* Cadillac's sales of 66,130 for 1941 outsold all makes of cars in both the medium- and high-price groups.

The 1941 Cadillac lineup comprised two differing styles: "aerodynamic" Series Sixty-One models and "popular Torpedo styling" Series Sixty-Two, Sixty-Three, and Sixty-Seven models. The ad displays the Cadillac Series Sixty-One aerodynamic five-passenger Coupe, Model 6127, priced at $1,345. (Fig. 310)

150   Chapter 4: 1941 Cadillac Factory Air-Conditioning

The Cadillac brochure displayed the aerodynamically styled De Luxe Touring Sedan, Model 6109D. (Fig. 311)

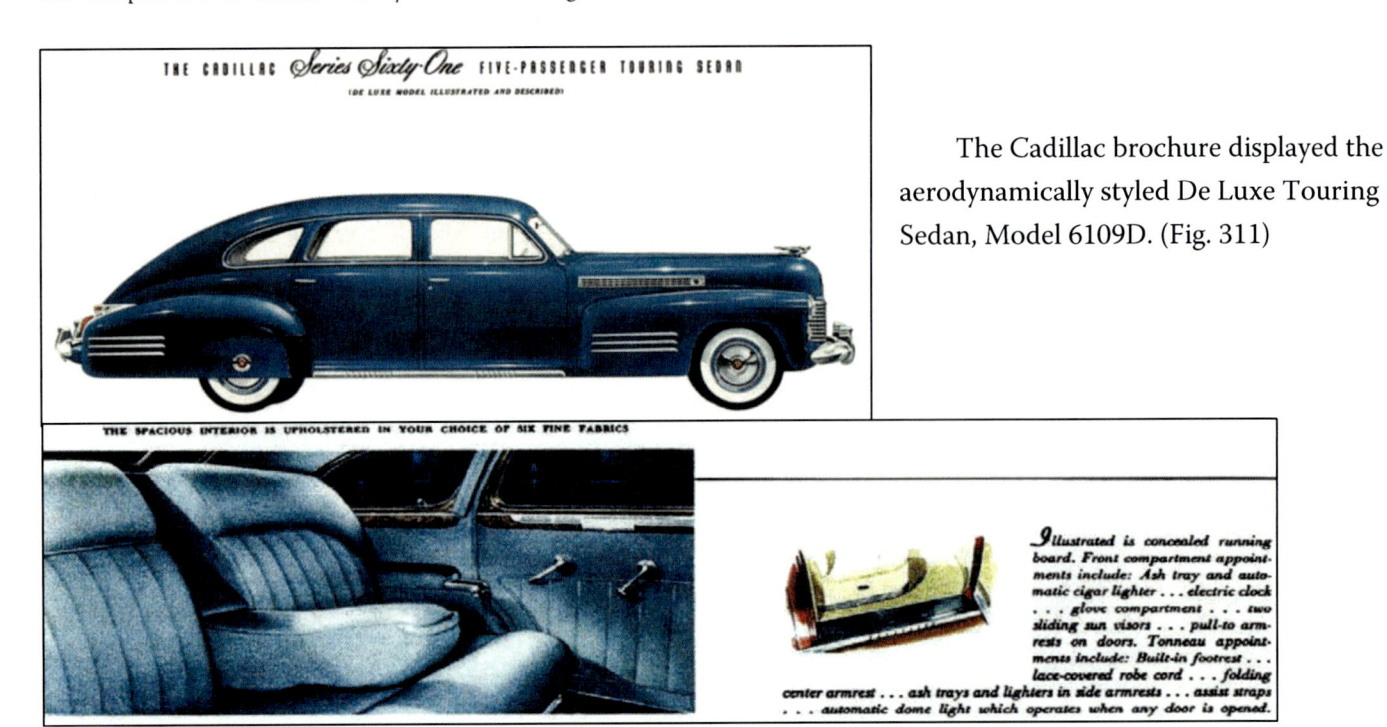

Figure 311. 1941 Cadillac Series Sixty-One Five-Passenger De Luxe Sedan, Model 6109D, and interior information   ABS

The brochure continues with the Cadillac Series Sixty-Two, shown in the popular torpedo styling, five-passenger De Luxe Touring Sedan, Model 6219D. (Fig. 312) The series included convertible styles of coupes and sedans.

Not shown: The Sixty-Two Series lineup also included a four-passenger De Luxe Coupe, Model 6227D. This car, owned by Mr. Doug Houston (1929–2016), is the subject of discussion in the "Spotlight" section of this chapter, with descriptions and photos of Cadillac's factory air-conditioning system. (Figs. 381–402)

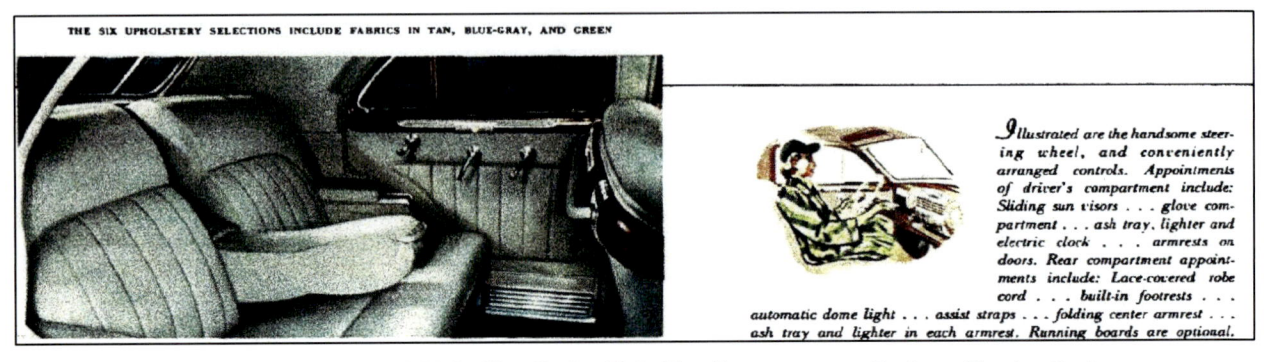

Figure 312. 1941 Cadillac Series Sixty-Two five-passenger De Luxe Touring Sedan, Model 6219, and interior information   ABS

The Cadillac body by Fleetwood presented two separate Cadillac models for 1941. The brochure displayed the Cadillac Fleetwood Series Sixty Special Five-Passenger Touring Sedan, Model 6019S. (Fig. 313) It also offered an optional Sunshine Turret Top sunroof, Model 6019S-A.

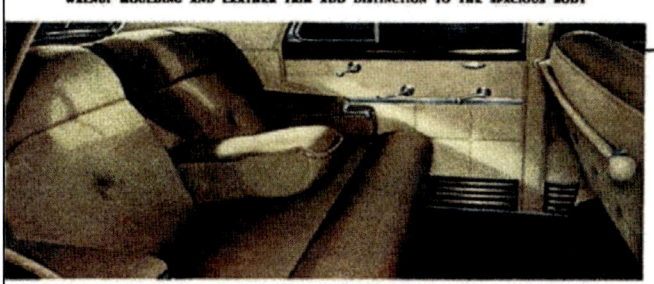

Figure 313. 1941 Cadillac Fleetwood Series Sixty Special five-passenger Touring Sedan, Model 6019S, and interior information   ABS

Later in this chapter, the "Spotlight" section presents images with a discussion of the air-conditioning system installed in Dr. Richard Zeiger's Model 6019S-A car. (Figs. 308, 342, 345–376, 380)

The Cadillac brochure displayed the Cadillac Fleetwood Series Seventy-Five Touring sedan with seating for five, Model 7519, or for seven passengers, Model 7523 as shown in Fig. 314.

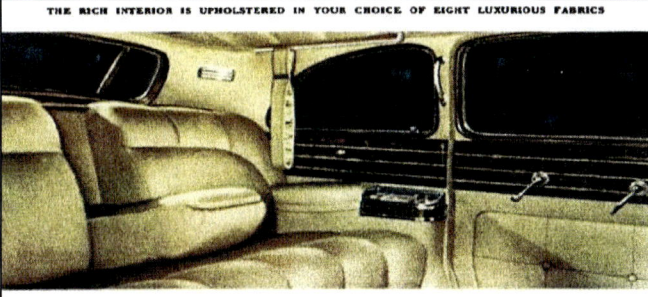

Figure 314. 1941 Cadillac Fleetwood Series Seventy-Five, Seven-Passenger Touring Sedan, Model 7523, and interior information   ABS

Each model offered an available electrically powered Imperial Division glass window for chauffeur driving. The "Spotlight" section features images and discusses Mr. Richard Kughn's formerly owned 1941 Cadillac Fleetwood Series Seventy-Five, Five-Passenger, Model 7519, equipped with air-conditioning. (Figs. 403–412)

152   Chapter 4: 1941 Cadillac Factory Air-Conditioning

Cadillac introduced model series changes for 1941, including its most powerful and economical 150-hp, 346-CID V-8 engine ever produced, standard directional signals, and a whimsical hidden gas filler cap under the left taillight. Its 1941 styling cues included headlight integration with the front fenders, the now-famous forward-thrusting Cadillac eggcrate-shaped grille, and three chromed horizontal speedlines on each fender, except for the Series Sixty Special.

Coincidentally, for 1941, Packard also integrated its first-ever headlights with the front fenders, while it featured "four sparkling speedline stripes" on the front and rear fenders instead of the three stripes on Cadillac's fenders.

Noteworthy is that both Cadillac and Packard eliminated the speedline stripes on their 1942 models.

Two other 1941 Cadillac mechanical features bear discussion. After Oldsmobile introduced the industry-first Hydra-Matic automatic transmission in 1940, Cadillac became the first luxury-priced car to introduce it in 1941. Cadillac drivers soon benefited from Hydra-Matic's technical achievement, eliminating the clutch pedal and the related gear shifting. (Fig. 315) Thirty percent of Cadillac owners selected the Hydra-Matic in 1941, listed at $125. Seemingly insignificantly priced now, its equivalent price in 2021 dollars is $2,318, according to the CPI.

Figure 315. 1941 Cadillac offered the Hydra-Matic Drive for $125, optional on all Cadillacs for an extra cost   ABS

Cadillac offered factory air-conditioning for 1941, one year after Packard. It listed air-conditioning as $364, according to 1941 Cadillac dealer invoices supplied by Mr. Doug Houston. (Fig. 386) The 1942 published retail price was $365 (Fig. 413), according to Mr. Roy A. Schneider's book, *Cadillacs of the Forties*, published in 1999. The 2021 equivalent price is $6,081, according to the CPI.

Refrigeration systems from the late 1800s to 1929 cooled with toxic gasses—namely ammonia, methyl chloride, and sulfur dioxide. During the 1920s, refrigerators leaking methyl chloride caused several fatal accidents. As a result, some homeowners moved their refrigerators outside for safety reasons. Three American corporations, Frigidaire, General Motors, and DuPont, joined to research a safer refrigeration coolant.

In 1928, Charles Kettering aided his top research scientist, Thomas Midgley, in Freon's invention, a "miracle compound." Freon, identified here as Freon-12, composed of chlorofluorocarbons, also known as CFCs. It proved odorless, colorless, nonflammable, and noncorrosive. On December 31, 1928, the Patent Office awarded Frigidaire its first patent for the CFC formula, Freon-12.

In 1930, General Motors and DuPont formed the Kinetic Chemicals Company that manufactured Freon-12. The same year, the General Motors Research Laboratories worked on a vapor compression system with Freon-12. On September 23, 1932, it proposed the development of an air-conditioning system for General Motors. The Cadillac Division expressed interest yet delayed work until the summer of 1933. In 1939, Cadillac installed a trunk-mounted prototype, a self-contained air-conditioning unit in a 1938 Cadillac. (Fig. 316)

Figure 316. 1939 Cadillac prototype trunk-mounted, self-contained air conditioner installed in a 1938 Cadillac @ASHRAE JOURNAL, September 1999

It was Packard, however, in cooperation with Bishop & Babcock Mfg. Co. of Cleveland, Ohio that eclipsed Cadillac's air-conditioning development in 1939 when it introduced a complete automotive air-conditioning system for the 1940 models. Packard's "Weather Conditioner" supplied refrigerated or heated air by switching dampers on the trunk-mounted evaporator. The 1940 Packard offered the combination system for $274. The equivalent price in January 2021 dollars is $5,156, according to the CPI.

> ### FRIGIDAIRE
> By 1935, *FRIGIDAIRE*, along with its competitors, sold eight million new Freon-12 cooled refrigerators in the United States. The "fridge," as it became known, represented the abbreviation of the *FRIGIDAIRE* brand refrigerator name.
>
> Similarly, the term "Frigidaire" became a generic household name for other brands of refrigerators.

154 Chapter 4: 1941 Cadillac Factory Air-Conditioning

No available advertising publicizes the 1941 Cadillac air-conditioning. This finding contrasts with Packard's extensive 1940–1942 advertising campaigns and Chrysler's full-color air-conditioning brochure in 1941, followed by a similar brochure for the 1942 DeSoto. Presented here are the three Cadillac Division publications, *sans* copies from national advertising.

Figure 317. 1941 Cadillac Data Book cover   GM Heritage Center

The 1942 *Cadillac Data Book* (Fig. 317) contains "Supplement Pages" titled "Cadillac Air-Conditioning," dated February 7, 1941. (Figs. 318–321) Cadillac identified the difference between other makes' air-cooling attempts and refrigerated air from Cadillac. It stated:

> The entire system was designed and built under the supervision of Cadillac engineers for Cadillac cars. The system must be installed at the factory. (It) filters and cools the car interior many degrees below that of outside temperatures.

Author note: The system design, involvement, and factory installation are qualified later in this chapter, p. 158.

1941 CADILLAC DATA BOOK AIR CONDITIONING SUPPLEMENT 2-7-41
ALLEN B. SIMONS -CLC#16572- 8-10-17 SUBMITTED BY WARREN RAUCH -CLC#4286-

CADILLAC *Air Conditioning*

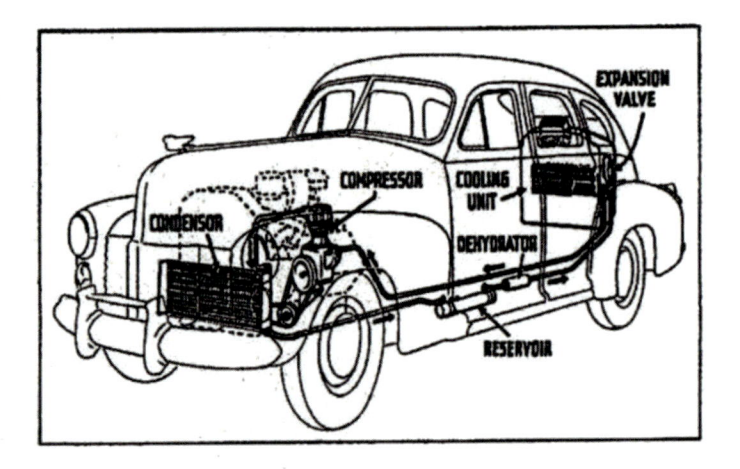

Cadillac Air Conditioning provides the automobile owner with the finest system of air cooling available. There are other designs whose names suggest air conditioning, but unless they actually provide refrigerated air, they do not secure the same results. The entire system was designed and built under the supervision of Cadillac engineers for Cadillac cars and provides a large volume of cool air that assures greater driving comfort in warm weather.

As it is part of the car, and not just an attachment, the Cadillac Air Conditioning system must be installed at the factory at the time the car is built. No mechanical knowledge is needed to operate Cadillac Air Conditioning. It is as simple as a heater. The refrigerant used is non-poisonous and odorless and under no condition can cause discomfort.

Several cars on the market offer a combination air filter and air heater. These do not in any way compare with Cadillac Air Conditioning. It operates on the same principle as home refrigerators and delivers filtered, cool air to the car interior, maintaining a temperature many degrees below that outside. With Cadillac Air Conditioning you shut discomfort out and ride in the height of motoring luxury.

2-7-41   AC-1   Printed in U. S. A.

*Figure 318. 1941 Cadillac Data Book air-conditioning supplement, p. AC-1
CLC (Cadillac & LaSalle Club member)   Warren Rauch*

Simple in Operation: (Fig. 319)
1. Compressor: Located on the driver's side of the engine, connected by two belts driven from a pulley on the crankshaft.
2. Condensor, [sic], now spelled "Condenser": It receives the heated, pressurized gas from the compressor and cools it to a liquid by the radiator fan airflow.
3. Reservoir: The receiver stores the cooled refrigerating liquid, Freon-12, for use as the expansion valve meters spray into the evaporator cooling coils.
4. Expansion Valve and Cooling Coils: Located next to the evaporator in the trunk's forward area. Liquid Freon-12, under pressure, enters the expansion valve, then the cooling coils inside the evaporator, and absorbs heat from the return air passing over the coils by a fan. The fan blows the cooled air upward through the cooling register behind the rear seat into the cabin. The heated Freon-12 gas completes the cycle as it returns to the compressor through copper tubing.

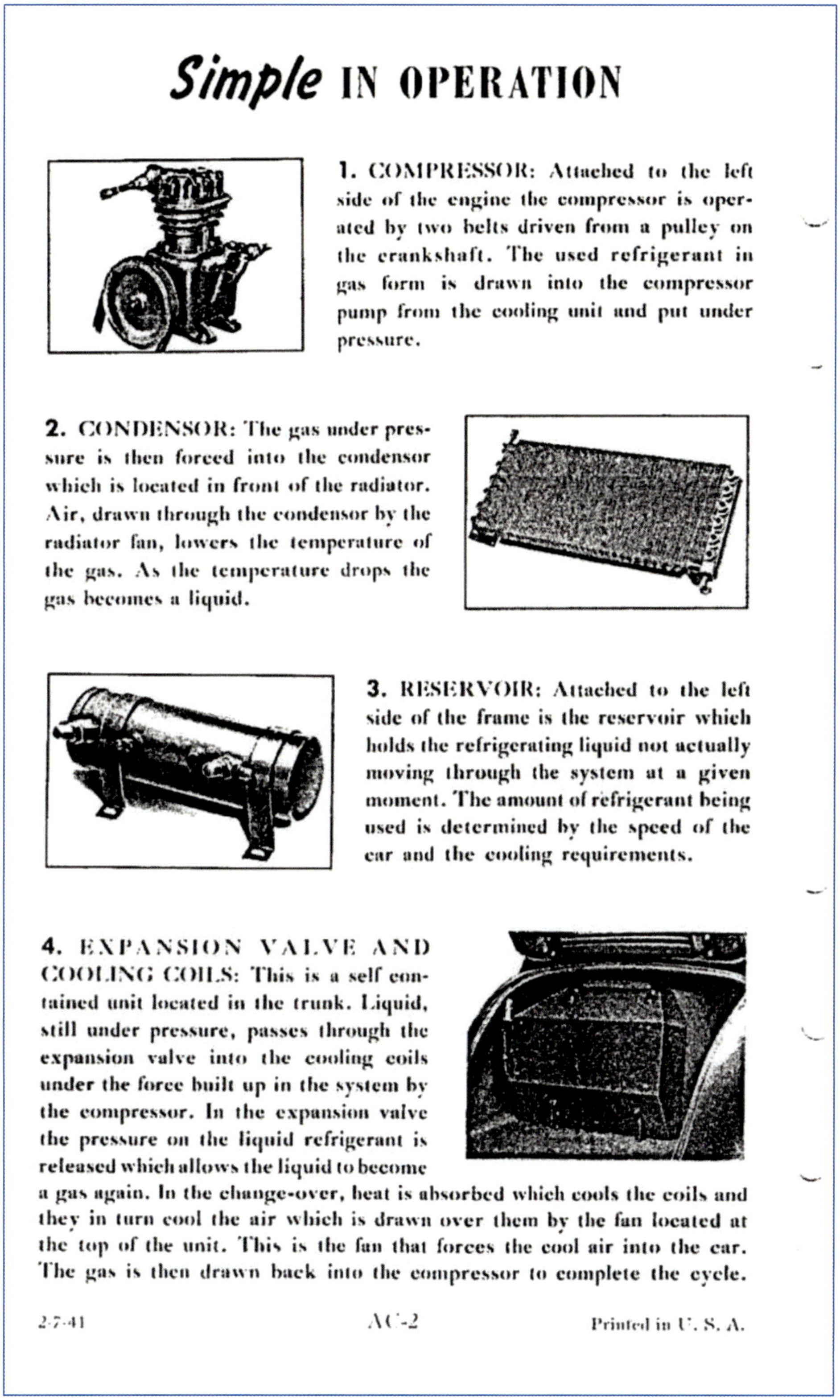

Figure 319. 1941 Cadillac Data Book air-conditioning supplement, p. AC-2
CLC Warren Rauch

Controlled Air Circulation: (Fig. 320)

    The fan moves the return air from the passenger cabin through filters that remove dust, dirt, and pollen, then blows over the refrigerating coils. The cooled airflow moves over the passenger's heads as it follows the headliner, settles to the floor, and returns under the seat. Moveable louvers in the cooling register direct the airflow, while the three-speed switch on the left side of the dash controls blower speeds and "Off."

    The supplement cautions the operator to keep doors, windows, and cowl ventilators closed to control the interior cooling. It recommends the driver open the cowl ventilator slightly with a full passenger load, inferring a smoke-filled cabin from the era's cigarette smokers. The system removed moisture and resisted frosting by thermostatic control of the expansion valve to a few degrees above freezing.

Controlled AIR CIRCULATION

Air in the car which has been cooled from passing over the refrigerating coils moves over the heads of the passengers by the force of the fan in the rear unit and gradually settles to the floor, the cool air being heavier than the warmer air in the car. The circuit is completed as the air is drawn back into the cooling unit. Filter elements in the return passage remove dust and dirt as well as reducing the amount of pollen. The direction of the current can be controlled by movable louvers in the outlet opening and further control of the air flow is possible as a three-way switch, located on the dash, controls the speed of the fan as well as turning it on and off.

A cooling system can of course, give maximum results only when the amount of air is controlled. Your home refrigerator does not work satisfactorily when the door is left open. In the same way, Cadillac Air Conditioning gives the best results when doors, windows and cowl ventilator are closed. An extra supply of air that might be desirable with a full passenger load can be secured by slightly opening the cowl ventilator.

The coils not only cool the air that passes over them but in humid weather also remove excess moisture. This gives the air greater cooling properties. Frosting of the coils during operation is prevented by thermostatic control of the expansion valve which maintains the coils at a temperature a few degrees above freezing.

2-7-41   AC-3   Printed in U. S. A.

Figure 320. 1941 Cadillac Data Book air-conditioning supplement, p. AC-3
CLC   Warren Rauch

Comfortable Interior Temperature (Fig. 321)

Cadillac's air-conditioning unit offers:

1. Large Capacity: Its capacity supplies almost twice as much cooling required to cool a large room as any available system. This capacity ensures comfort for all passengers.

2. Adequate Insulation: Cadillac confirms that the standard body insulation sufficiently prevents heat incursion from the "top, sides, and floor," and that competitors must add extra insulation to their body when equipped with air-conditioning.

Author note: The *Packard Data Book for 1941* confirmed adding "special insulation" to those areas before shipping the car to its contractor.

3. Immediate Operation: Cadillac's air-conditioning system runs continually, offering comfort for passengers on short and long trips.

The dash switch controls the air volume so that passengers will be comfortable at any speed.

An in-line dehydrator filters out moisture and foreign matter in the Freon-12 to prevent freeze-ups and other problems.

Comfortable INTERIOR TEMPERATURE

Maintenance of desired interior temperatures is assured by three factors:

1. LARGE CAPACITY

The Cadillac Air Conditioning unit produces larger quantities of cool air than any system now available. It supplies almost twice as much as is required to cool a large room. Adequate capacity means real comfort for all passengers.

2. ADEQUATE INSULATION

No special insulation is necessary to give the desired results. Standard Cadillac construction gives the necessary protection which is not the case with other cars offering this type of air conditioning. Competitive makes require special insulation for the top, sides and floor to keep outside heat from interfering with the cooling system.

3. IMMEDIATE OPERATION

As the system is directly connected with the engine it goes into operation as soon as the car is started. Because of this, passengers ride in comfort whether on a short or a long trip. Since the amount of cool air is controlled by a dash switch, interior temperatures will be just right at any reasonable speed.

Continuously satisfactory operation is insured by a special Cadillac feature. A dehydrator is fitted into the line between the reservoir and the cooling coils. It takes out all traces of moisture which would freeze in the valves and any foreign matter that might cause other trouble.

2-7-41　　　　　　　　　　AC-4　　　　　　　　Printed in U. S. A.

Figure 321. 1941 Cadillac Data Book air-conditioning supplement, p. AC-4
CLC Warren Rauch

158 Chapter 4: 1941 Cadillac Factory Air-Conditioning

While General Motors, Frigidaire, and DuPont invented Freon, later known as Freon-12, the 1940 Packard contracted with the Bishop & Babcock Mfg. Co. of Cleveland, Ohio, for the first mass-produced automotive air-conditioning system installation. Bishop & Babcock Mfg. Co. sold a variety of automotive products, including fans, heaters, thermostats, and radiators. It explored automotive air-conditioning, selected a Servel commercial refrigeration compressor, and modified it for automotive use. Working together, Packard introduced its "Weather Conditioner" in late 1939, in time for the 1940 New York, Chicago, and San Francisco auto shows.

Cadillac, for 1941, stretched their Cadillac designing and engineering involvement, p. 154, as it also contracted with Bishop & Babcock for its air-conditioning system, verified by Servel compressor images. (Figs. 332, 348–352, 387, 390, 398, 406–408, 474) Both Cadillac and Packard advertised a "factory-installed system," with statements of assembly completion in their respective factories. Cadillac's obfuscation appeared in the 1941 *Cadillac Air-Conditioning Manual*, p. 5: (Fig. 331)

> The air-conditioning system is available at extra cost on all 41-Series Cadillac cars, but the *installation must be made at the Cadillac factory at the time the car is being assembled*. (emphasis by author)

Neither Cadillac nor Packard's literature acknowledged the contractor involvement or location of the air-conditioning installation. At least one writer believes Packard shipped the selected, nearly completely assembled cars to the Bishop & Babcock Mfg. Co. in Cleveland for the air-conditioning componentry installation. Without citing a reference, author L. Morgan Yost stated the following on p. 365, from *Packard: A History of the Motor Car and the Company*:

> Special insulation for cars so equipped was provided at the factory, but, although noted in Packard literature as "factory installed," the refrigerating unit itself was actually fitted into the car by Bishop & Babcock in Cleveland, and in theory, vehicles ordered with it were shipped there, the unit was installed and then the cars were forwarded to the dealers.

Upon delivery, the owner of an air-conditioned 1941 Cadillac received a seven-page owner's manual titled, *"Regarding the Air-Conditioning System on your Cadillac."* (Figs. 322–328)

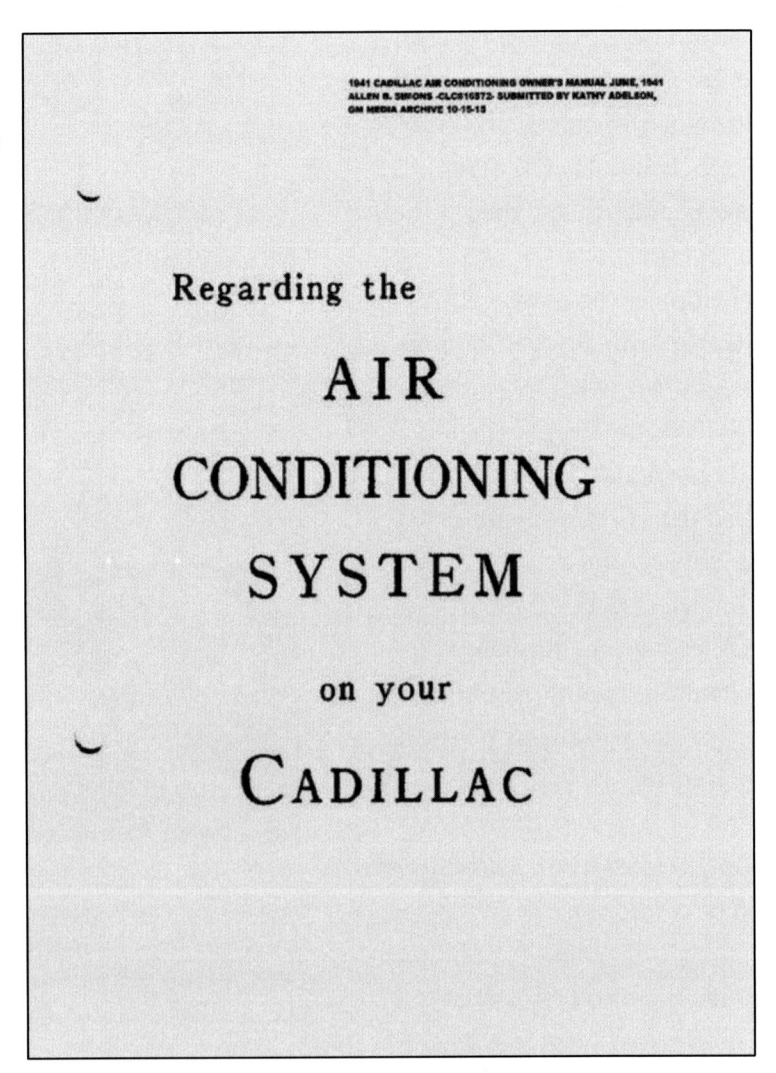

Figure 322. 1941 Cadillac air-conditioning owner's manual, Regarding The Air-Conditioning System On Your Cadillac
GM MEDIA ARCHIVE   Kathy Adelson

Cadillac supplied information about passenger comfort through proper use and maintenance.

Controls: A dash diagram showed the single, three-speed, and "Off" air-conditioning blower switch under the dash's left side. (Fig. 323) Unexpectedly, Cadillac chose not to inscribe the knob with a letter such as "A" or "AC," like the "B" initial on the back-up light knob next to it in Dr. Zeiger's car. (Fig. 323 and Dr. Zeiger's "B" knob, Figs. 353, 354)

Cadillac Air Conditioning

The Cadillac Air Conditioning System with which your car is equipped is designed to make the interior of your car more comfortable under very adverse conditions of heat and high humidity, provided it is properly used and properly maintained.

Controls

The driver has only one control to operate, and that is the switch for the blower. This switch controls both the circulation of air and the cooling action of the system. It has three speeds, as well as "Off".

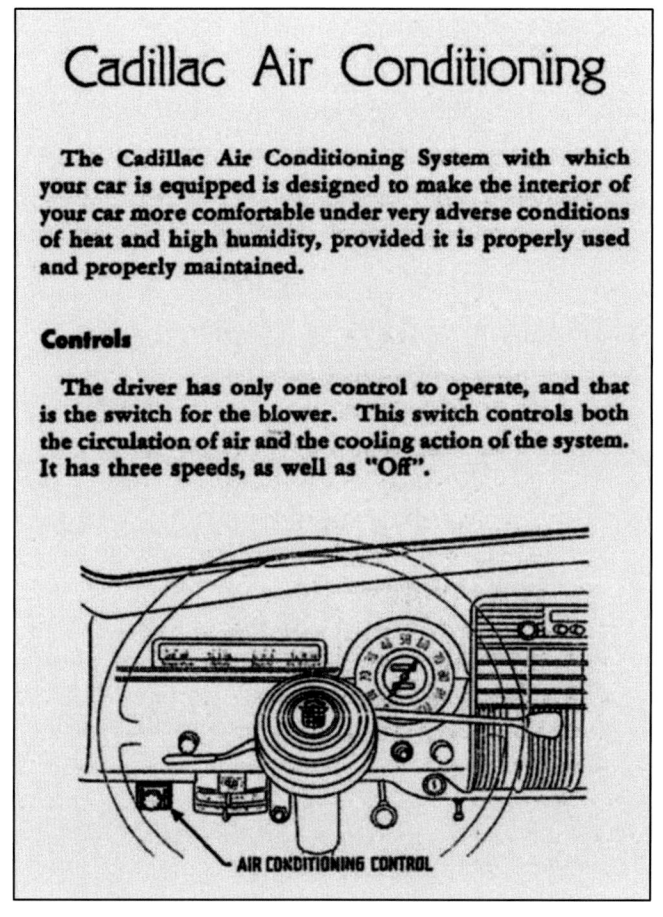

Figure 323. 1941 Cadillac air-conditioning owner's manual, p. 2   GM MEDIA ARCHIVE   Kathy Adelson

• CADILLAC AIR CONDITIONING •

Action of System

The air conditioning system does three things, all of which help to make the car interior more comfortable:

It cools the air
It de-humidifies
It filters

The cooling action of the system is not intended to achieve refrigerator temperatures, as too great a spread between interior and exterior temperatures is unhealthful and uncomfortable. It is rather designed to provide a temperature drop sufficient for increased comfort under a variety of conditions.

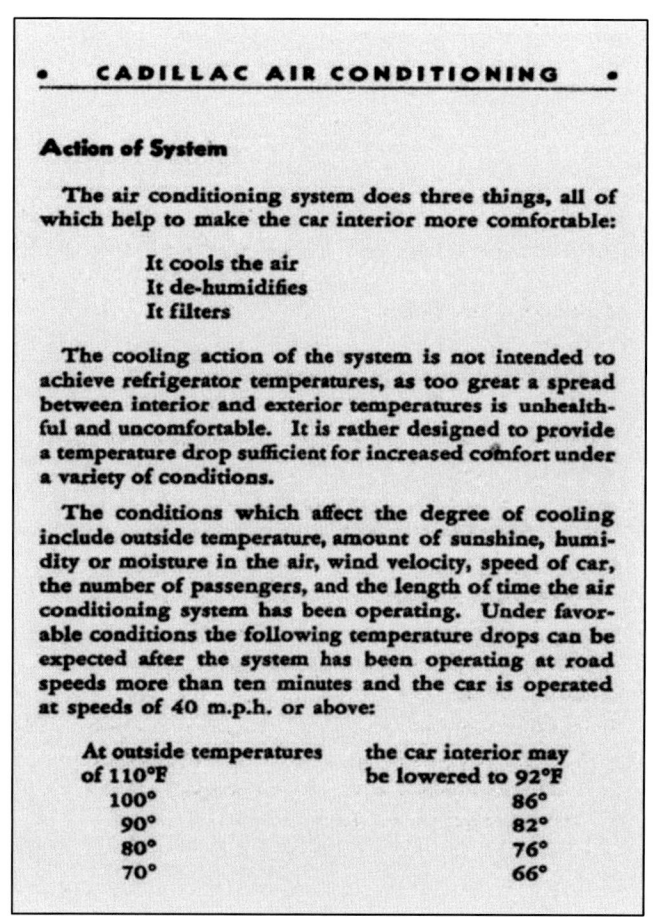

The conditions which affect the degree of cooling include outside temperature, amount of sunshine, humidity or moisture in the air, wind velocity, speed of car, the number of passengers, and the length of time the air conditioning system has been operating. Under favorable conditions the following temperature drops can be expected after the system has been operating at road speeds more than ten minutes and the car is operated at speeds of 40 m.p.h. or above:

At outside temperatures	the car interior may
of 110°F	be lowered to 92°F
100°	86°
90°	82°
80°	76°
70°	66°

Figure 324. 1941 Cadillac air-conditioning owner's manual, p. 3   GM MEDIA ARCHIVE   Kathy Adelson

Action of System: The air conditioner cooled, dehumidified, and filtered the passenger cabin's air. (Fig. 324) Importantly, Cadillac addressed the interior temperature and passenger's expectations, stating:

> The cooling action of the system is not intended to achieve refrigerator temperatures, as too great a spread between interior and exterior temperatures is unhealthful and uncomfortable.

Author note: M. S. Bhatti, Ph.D., wrote in his 1999 ASHRAE report:

> At that time, it was believed that if the difference between the outside air and the conditioned air exceeded 10°F, the occupant of the conditioned space could experience a thermal shock upon emerging into the outside air!

Interior cooling was a function of several outside and inside influences. (Fig. 324) The system achieved optimum interior temperatures based on 40 MPH or higher speed for ten minutes, ranging from 92°F down to 66°F as outside temperatures ranged from 110°F to 70°F.

160   Chapter 4: 1941 Cadillac Factory Air-Conditioning

> • CADILLAC AIR CONDITIONING •
>
> When outside temperatures are very high, the system can achieve a temperature difference of as much as 10 to 15°, whereas at temperatures of 80° or below the difference will be very slight.
>
> At these lower temperatures, however, the system is most likely to be used because of damp and sultry conditions, and the *de-humidifying* action makes the car interior more comfortable than the *temperature* difference might seem to indicate.
>
> An important advantage of the system is the filtering and recirculating action. Even in cool weather this can be used effectively in traveling dusty roads. The system should then be turned on high speed, and all windows and ventilators tightly closed.
>
> **Operation of System**
>
> The air conditioning system operates only when the engine of the car is running, and does not give the same cooling effect at idling speeds as when the car is driven at normal road speeds.
>
> When the blower is turned on, air is drawn from the car interior, under the rear seat, through an air filter and over the cooling coils, and is then discharged through louvers that circulate it along the roof and through the car.

Figure 325. 1941 Cadillac air-conditioning owner's manual, p. 4   GM MEDIA ARCHIVE   Kathy Adelson

Switched on, the air-conditioning system draws air under the rear seat, filters it, and passes over the cooling coils, whereby the blower discharges cool air upward through the louvered vent that circulates it towards the headliner and throughout the car. Closed windows provide the best operation and maximum efficiency. (Fig. 326) The fresh air inlet from the heater or cowl ventilator supplies fresh air as desired. Smoking occupants in a crowded car add an abnormal load to the system, especially in humid weather.

Refrigeration Cycle: A reservoir supplies the Freon-12 refrigerant under pressure. It is a non-inflammable (*sic*), nontoxic, non-irritating, and nearly odorless substance.

The manual explained that the system cooling worked best at higher outside temperatures, *lowering inside temperatures "as much as 10°F to 15°F"* (emphasis by author), whereas the cooling would not be as effective for temperatures of 80°F or lower. (Fig. 325)

Under "damp and sultry conditions," the system's dehumidification qualities provided more comfort for passengers than just the temperature decrease; also, the air filtration and use of recirculated air insulated passengers on dusty roads.

Operation of System: The air-conditioning system operated whenever the engine ran. It cautioned that cooling at idle speed was not as effective as at normal road speeds.

> • CADILLAC AIR CONDITIONING •
>
> All windows must be closed for the system to function and for maximum efficiency. All window, door and cowl ventilators should be closed also. Ventilation, if desired, however, can be secured either by opening the fresh air inlet for the heater (if a ventilating heater has been installed) or by opening the cowl ventilator not more than one notch.
>
> The ventilation requirements for a crowded car may impose an abnormal load on the system, especially if the occupants of the car are smoking. This is true particularly in hot, moist weather.
>
> **Refrigeration Cycle**
>
> The refrigerant is Freon, a non-inflammable, non-toxic, non-irritating and practically odorless substance. A supply is carried, in liquid form under pressure, in a reservoir attached to the frame of the car.

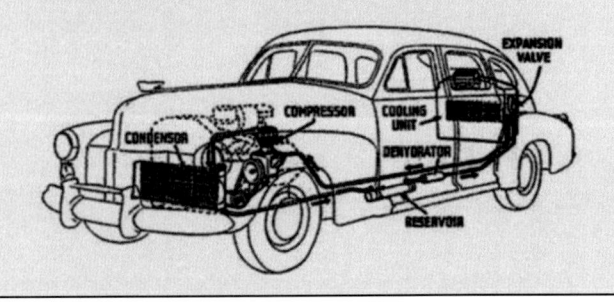

Figure 326. 1941 Cadillac air-conditioning owner's manual, p. 5   GM MEDIA ARCHIVE   Kathy Adelson

The Freon-12 refrigerant continues through the expansion valve and enters the cooling coils. It vaporizes rapidly as the warm air passes over the cooling coils. (Fig. 327) The fan discharges the now-cooled air upwards along the headliner into the passenger cabin through heat exchange. Excess humidity condenses on the cooling coils.

From the evaporator coils, the vaporized Freon-12 continues to the compressor, then travels to the condenser in front of the radiator. By the car's forward movement and the radiator fan's use, heat dissipates from the Freon-12 and condenses. The liquid refrigerant flows into the reservoir.

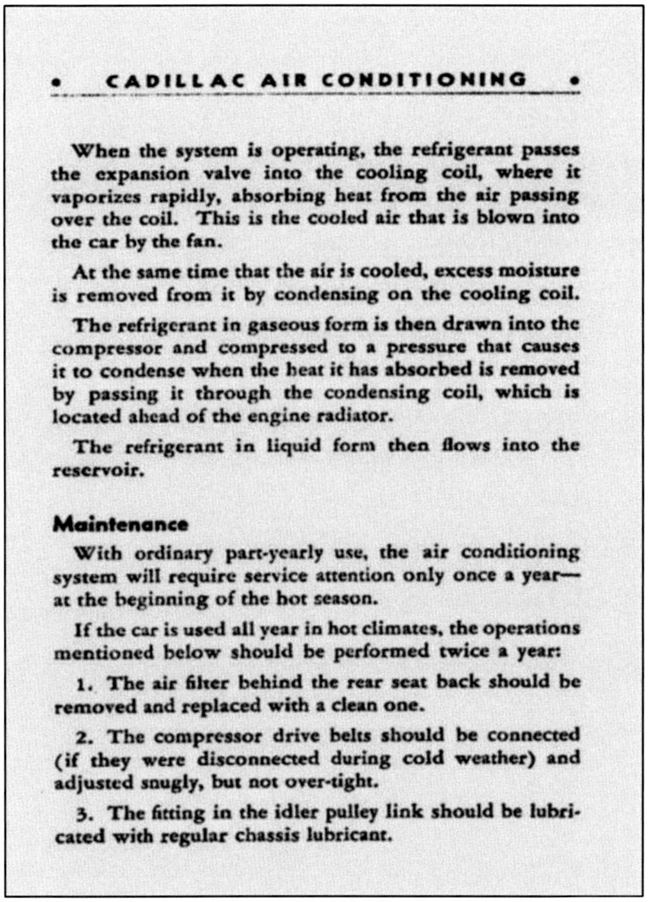

• CADILLAC AIR CONDITIONING •

When the system is operating, the refrigerant passes the expansion valve into the cooling coil, where it vaporizes rapidly, absorbing heat from the air passing over the coil. This is the cooled air that is blown into the car by the fan.

At the same time that the air is cooled, excess moisture is removed from it by condensing on the cooling coil.

The refrigerant in gaseous form is then drawn into the compressor and compressed to a pressure that causes it to condense when the heat it has absorbed is removed by passing it through the condensing coil, which is located ahead of the engine radiator.

The refrigerant in liquid form then flows into the reservoir.

Maintenance

With ordinary part-yearly use, the air conditioning system will require service attention only once a year—at the beginning of the hot season.

If the car is used all year in hot climates, the operations mentioned below should be performed twice a year:

1. The air filter behind the rear seat back should be removed and replaced with a clean one.

2. The compressor drive belts should be connected (if they were disconnected during cold weather) and adjusted snugly, but not over-tight.

3. The fitting in the idler pulley link should be lubricated with regular chassis lubricant.

Figure 327. 1941 Cadillac air-conditioning owner's manual, p. 6 GM MEDIA ARCHIVE Kathy Adelson

Maintenance: The system requires maintenance once or twice yearly when used all year round: (Fig. 327)

1. Remove and replace the filter behind the seat.
2. Reconnect compressor and pulley drive belts if removed during winter operation.
3. Lubricate the pulley bearing.
4. Test the refrigerant level (Fig. 328), according to the separate publication, the *Cadillac Air-Conditioning Manual.* (Fig. 329)
5. Clean and deodorize the evaporator cooling coil.

Separately, it recommended drive belt removal "at the end of the warm season."

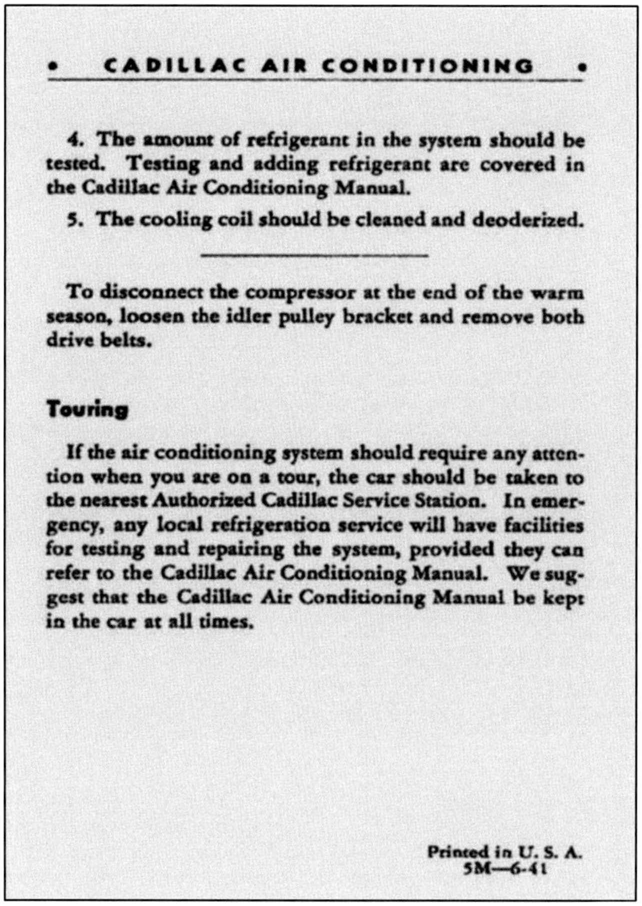

• CADILLAC AIR CONDITIONING •

4. The amount of refrigerant in the system should be tested. Testing and adding refrigerant are covered in the Cadillac Air Conditioning Manual.

5. The cooling coil should be cleaned and deoderized.

To disconnect the compressor at the end of the warm season, loosen the idler pulley bracket and remove both drive belts.

Touring

If the air conditioning system should require any attention when you are on a tour, the car should be taken to the nearest Authorized Cadillac Service Station. In emergency, any local refrigeration service will have facilities for testing and repairing the system, provided they can refer to the Cadillac Air Conditioning Manual. We suggest that the Cadillac Air Conditioning Manual be kept in the car at all times.

Printed in U. S. A.
5M—6-41

Figure 328. 1941 Cadillac air-conditioning owner's manual, p. 7 GM MEDIA ARCHIVE Kathy Adelson

Touring: For repairs, visit a Cadillac dealer or local refrigeration service and refer to the separate publication stored in the car, the *Cadillac Air-Conditioning Manual.*

The back page showed, "Printed in U. S. A. 5M—6-41." The 6-41 referred to the print date of June 1941. (Fig. 328)

For 1941, the *Cadillac Air-Conditioning Manual* publication discussed servicing the system in a 15-page format. (Figs. 329–341. Page 2 and 4 are blank.)

The brevity of the 1941 manual left many questions and issues unanswered:

1. The Servel air-conditioning compressor's specifications, tolerances, and manufacturer's information.

2. The omission of on-car photos of air-conditioning componentry, such as the compressor, condenser, hoses, tubing, framing, receiver, bracketing, louvered cooling register, evaporator with/without the inspection panel to show the cooling coils, expansion valve; a technician holding and reading the testing gauges during Freon-12 measurement; illustrations/photos for air filter replacement, removal, and installation of the compressor belt; more comprehensive outside temperature/inside cabin temperature charts under varying outside conditions; and the CFM, cubic feet per minute airflow from the cooling register.

The lack of Cadillac service manual photos is surprising considering their contractor, Bishop & Babcock, delayed its service manual publication until the 1941 model year, especially since the 1940 Packard introduced its Weather Conditioner air-conditioning system a full model year before Cadillac.

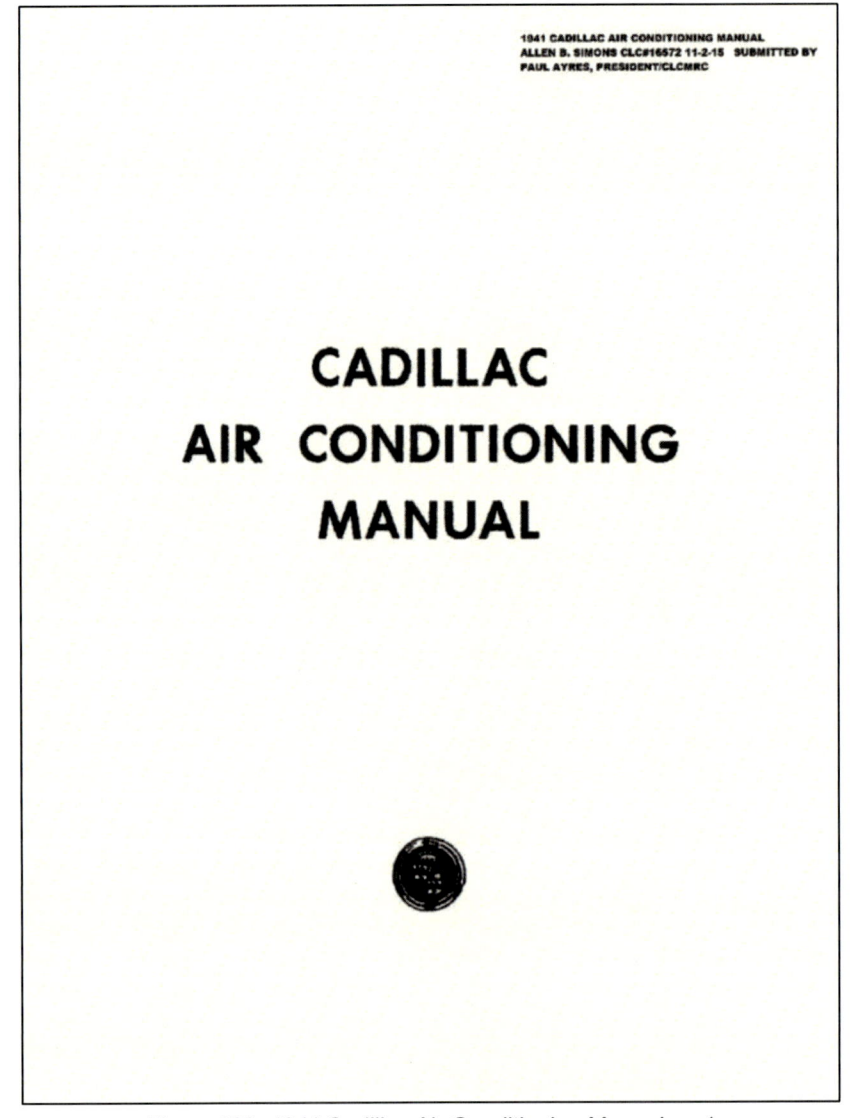

Figure 329. 1941 Cadillac Air-Conditioning Manual, p. 1
CLCMRC (Cadillac & LaSalle Club Museum and Research Center)   Paul Ayres

CONTENTS

General Description   Page

    Controls 5
    Air Circulation 5
    Refrigeration Cycle 5
    Expansion Valve 6
    Compressor and Drive 6
    Dehydrator 7
    Air Filter 7

Service Operations

    Periodic Items 7
        Air Filter Replacement 7
        Compressor Drive Adjustment 7
        Testing Refrigerant 7
        Cleaning Evaporator Coil 8
    Operation of Service and Shut-Off Valves 8
    Testing System for Leakage 8
    Removing and Installing Expansion Valve 9
    Removing and Installing Condenser 10
    Removing and Installing Compressor 10
    Removing and Installing Dehydrator 11
    Adding Refrigerant to System 11
        Recharging System Completely 12
        Fusible Plug 12
    Checking Compressor Oil Level 12

Diagnosis Chart 14

Figure 330. 1941 Cadillac Air-Conditioning Manual, p. 3 (p. 2 left blank)   CLCMRC   Paul Ayres

CADILLAC AIR CONDITIONING MANUAL

GENERAL DESCRIPTION

The Cadillac air-conditioning system provides a means of cooling, dehumidifying, filtering and recirculating the air in the interior of the car, thereby providing a comfortable car interior during extremely warm weather.

The air-conditioning system is available at extra cost on all 41-Series Cadillac cars, but the installation must be made at the Cadillac factory at the time the car is being assembled.

1. Controls

The only control which the driver needs to manipulate is the switch for the evaporator fan, located on the flange of the instrument panel. This switch has three speeds, as well as "off," to permit the desired degree of cooling to be selected.

Continuous cooling is possible only by keeping the engine of the car running.

When the car is to be operated for some time without need of the air-conditioning system, the system can be disconnected by removing the compressor drive belt.

2. Air Circulation

When the evaporator fan is turned on, air is drawn from the car interior under the rear seat and through the air filter into the cooling or "evaporator" coil, where it is cooled and dehumidified. It is then discharged by the blower through the directional louvers so that it follows the contour of the roof and circulates through the car. (See Figure 1.)

The system will function most efficiently if all windows and ventilators are closed. Ventilation, if desirable, can be secured either by opening the fresh air inlet for the heater (if a ventilating heater has been installed) or by opening the cowl ventilator not more than one notch.

3. Refrigeration Cycle

The refrigerant used is Freon (F-12), a non-toxic, non-corrosive, non-inflammable, non-irritating, and practically odorless gas which has a boiling point of -21°F., at atmospheric pressure.

The refrigerant is admitted through the expansion valve to the cooling or evaporator coil. Before passing through the valve,

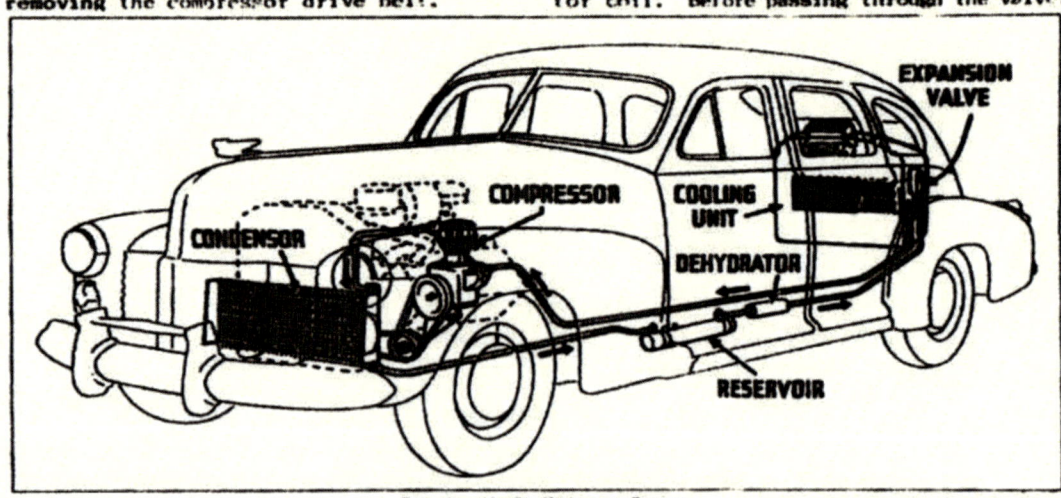

Fig. 1. Air Conditioning System

5

Figure 331. 1941 Cadillac Air-Conditioning Manual, p. 5 (p. 4 left blank)   CLCMRC   Paul Ayres

CADILLAC AIR CONDITIONING MANUAL

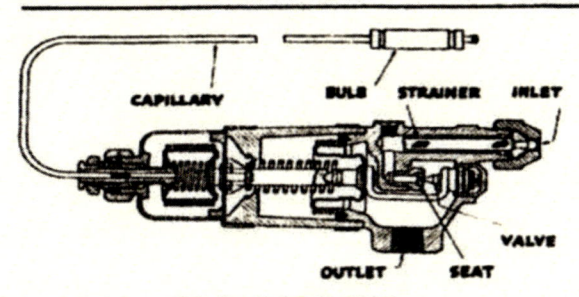

Fig. 2. Expansion Valve

the refrigerant is in liquid form at high pressure. The low pressure maintained in the cooling coil by the compressor causes the liquid refrigerant to vaporize or evaporate after it passes through the valve, and when this change from a liquid to a vapor occurs, there must be an absorption of heat from some source. As the only source from which heat can be absorbed is the air passing over the cooling coil, this air is consequently reduced in temperature. This is the cooled air which is introduced into the car by the fan. At the same time the air is reduced in temperature, the excess moisture is removed by condensing on the cooling coil.

The refrigerant in gaseous form, laden with heat absorbed in the evaporator coil, is then drawn into the compressor and compressed to a pressure that will cause it to return to a liquid when the heat it has absorbed is removed by passing the refrigerant through a condensing coil that is exposed to atmospheric temperatures.

The condensing coil, or condenser, is mounted in the space between the car radiator and the radiator grille, so that there is a continuous flow of air through it, induced by fan suction when the car is standing still or by combined fan suction and car velocity when the car is in motion.

The refrigerant in liquid form then flows into the liquid reservoir, which is a storage tank for a small surplus of refrigerant.

4. Expansion Valve

The expansion valve (Figure 2), which is mounted at the inlet of the evaporator coil, regulates the flow of liquid refrigerant into the coil, permitting passage of the maximum amount that will evaporate completely. This flow of refrigerant is controlled by a thermostatic bulb clamped to the low pressure gas line at the evaporator coil outlet and connected to the valve through a capillary. The action is as follows:

If too much refrigerant passes into the coil, it will not evaporate completely and some liquid will flow into the low pressure gas line. This liquid will cool the bulb, and this in turn will close the valve partially and reduce the amount of liquid entering the coil. If not enough liquid is entering the coil, the cooling action will be inadequate and the bulb will warm up, opening the valve wider to let more liquid enter the coil.

The expansion valve has a 3/8" S.A.E. inlet connection, a 1/2" S.A.E. outlet, a 5/32" orifice, a 55-pound Freon, gas-charged thermostat bellows and a 24" capillary.

5. Compressor and Drive

The compressor is a two-cylinder, high speed, air cooled unit designed especially

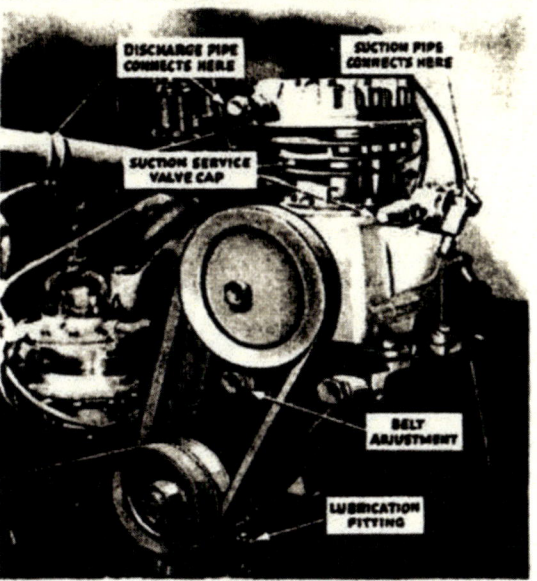

Fig. 3. Compressor and Drive.

6

Figure 332. 1941 Cadillac Air-Conditioning Manual, p. 6   CLCMRC   Paul Ayres

CADILLAC AIR CONDITIONING MANUAL

for automotive use. It is mounted on the engine at the front of the left cylinder block. It is lubricated by a special compressor oil which is added to the crankcase of the compressor, but which mixes readily with the Freon and is present in all parts of the system.

The compressor is driven from the engine crankshaft by two short belts, one from crankshaft to idler pulley assembly, and the second from idler pulley to compressor. The idler pulley assembly is mounted on an adjustable support arm, by means of which the correct belt tension is obtained. This support also incorporates an equalizing link to equalize the tension of the two belts. (See Figure 3.)

6. Dehydrator

In order to remove any slight amount of moisture that might find its way into the system, a dehydrator is installed in the line between the reservoir and the expansion valve. This is merely a cylinder which contains a chemical which absorbs moisture from the Freon as it passes through.

7. Air Filter

The air filter is a replacement type filter constructed of corrugated fiber board arranged to form cellular passages. This fiber board is coated with a special compound, a viscous "tacky" oil, which is odorless, and which remains "tacky" over a wide temperature range.

Dust laden air entering the filter immediately changes its direction at a 45° angle, thus causing a scrubbing along the sides of the cellular passages. As dust impinges against the passages, oil is absorbed into each dust particle by capillary attraction, and that dust particle then becomes the medium for catching other dust particles. This process continues until the filter becomes completely dust laden.

SERVICE OPERATIONS

8. Periodic Items

Under ordinary part-season use, the air conditioning system will require service attention only once a year--at the beginning of the hot season. If the car is used year-round in hot climates, the operations listed below should be performed twice a year:

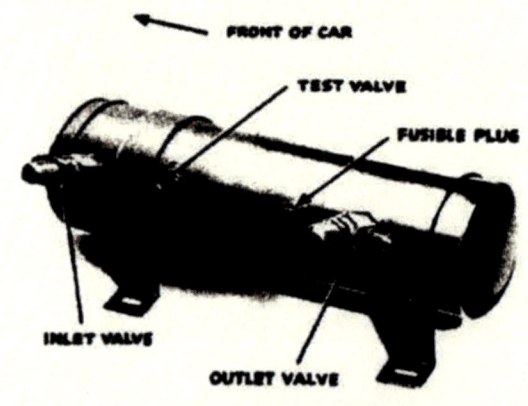

Fig. 4. Liquid Reservoir

Air Filter--The first regular maintenance operation is replacement of the air filter. This is necessary both to assure the circulation of clean air and to permit the air conditioning system to operate at maximum efficiency.

To replace the filter, remove the rear seat cushion, the rear seat back, and the metal pan over the filter. Then slide the old filter out of position and install a new one.

Compressor Drive--Reconnect the compressor drive belts if they were disconnected for the winter driving season. Readjust the idler pulley support to bring both belts up snug, but not over-tight. Lubricate the fitting in the link with regular chassis lubricant.

To disconnect the compressor at the end of the warm season, loosen the idler pulley bracket and remove both drive belts.

Testing Refrigerant--Test the operation of the system to determine whether or not it contains a sufficient amount of Freon for normal operation. To do this:

Figure 333. 1941 Cadillac Air-Conditioning Manual, p. 7   CLCMRC   Paul Ayres

CADILLAC AIR CONDITIONING MANUAL

1. Start the engine of the car and turn on the fan for the air conditioning system.

2. Open slightly the test valve at the top of the liquid reservoir (See Figure 4).

3. If there is liquid refrigerant up to the correct level, liquid Freon will come out in a milky white flow. In this case, shut off the test valve at once.

4. If only gas blows out, this indicates a shortage of refrigerant, and more should be added, in accordance with the procedure given in Note 15.

Evaporator Coil--The evaporator coil also should be cleaned at this time, by removing the cover plate on the back of the evaporator housing and spraying the coil with a solution of Diversal cleaner and water. This will remove all dirt which may be clinging to the coil, washing it down into the return duct and out the drain tube.

9. Operation of Service and Shut-Off Valves

The flow of refrigerant in the system can be controlled and localized for service operations by means of four valves: a "service" valve at the suction side of the compressor, a service valve at the discharge side of the compressor, a shut-off valve at the reservoir inlet, and a shut-off valve at the reservoir outlet.

The "service" valves are illustrated in Figure 5. They are of the double seating type, which seat or close when screwed all the way in or all the way out. When a valve is in the "in" position, the line to the compressor is shut off. When it is in the center position, the line is open. When it is in the "out" position (which is also called "back-seated"), the service plug may be removed for installation of a gauge.

The compressor suction service valve (inlet) is threaded for 5/8-inch tubing; the compressor discharge service valve (outlet) for 1/2-inch tubing.

The reservoir shut-off valves are of the single seating type shown in Figure 6. The outlet valve (3/8" SAE flare to 3/8" MPT) is closed when all of the refrigerant is to be pumped into the reservoir to permit removal of some part of the system. The inlet valve (1/2" SAE flare to 3/8" MPT) is then used to retain the Freon in the receiver.

10. Testing System for Leakage

It is possible to check for severe leaks by visual inspection. Oil from the compressor mixes with the Freon and travels throughout the entire system, and consequently oil leaks at any point in the system indicate gas leaks at the same point. This method, however, is practical only for very bad leaks.

An entirely different method must be used to detect slight leaks. This requires the use of a Halide Gas Leak Detector, which is a small alcohol torch with a long rubber tube that feeds air to the flame.

To use this torch, pass the open end of the tube around the joints and connections to be tested, and particularly just below the joints, as Freon is heavy enough to

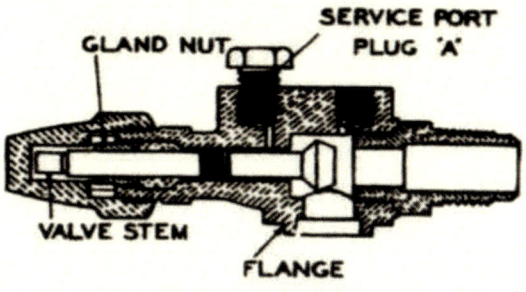

Fig. 5. Service Valve

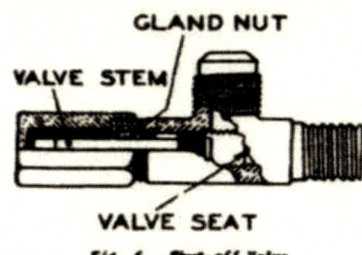

Fig. 6. Shut-off Valve

Figure 334. 1941 Cadillac Air-Conditioning Manual, p. 8   CLCMRC   Paul Ayres

CADILLAC AIR CONDITIONING MANUAL

settle slowly. If a leak exists, the flame of the torch will turn to a brilliant green color.

If a leak is detected in a flare connection, draw up the flare nut tightly. If the leak still exists, the flare on the tubing is probably defective. In this case, the gas in that part of the system should be pumped into the receiver, as described in Note 11, the tubing disconnected, and a new flare made in the tubing.

If a leak is detected in a soldered joint, relieve the pressure in that part of the system down to zero on the gauge, either by pumping the refrigerant into another part of the system or, in the case of the reservoir, by purging the gas out of the system into the air or into a small refrigerant service drum. Do not attempt to solder the joint while there is pressure in that part of the line.

11. **Removing and Installing Expansion Valve**

   1. "Backseat" compressor suction service valve by removing cap and turning valve stem counter-clockwise as far as possible with a 1/4 inch valve stem wrench.

   2. Remove service plug from port in valve and attach compound (pressure and vacuum) gauge to this port.

   3. Attach high pressure gauge (reading up to 300 pounds) to service port of compressor discharge service valve in the same manner.

   4. Open suction and discharge service valves one turn only by turning stem clockwise.

   5. Remove cap from reservoir outlet valve and close valve by turning valve stem clockwise as far as it will go.

   6. Start engine and run at slow speed. Compressor will now pump all gas from 3/8" liquid line, expansion valve, evaporator coil, and 5/8" low pressure gas line (suction line) and store it in the reservoir.

Watch both gauges. If high pressure gauge goes to limit of gauge dial, stop engine, as either the compressor has been run at too high a speed or the installation received too much of a Freon gas charge when initially charged. If pressure is due to an excessive compressor speed, it can be reduced by allowing compressor to remain stopped for a few minutes until pressure drops to normal, when engine can be restarted.

If there has been too much Freon charged initially, the pressure will not drop when the engine is stopped, and it will be necessary to vent the excess gas by loosening the flare nuts at the expansion valve inlet and outlet, and allowing the gas to escape slowly to the atmosphere until the lines are empty. The correct Freon charge should be retained in the liquid receiver, as it is closed off from the rest of the system by the shut-off valve and the discharge valves of the compressor.

Run engine until compound gauge reads 2" of vacuum. Then stop engine. It is desirable not to open the refrigerant lines while the compound gauge registers a vacuum and it may be necessary to "crack" the liquid reservoir shut-off valve until the compound gauge reads zero or slightly above. The lines may then be opened.

   7. Remove expansion valve by loosening inlet and outlet flare nuts. Some gas may escape, but the loss should be negligible if reservoir shut-off valve is closed tightly.

   NOTE: When expansion valve is removed or at any time when lines are open, care must be taken not to start engine unless compressor has been unbelted.

   8. Remove capillary bulb from its clamp on the 5/8" line leaving evaporator.

   9. Install new valve and pull flare nuts tight.

   10. Clamp capillary bulb to 5/8" line as before.

9

Figure 335. 1941 Cadillac Air-Conditioning Manual, p. 9   CLCMRC   Paul Ayres

CADILLAC AIR CONDITIONING MANUAL

11. "Crack" receiver shut-off valve until compound gauge on compressor reads 30 to 40 pounds; then close.

12. Test around expansion valve nuts for leaks.

13. Open reservoir shut-off valve by turning valve stem counter-clockwise three or four turns. Tighten gland nut around valve stem. Replace cap tightly.

14. Start engine and watch gauges. Compound gauge should read 20 to 40 pounds, and high pressure gauge 140 pounds to 190 pounds, after engine runs at slow speed for a few minutes (with blower running).

15. "Backseat" compressor discharge and suction service valves by turning valve stems counter-clockwise as far as they will go. Remove gauges, replace plugs, tighten gland nuts and replace caps tightly.

16. Recheck level of refrigerant (Note 8) to make certain that system has been properly refilled.

12. **Removing and Installing Condenser**

1. Remove radiator grille from car.

2. With blower running, start engine and run at slow speed until condenser is warm.

3. Stop engine, remove caps from compressor discharge service valve and reservoir inlet valve and turn stems clockwise as far as they will go.

4. Loosen flare nut on compressor discharge service valve and allow gas in condenser to escape slowly.

5. Unscrew flare nut after all gas has escaped, and unscrew flare nut on condenser outlet connection.

6. Remove bolts securing condenser to the radiator dust shield and take off condenser. Cover inlet and outlet to keep damp air and dirt out of condenser.

7. To install condenser, put it in place and attach bolts to radiator dust shield.

8. Connect flare nut to condenser outlet connection and tighten.

9. Connect flare nut to compressor discharge service valve, but do not tighten.

10. "Crack" (open slightly) reservoir inlet valve. This will allow gas from the 1/2" line to the reservoir to pass up through the condenser and out at the loose nut on the upper valve, thus purging air out of condenser. It is necessary to purge only a few seconds.

11. Tighten flare nut on compressor discharge valve, open this valve and reservoir inlet valve fully (counter-clockwise), tighten gland nuts and put on and tighten valve caps.

12. Test joints and connections at both valves for leaks. If no leaks are found, radiator grille may be reinstalled.

13. **Removing and Installing Compressor**

1. Run engine at slow speed for a few minutes until compressor is warm.

2. Stop engine, remove caps from compressor discharge and suction service valves and close both valves by turning stems clockwise as far as they will go.

3. Loosen plug (Fig. 5) in service port of suction service valve and allow gas in the compressor to escape slowly. Also remove plug in service port of discharge service valve and allow gas in compressor head to escape.

4. Remove cap screws that hold compressor service valves to the compressor and carefully lift valves away from compressor.

5. Loosen bolts holding compressor to the compressor base, and remove idler-to-compressor drive belt.

10

Figure 336. 1941 Cadillac Air-Conditioning Manual, p. 10   CLCMRC   Paul Ayres

CADILLAC AIR CONDITIONING MANUAL

6. Remove bolts and compressor from base.

7. Take off nut holding compressor pulley to shaft, remove set screw on compressor side of pulley and remove pulley.

8. See that Woodruff key is in place on replacement compressor and transfer pulley to replacement compressor. Slide it in place and tighten shaft nut. Tighten set screw.

9. Put replacement compressor on base, put in bolts loosely, put on belt and pull up bracket until belts are tight--not too tight, just enough to eliminate slip. Also see that belts and pulleys are lined up. Tighten bolts.

10. Put in new copper flange gaskets between the service valves and compressor. Put valve flanges down against copper gaskets, run in cap screws finger tight. Tighten cap screws evenly so that flange fits flat against the gasket.

11. Replace plug in service port of suction service valve, and crack this valve to blow gas from the suction line up through the compressor and out service port of discharge service valve. Turn compressor pulley over by hand to assist in purging air from compressor.

12. Install and tighten plug in service port of discharge service valve, open valve, tighten gland nut, install and tighten cap.

13. Open suction service valve, tighten gland nut, install and tighten valve cap.

14. Test for leaks around service valve flanges, service ports, and caps. If there are no leaks, the compressor is ready for use.

14. Removing and Installing Dehydrator

The chemical in the dehydrator acts continuously to absorb moisture from the Freon passing through it and in time becomes completely saturated. For this reason, replacement of the dehydrator is recommended whenever the system is opened up for any service work where the loss of the charge of Freon was involved.

To make the replacement--

1. Exhaust the Freon from the reservoir-to-compressor line, as explained in Note 11.

2. Remove old dehydrator.

3. Connect new dehydrator to tube from reservoir.

4. Crack reservoir shut-off valve to purge air from dehydrator.

5. Connect other end of dehydrator to expansion valve line.

6. Open reservoir shut-off valve.

15. Adding Refrigerant to System

1. Install gauges in compressor service valves as described in Note 11, except that a tee fitting should be used between the compound gauge and the suction valve.

2. Connect a Freon drum to this tee by means of a 1/4-inch copper tubing with flared end.

3. Purge this line of air by "cracking" valve on Freon drum with flare nut loose. Tighten flare nut.

4. Start engine and run it at slow speed.

5. Turn suction service valve to middle position, and open valve on Freon drum until compound gauge reads 55 to 60 pounds. Compressor is then drawing gaseous Freon from drum and delivering it to condenser where it condenses into a liquid and goes on to reservoir.

NOTE: Stand Freon drum with valve end up, so that only Freon gas is drawn into compressor. Do not invert drum or lay it on its side, as

11

Figure 337. 1941 Cadillac Air-Conditioning Manual, p. 11   CLCMRC Paul Ayres

CADILLAC AIR CONDITIONING MANUAL

this will allow liquid Freon to enter the compressor and perhaps damage the valves.

If drum gets cold, set it in a pail of warm water to hasten the vaporizing.

6. Keep trying the liquid tester. When liquid Freon can be drawn from it, shut off Freon valve and stop engine.

7. Turn suction service valve all the way out; remove charging line, tee and gauge; replace plug in service port; open valve, tighten gland nut, replace and tighten valve cap.

8. Turn discharge service valve all the way out, remove high pressure gauge from discharge valve service port; replace plug; tighten gland nut; replace and tighten cap.

Complete Recharge of System—

1. If no liquid or gas comes out when liquid tester is opened, put gauges on compressor (as described in Note 11). If no pressure is registered on either gauge, it indicates that all of the Freon has been lost from the system and that it must be completely recharged.

2. Reconnect compound gauge to compressor suction service valve with tee, charging line and Freon drum, as just described.

3. Close compressor discharge service valve, and remove gauge from service port.

4. With blower running, start engine and run slowly. Compressor is now pumping out any slight amount of gas (or air) from entire system. As compressor continues to pump, compound gauge should read lower and lower into a vacuum.

5. If only a slight vacuum can be obtained, air is coming into system, probably at the place where the Freon escaped. In such a case, stop engine and open valve on Freon drum to get a reading of 50 to 60 pounds on compound gauge. Then go over entire system, find leak and repair it.

6. When a vacuum of from 25 to 29 inches has been obtained (depending on altitude and barometric pressure), all air has been expelled from system.

7. Stop engine, and immediately open valve on the Freon drum, but close it when compound gauge reads zero and gas starts to escape from service port of discharge service valve.

8. Put high pressure gauge back in service port of discharge service valve and open this valve so that gauge will indicate.

9. Open valve on Freon drum to obtain 50 to 60 pounds on compound gauge and test entire system for leaks. Take no joint for granted.

10. Start engine and run slowly and add refrigerant to normal amount as described in previous note.

Fusible Plug—The liquid receiver is equipped with a fusible plug set to discharge at 212°F. This is a safety device to prevent excessive pressure in the event the system is overcharged with Freon gas. When an excessive pressure is reached, the fusible plug will melt, allowing all of the charge to escape. It is then necessary to replace this plug with a new plug having the same temperature setting (212°F.) before recharging. In replacing the plug, make sure it is screwed in tightly.

16. **Checking Oil Level in Compressor**

If there has been a loss of liquid Freon from the system, some of the oil (which mixes readily with Freon and is present in all parts of the system) may also have been lost. The level in the compressor may be checked as follows:

12

Figure 338. 1941 Cadillac Air-Conditioning Manual, p. 12   CLCMRC   Paul Ayres

CADILLAC AIR CONDITIONING MANUAL

1. With blower running, start engine and run at slow speed for a few minutes until compressor crankcase is warm. Then stop engine.

2. Remove cap and turn compressor suction service valve clockwise as far as it will go. Loosen plug in service port and allow the gas in the compressor to escape slowly.

3. When there is no further escape of gas, remove the oil filler plug, a hex head plug on the side of the compressor crankcase.

4. Insert a clean rod to bottom of compressor crankcase, and measure height of oil level. It should be up to, or within 3/8" below center line of compressor shaft.

5. If oil level is low, add oil as necessary. **USE SPECIAL CADILLAC COMPRESSOR OIL ONLY.**

6. Replace oil filler plug, tighten plug in service port on suction service valve, open valve, tighten gland nut and replace, and tighten valve cap. Test for leaks. No purging of compressor is necessary for this operation.

Figure 339. 1941 Cadillac Air-Conditioning Manual, p. 13   CLCMRC   Paul Ayres

DIAGNOSIS CHART

SYMPTOM	CAUSE	TEST	REMEDY
Air Evaporator Not Cold	1. Compressor not running, or running slowly. Belt broken or loose and slipping.	1. Obvious upon inspection.	1. Tighten belts; replace if necessary. Be sure to line up pulleys properly.
	2. Loss of refrigerant	2. Open liquid tester on reservoir. If gas only comes out, some Freon has been lost.	2. Test entire system for leaks thoroughly. Find and repair leaking joint, then add refrigerant as described in Note 15.
	3. Expansion valve stopped up with foreign matter.	3. Low suction pressure; evaporator coil not active; suction tube out of evaporator warm.	3. Stoppage will probably be found at screen at inlet of valve. Remove and wash in clean naphtha.
	4. Moisture in expansion valve.	4. Same as 3 above.	4. Same as 3 above and also install new dehydrator.
	5. Condenser plugged with dirt, bugs or lint.	5. High compressor head pressure. Gas line from compressor extra hot.	5. Remove radiator grille and clean condenser thoroughly.
Not Enough Air From Blower	1. Air filter stopped up.	1. Low suction pressure. Suction line cold.	1. Remove air filter and replace with a new one.
	2. Blower running under speed—Loose or corroded connections, rheostat switch broken battery charge low.	2. Insufficient circulation.	2. Trace circuits for bad connections. Check rheostat switch and replace if necessary. Check battery and recharge if low.
	3. Evaporator coil stopped with dirt or lint.	3. Low suction pressure, Suction line cold.	3. Cleanse with spray gun and a solution of Diversal cleaner and water.
Interior of Car not Cool, but Normal Amount of Cold Air From Blower.	1. Windows of ventilator open. Doors opened too much of time.	1. Obvious. NOTE: Ventilation requirements for crowded car especially if occupants are smoking may impose abnormal load particularly in hot, moist weather.	1. Use care in regulating ventilation for smoke and in leaving doors open.
	2. Car engine idling or running very slowly large portion of time.	2. High suction pressure, high discharge pressure.	2. Set engine idling speed somewhat higher, and run blower full speed.
	3. Extra Moist Weather perhaps raining	3. The inside of the car will be more comfortable than an unconditioned car, as the unit removes surplus moisture from the air. However, the dry bulb temperature (as read on an ordinary thermometer) may not read any lower than a shaded thermometer outside the car.	3. Use very minimum of ventilation and operate blower at maximum speed.
Compressor Unit Noisy	1. Loose drive pulley or compressor pulley.	1. Nut on compressor shaft or key from shaft to pulley may be loose.	1. Tighten shaft out and replace key if loose.
	2. Squeaky drive belt. Belt loose or greasy.	2. Belt should not have more than 3/8" slack. Inspect for oil on belt.	2. Tighten belt if loose. If greasy, wipe clean with naphtha and coat with powdered talc.

14

Figure 340. 1941 Cadillac Air-Conditioning Manual, p. 14   CLCMRC   Paul Ayres

DIAGNOSIS CHART

SYMPTOM	CAUSE	TEST	REMEDY
Compressor Unit Noisy (Cont)	3. Liquid Freon instead of gas being returned to compressor through suction line. This causes oil pumping and inadequate lubrication to the compressor. Usually caused by capillary bulb loose on suction line.	3. Wet and cold suction line—even the suction service valve and the compressor cylinder housing and crankcase may be cold although the engine and compressor are running—Causes knocking sound similar to loose bearings.	3. Clamp capillary bulb tightly to the suction line.
	4. Compressor loose on base.	4. Obvious upon inspection.	4. Tighten compressor hold-down bolts.
	5. High discharge pressure: Air in system. Stoppage in valves or lines. Overcharge of Freon.	5. Check with gauge in discharge service valve. Causes pounding, laboring sound.	5. Correction depends on cause of the high discharge pressure. Air in system; purge air at discharge system and entirely recharge. Stoppage: remove stoppage. Overcharge of refrigerant: Purge Freon to tester level.
	6. Too much oil in compressor crankcase.	6. Evidenced by a dull thumping sound.	6. Check at compressor oil filler plug as in Note 16.
	7. Not enough oil in compressor crankcase.	7. Usually not evidenced until bearings and pistons are worn and knocking or seal leaking. Denoted by very hot crankcase.	7. This condition is frequently the result of leakage of oil from the system and if noticed and caught soon enough may be corrected by adding oil to the compressor but if damage has already resulted, it will be necessary to replace the compressor.
	8. Solder or foreign particles in cylinders.	8. Evidenced by a knocking sound similar to loose bearings.	8. Open compressor and clean.
	9. Broken parts in compressor.	9. Same as 8 above.	9. Replace entire compressor.
Noisy Operation Other Than Compressor	1. Loose tubing, straps, brackets or sheet metal parts.	1. If any question as to whether the compressor is responsible for the noise, operate with belt removed.	1. Trace sound for location and cause and repair as may be required.
	2. Loose bearings of blower motor or fan.	2. Noticeable at starting of blower.	2. Remove blower motor and fan and replace motor.
	3. Hissing sound at expansion valve.	3. May be no more than normal, but if excessive may indicate low Freon charge.—Loss of Freon.	3. Check liquid tester on receiver.
	4. Gurgling sound in evaporator.	4. Not audible except under very quiet conditions.	4. Entirely normal—merely Freon passing through evaporator. No correction required.

Lithographed in U.S.A.

15

Figure 341. 1941 Cadillac Air-Conditioning Manual, p. 15   CLCMRC   Paul Ayres

The relatively few owners of the air-conditioned 1941 Cadillac primarily lived in the Dallas-Ft. Worth, Texas area, according to Mr. Rod Barclay, author of *BOY! That Air Feels Good!* He reported:

Cadillac rushed production of 300 Bishop & Babcock air-conditioned models for 1941, with the majority shipped to Lone Star Cadillac of Dallas, Texas. (It) served as the predominant retailer of the 1941 Cadillac air-conditioned cars.

The Bishop & Babcock system set the standard for many years. This was the system that most DIYs (Do-it-yourself-type persons) emulated, and most start-ups like ARA copied in one way or another. "Why Dallas?" you may ask. The three-word answer is "heat, humidity, and money."

Cadillac offered the Bishop & Babcock-supplied air-conditioning unit during the 1941 model year. The reasoning behind its discontinuance for the 1942 model year is unknown. As the United States entered WWII the day after the Pearl Harbor bombing on December 7, 1941, all automobile production soon halted. By February 9, 1942, the plants ceased automobile production and converted to war materiel production.

Near the end of the 1940s, the men of Lone Star Cadillac of Dallas, Texas, at the request of their Cadillac customers, joined to assemble trunk-mounted air-conditioning units copied from the Bishop & Babcock system.

Figure 342. 1941 Cadillac Fleetwood Series Sixty Special, Model 6019S-A Owned by Dr. Richard Zeiger Ronald Verschoor photo

Of the 66,130 Cadillac automobiles produced in 1941, it equipped 300 cars with air-conditioning, a ratio to total sales of .45%.

Author note: Only three air-conditioned 1941 Cadillac automobiles exist today. This chapter features all three 1941 Cadillacs equipped with air-conditioning.

The first of three "Spotlight" sections displays Dr. Richard Zeiger's air-conditioned 1941 Cadillac Fleetwood Series Sixty Special Five-Passenger Touring Sedan, Model 6019S-A. (Fig. 342) This Series Sixty Special was one of 185 "Sunshine Turret Top" sunroof-equipped sedans produced of the 4,098 Fleetwood models sold.

Figure 343. 1941 Cadillac Series Sixty-Two De Luxe Coupe, Model 6227D Former owner and photo by Doug Houston

The second "Spotlight" describes the 1941 Cadillac Series Sixty-Two De Luxe Coupe, Model 6227D, one of 1,900 produced, formerly owned by Mr. Doug Houston and sold at his estate auction. (Fig. 343)

The third "Spotlight" presents the 1941 Cadillac Series Seventy-Five, Five-Passenger Touring Sedan, Model 7519, one of 422 produced, formerly owned, and sold at auction by Mr. Richard Kughn. (Fig. 344)

Figure 344. 1941 Cadillac Fleetwood Series Seventy-Five, Model 7519 Former owner Richard Kughn   Rex Roy photo

176 Chapter 4: 1941 Cadillac Factory Air-Conditioning

Spotlight on Cadillac, Featured Car Number 1

1941 Cadillac Fleetwood Series Sixty Special, Five-Passenger Touring Sedan, Model 6019S-A, Equipped with Factory Air-Conditioning

Owned by Dr. Richard Zeiger

Photos by Ronald Verschoor

Figure 345. 1941 Cadillac Fleetwood Series Sixty Special Five-Passenger Touring Sedan, equipped with factory air-conditioning, 1 of 300 produced. The "Sunshine Turret Top" sunroof, shown open, was optionally available only on this Series Sixty Special, identified as Model 6019S-A. Owner Dr. Richard Zeiger   Ronald Verschoor photo

Photographed up close, the 1941 Series Sixty Special's coffin-nosed hood design incorporates small block letters above the grille.

Cadillac's new crest with upswept wings leads to the "Flying Lady" dual-purpose hood ornament/latch handle. (Figs. 308, 345, 346)

Figure 346. 1941 Cadillac Series Sixty Special close-up view of block Cadillac letters and Cadillac Crest with upswept wings lead to the Flying Lady combination hood ornament and latch handle. Owner Dr. Richard Zeiger   Ronald Verschoor photo

The two former separate hood louvers in the 1940 Cadillac expanded into one long, ventilated hood louver with vertical grille bars, complemented by a Cadillac crest. (Fig. 347)

Figure 347. 1941 Cadillac Series Sixty Special ventilated hood louvers and Cadillac crest Owner Dr. Richard Zeiger   Ronald Verschoor photo

178   Chapter 4: 1941 Cadillac Factory Air-Conditioning

The Bishop & Babcock-based Servel compressor rests on the driver's side of the 150-hp Cadillac 346-CID V-8 engine. (Fig. 348) Shown operating, the size of the factory green-painted compressor does not crowd the engine bay. The constantly run compressor pulley connects below with a two-belt idler pulley to the crankshaft.

Author note: Although the control panel knob featured an "OFF" position, the compressor ran continuously, potentially leading to a mechanical breakdown far sooner than the expected design lifespan.

Figure 348. 1941 Cadillac Series Sixty Special air-conditioning compressor while operational.   Owner Dr. Richard Zeiger   Ronald Verschoor photo

Far into the future, the reintroduced 1953 air-conditioning technology remained comparable to the prewar systems. The 1954 Cadillac, Oldsmobile, Buick, Pontiac, and Nash cars and the 1955 and 1956 Packard cars achieved a technological milestone by introducing an electromagnetic clutch that disengaged the compressor when the air-conditioning turned off, potentially extending the compressor's service life while lowering gasoline consumption.

In a profile view, the compressor cylinder head reveals the two-cylinder configuration, the high-pressure copper tubing on the left, leading to the condenser in front of the radiator, and the low-pressure copper tubing on the right, extending from the trunk-mounted evaporator. (Figs. 349–352) The radiator fan blurs during operation.

Figure 349. 1941 Cadillac Series Sixty Special Servel air-conditioning compressor profile view of high and low-pressure copper tubing while operational. Owner Dr. Richard Zeiger   Ronald Verschoor photo

Factory Air: Cool Cars in Cooler Comfort 179

Figure 350. 1941 Cadillac Series Sixty Special Servel air-conditioning compressor and idler pulley frontal view. Owner Dr. Richard Zeiger   Ronald Verschoor photo

Figure 351. 1941 Cadillac Series Sixty Special Servel air-conditioning compressor frontal view and high-pressure copper tubing leading to the condenser hidden under shielding.   Owner Dr. Richard Zeiger Ronald Verschoor photo

Figure 352. 1941 Cadillac Series Sixty Special elevated engine bay view of Servel air-conditioning compressor. Note bottom, center view of thermostatically controlled connection rod for the radiator's automatic shutter opener. Owner Dr. Richard Zeiger   Ronald Verschoor photo

180 Chapter 4: 1941 Cadillac Factory Air-Conditioning

An interior illuminated by opening the "Sunshine Turret Top" sunroof highlights the symmetrically designed instrument panel. (Fig. 353) The air-conditioning knob (not shown in Fig. 353) rests to the knob's left, labeled with a "B" (yellow arrow) for the optional single backup light illumination.

The lower dash view shows the unmarked air-conditioning knob (yellow arrow, Fig. 354), a blank space for an optional "F" control knob for fog lights, the "B" knob for back-up light illumination, and the heater-defroster controls. Inexplicably, no inscription identified the blank control knob's purpose, in contrast to the 1941 and 1942 Packard lettering placed on the control panel face.

Figure 353. 1941 Cadillac Series Sixty Special instrument panel layout. Illumination provided through opened. "Sunshine Turret Top" sunroof. Air-conditioning control knob (not shown), on left of knob with letter "B," (yellow arrow).
Owner Dr. Richard Zeiger   Ronald Verschoor photo

Figure 354. 1941 Cadillac Series Sixty Special lower left dash view of air-conditioning control knob (yellow arrow), backup light knob with letter "B," and heater-defroster controls.
Owner Dr. Richard Zeiger   Ronald Verschoor photo

The cooling register on the rear seat's package shelf uses a fan-forced, three-speed blower to circulate the refrigerated air. (Fig. 355, yellow arrow)

The cooled air flowed upward and followed the headliner toward the front seat passengers. The fan recirculated the air from the front seat and pulled the cool air past the rear seat passengers, whereby it drew the air underneath the rear seat and filters, then flowed upward and through the cooling coils in the trunk-mounted evaporator.

Figure 355. 1941 Cadillac Series Sixty Special view of air-conditioning cooling register. (yellow arrow) Optional equipment includes Cadillac Venetian blinds accessory on rear window and opened, partial view of the "Sunshine Turret Top" sunroof in headliner.
Owner Dr. Richard Zeiger   Ronald Verschoor photo

Close-up images reveal the air-conditioning register, louvers, and the handle for louver adjustment of the airflow, a new feature added by Bishop & Babcock for 1941. (Figs. 356, 357)

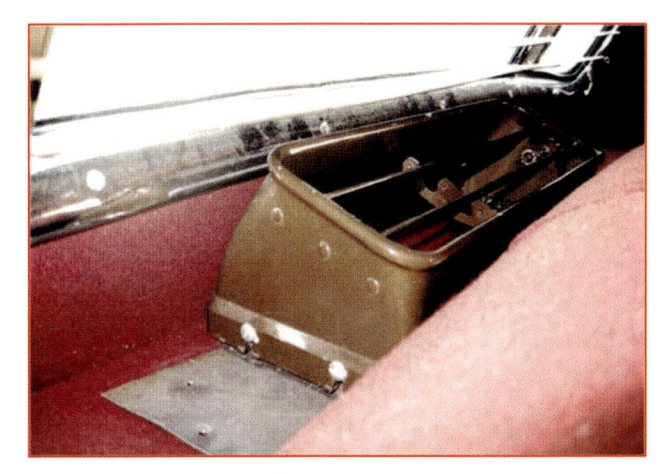

Figure 356. 1941 Cadillac Series Sixty Special air-conditioning cooling register and adjustable louvers
Owner Dr. Richard Zeiger   Ronald Verschoor photo

Figure 357. 1941 Cadillac Series Sixty Special air-conditioning cooling register showing louver adjustment handle
Owner Dr. Richard Zeiger   Ronald Verschoor photo

Dr. Zeiger commented about the optional Venetian blinds: (Fig. 358)

> It reduced the amount of direct sunlight entering the passenger compartment and helped direct the cool air from the blower into the passenger compartment rather than blowing directly on the rear backlight.

His statement compares with Mr. Houston's comments about the cool air effect on his three-window backlight not equipped with Venetian blinds. (Fig. 385) Dr. Zeiger reported that no insulation existed in the body during restoration, contrary to the *Cadillac Data Book* information. When asked how well the air conditioner cools, he reported:

Figure 358. 1941 Cadillac Series Sixty Special view of Venetian blinds, air-conditioning cooling register louvers, and adjustment handle
Owner Dr. Richard Zeiger   Ronald Verschoor photo

> The Cadillac remains cool in city and highway driving, due in part to the extensive insulation added to this car. The restorers applied insulation on the firewall and on the floorboard. As an added benefit, the insulation makes the car quieter than most Cadillacs of this year.

Figure 359. 1941 Cadillac Series Sixty Special trunk-mounted air-conditioning evaporator
Owner Dr. Richard Zeiger   Ronald Verschoor photo

The evaporator housing containing the cooling coils occupies the trunk's forward ledge space. Cadillac's trunk design allows vertical spare tire storage. (Fig. 359) Close-up views of the evaporator reveal the "Cadillac Air Conditioner" nameplate and the expansion valve's external location leading to the evaporator cooling coils. (Figs. 360, 361)

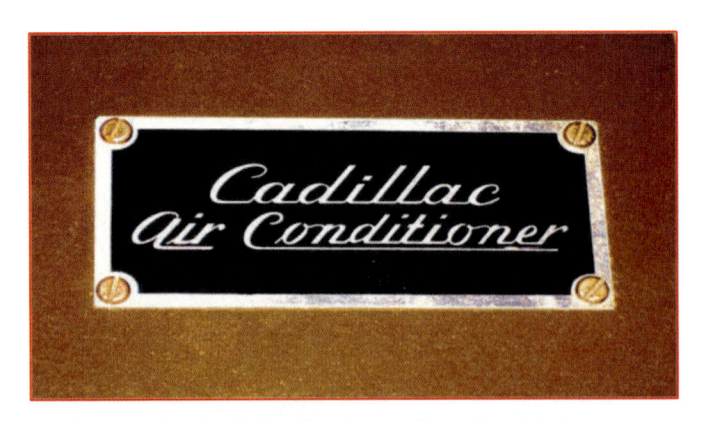

Figure 360. 1941 Cadillac Series Sixty Special close-up view of "Cadillac Air Conditioner" nameplate affixed to the air-conditioning evaporator
Owner Dr. Richard Zeiger
Ronald Verschoor photo

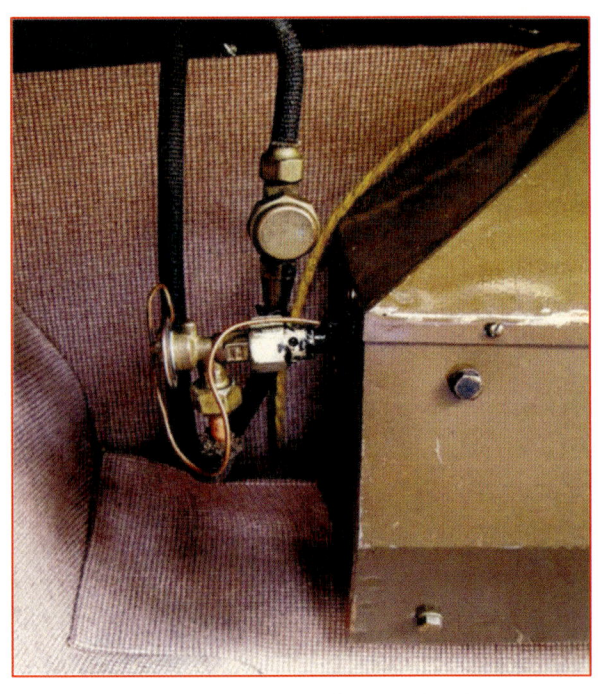

Figure 361. 1941 Cadillac Series Sixty Special's air-conditioning expansion valve connected to cooling coils inside evaporator housing
Owner Dr. Richard Zeiger   Ronald Verschoor photo

Factory Air: Cool Cars in Cooler Comfort 183

Other views include unique Fleetwood touches that made this Cadillac a "Standard of the World." (Figs. 362–377)

Figure 362. 1941 Cadillac Series Sixty Special features non-decorated fender-wheel shields extending toward the rear
Owner Dr. Richard Zeiger   Ronald Verschoor photo

Figure 363. 1941 Cadillac Series Sixty Special displays the three-piece backlight
Owner Dr. Richard Zeiger   Ronald Verschoor photo

184   Chapter 4: 1941 Cadillac Factory Air-Conditioning

Figure 364. 1941 Cadillac Series Sixty Special front fender extends into front door and features a chromed "Fleetwood" script
Owner Dr. Richard Zeiger   Ronald Verschoor photo

Figure 365. 1941 Cadillac Series Sixty Special "Fleetwood" logo on door sill
Owner Dr. Richard Zeiger   Ronald Verschoor photo

Figure 366. 1941 Cadillac Series Sixty Special jewel-toned Cadillac crest affixed to trunk lid
Owner Dr. Richard Zeiger   Ronald Verschoor photo

Figure 367. The 1941 Cadillac wheel cover with a red Cadillac-crested medallion accentuates the wide whitewall tires
Owner Dr. Richard Zeiger   Ronald Verschoor photo

Factory Air: Cool Cars in Cooler Comfort 185

Figure 368. 1941 Cadillac Series Sixty Special passenger side dash view and deluxe Fleetwood door appointments
Owner Dr. Richard Zeiger   Ronald Verschoor photo

Figure 369. 1941 Cadillac Series Sixty Special view of dash painted in Madeira Maroon, a special order that replaced the standard burled woodgrain finish
Owner Dr. Richard Zeiger   Ronald Verschoor photo

Figure 370. 1941 Cadillac Series Sixty Special, first year of concealed gas filler cap under left taillight, in closed position
Owner Dr. Richard Zeiger   Ronald Verschoor photo

Figure 371. 1941 Cadillac Series Sixty Special, first year of concealed gas filler cap under left taillight, in open position
Owner Dr. Richard Zeiger
Ronald Verschoor photo

Figure 372. 1941 Cadillac Series Sixty Special features built-in footrest, folding center armrest, robe cord with Pom-Pom ends, and a "smoking case" in side armrest, shown in closed position.
Owner Dr. Richard Zeiger   Ronald Verschoor photo

Figure 373. 1941 Cadillac Series Sixty Special's view of opened "smoking case" in side armrest
Owner Dr. Richard Zeiger   Ronald Verschoor photo

Figure 374. 1941 Cadillac Series Sixty Special's courtesy lighting provides illuminated entry for rear seat passenger.
Owner Dr. Richard Zeiger   Ronald Verschoor photo

Factory Air: Cool Cars in Cooler Comfort 187

Figure 375. 1941 Cadillac Series Sixty Special Optional "Sunshine Turret Top" sunroof, shown in open position. Available only in this Model 6019S-A, one of 185 sold in 1941 for $85. Owner Dr. Richard Zeiger   Ronald Verschoor photo

Figure 376. 1941 Cadillac Series Sixty Special exterior view of optional "Sunshine Turret Top" sunroof, shown in open position. Available only in this Model 6019S-A, one of 185 sold in 1941 for $85. Owner Dr. Richard Zeiger   Ronald Verschoor photo

A Sunshine Turret Top is also available on the Sixty Special sedan without division at small additional charge. This design combines the open-air features of a convertible type with the safety of a steel roof. The easily operated sliding panel is effectively sealed from rain and draft and may be locked in any desired position.

Figure 377. 1941 Cadillac Series Sixty Special optional "Sunshine Turret Top" sunroof   1941 Cadillac Data Book, 9-13-40, p. 58 GM Heritage Center

188   Chapter 4: 1941 Cadillac Factory Air-Conditioning

The 1941 Cadillac accessory brochure listed the "Sunshine Turret Top," Style No. 6019, for $85. (Figs. 378, 379, and yellow arrow) According to the CPI, the equivalent price in 2021 for the "Sunshine Turret Top" sunroof is $1,576.

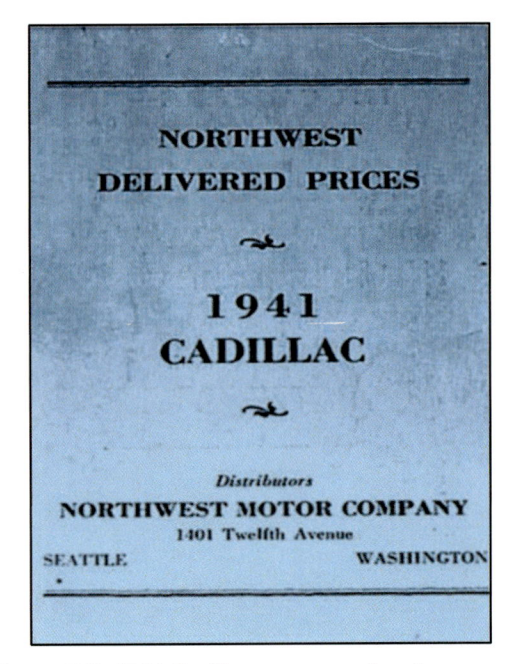

Figure 378. 1941 Cadillac accessory brochure cover Northwest Motor Company, Seattle, Washington

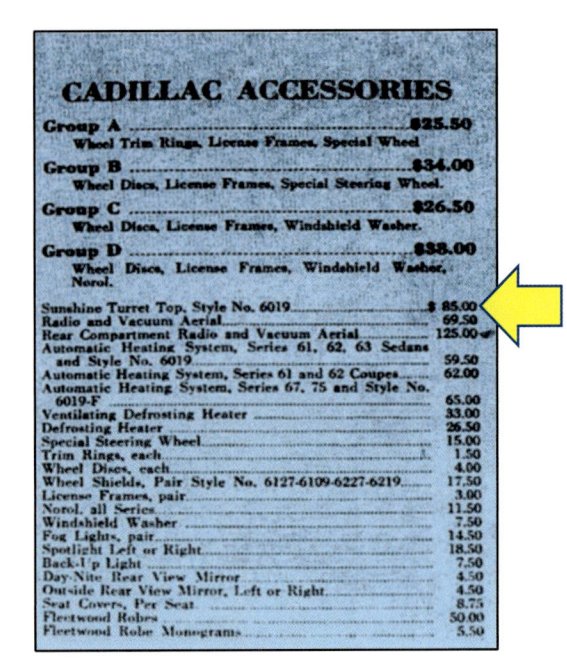

Figure 379. 1941 Cadillac accessory brochure pricing Northwest Motor Company, Seattle, Washington

Owner Dr. Richard Zeiger and photographer Ronald Verschoor stand beside the beautifully restored 1941 Cadillac Series Sixty Special, equipped with factory air-conditioning (one of 300 produced) and the optional "Sunshine Turret Top" sunroof (one of 185 produced). (Fig. 380)

The exceptional Madeira Maroon 1941 Cadillac won the Classic Car Club of America's (CCCA) first-place award in March 2013, scoring 98.5 points out of 100. In September 2013, Dr. Zeiger completed a 1,000-mile driving tour in Idaho without incident.

Figure 380. Dr. Richard Zeiger, owner (left), and Ronald Verschoor, photographer, with 1941 Cadillac Series Sixty Special, equipped with factory air-conditioning and a "Sunshine Turret Top" sun roof.

Spotlight on Cadillac, Featured Car Number 2
1941 Cadillac Series Sixty-Two De Luxe Coupe, Model 6227D, Equipped with Factory Air-Conditioning

After 50 years as a Cadillac & LaSalle Club member, Mr. C. Douglas "Doug" Houston, Jr. passed away on February 9, 2016. He served as a trusted knowledge base for the 1941 Cadillac factory air-conditioning system, later year V-16 Cadillacs, and Cadillac radios.

Figure 381. 1941 Cadillac Series Sixty-Two Coupe, Model 6227D, equipped with factory air-conditioning Formerly owned and photo by Doug Houston   CLC

Mr. Houston owned this two-toned Rivermist Gray over Dusty Gray 1941 Cadillac Series Sixty-Two De Luxe Coupe, Model 6227D, equipped with factory air-conditioning. (Fig. 381)

The *Original Cadillac Database* identified his coupe as one of only three surviving 1941 Cadillacs with air-conditioning. (Fig. 382)

1941 Cadillac photo page

Coupe interior (catalog illustration)

Once again, a De Luxe version of the Series 62 coupe was available.
It was termed style #6227D. Production totaled 1900 units; each one cost $1510.

This fine survivor - one of only three surviving 1941 Cadillacs with air-conditioning! -
belongs to enthusiast Doug Houston of Detroit, MI. The picture was taken at the
Grosse Pointe War Memorial , in Detroit, on the circular entrance drive

http://www.car-nection.com/yann/Dbas_txt/Phocad41.htm

Figure 382. 1941 Cadillac Series Sixty-Two De Luxe Coupe, equipped with factory air-conditioning   car-nection.com   Yann Sanders

190   Chapter 4: 1941 Cadillac Factory Air-Conditioning

In 1991, on the 50th anniversary of the 1941 Cadillac, the Cadillac-La Salle Club published a tribute to the car in *The Cadillac-La Salle Self-Starter Annual,* Volume XVII. (Fig. 383) It contains Mr. Doug Houston's article about the 1941 Cadillac, "Cadillac's Air Conditioner." (Figs. 384–387)

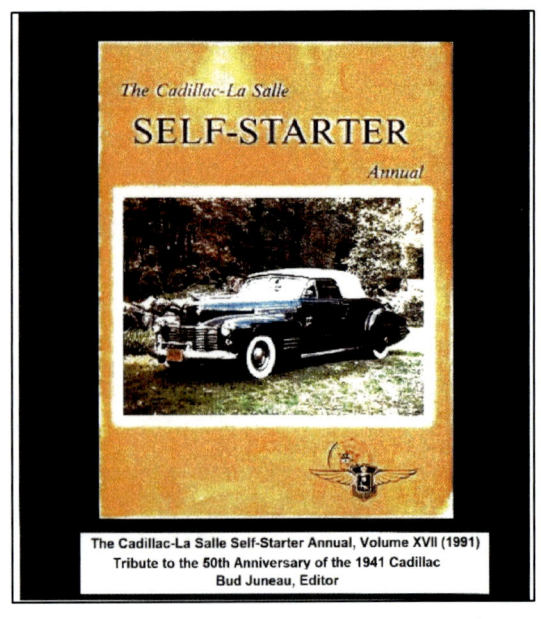

Figure 383. The Cadillac-LaSalle Self-Starter Annual, Volume XVII (1991), "Tribute to the 1941 Cadillac"

CADILLAC'S AIR CONDITIONER
By Doug Houston

Automotive technology had progressed so rapidly through the 1930's that any one aspect of it was difficult to follow, in many instances! This was the era when there were many car manufacturers competing with each other for excellence in engineering and overall design. The common goal was superior roadworthiness and customer appeal, all consistent with reliability. In Pennsylvania, the Highway of the future opened just in time for the 1941 model year. This was the dawn of an era, where a driver could travel non-stop for hours at high, safe speeds.

But travel through the hot, humid summer months hindered the enjoyment of travel for long stretches. Something was needed to overcome the fatigue that was the result. The answer to this problem was a refrigeration type cooler for the car's interior. Even if the air were to be dehumidified, driving fatigue could be drastically reduced.. Such a refrigerating type of air conditioner would be necessarily expensive, so the cost of such a system drove it into cars of higher stature, where customers were able and willing to buy it.

Both Packard and Cadillac arranged with Bishop and Babcock, a Cleveland firm, to design and supply a system, which could be installed on a passenger car. This was very much a pioneering situation. While the other elements of a refrigeration system were simple enough to fabricate, there had never been a refrigeration compressor that was designed for automotive application.

The system, as finally installed, employed a stationary commercial compressor, mounted to the car's engine, with four longer head bolts holding special bracketry, and belt driven from the front of the crankshaft, as shown. A clutching pulley was not available, for uncoupling the compressor when not desired, as on later systems, such as we have today, so the compressor remained fully operational at all times that the engine was running and kept the evaporator coils in the trunk cold. For this reason, some icing could occur and it was recommended that the fan be left on low speed. For Winter months, Cadillac recommended that the drive belt from the crankshaft be disconnected. Easier said than done, for the main drive belt from the crankshaft to the idler pulley, is to the inside of the regular fan belt and generator belt. The crankshaft pulley, by the way, on the air-conditioned cars is different, with room for the three belts, and with a smaller vibrational damper between the second and third pulleys. The timing pointer is different than regular production, since it has to be longer to span the inner pulley and still point to the timing marks on the damper, and it is held on with cap screws so it can be removed to allow removal of the inner air-conditioning belt. Regular production pointers are permanently attached.

The only control for the air-conditioning is the fan motor control, located on the bottom edge of the dash, to the left of the steering column, resembling the backing light or fog light switches, but bearing no letter markings.

Figure 384. "Cadillac's Air Conditioner," (1941) p. 30, by Doug Houston
The Cadillac-LaSalle Self-Starter Annual, Volume XVII (1991)   CLC   Terry Wenger, Sr.

> There are four positions on the switch; Off, Low, Medium, & High. As on the Hydra-Matic equipped cars, Cadillac had no exterior medallions or notation to indicate that this equipment was installed.
>
> The condenser was placed in front of the radiator, as is still the practice. Horns were relocated upward inside the sheet metal shroud. The radiator was unchanged for this system, but still cools sufficiently. The evaporator is mounted in the trunk compartment, as shown in the space that was normally the tool compartment. Drains for water condensate drained through the floor pan at the bottom of the evaporator housing. The air outlet is on the shelf, behind the rear seat backrest. Adjustable vanes direct the air through the passenger compartment as desired.
>
> Cadillac produced 300 cars with this air conditioner in the 1941 model year, and published a special service manual for the system. Of these 300 jobs, two are still known to exist. One, a series 75, is owned by Richard P. Kughn, of Detroit, and the other by this writer, a series 62 club coupe. The system in the series 75 car has been serviced and fine-tuned to perfect operation, and cools the car beautifully, while my own is not up to these standards, but is due for extensive work in the very near future. In the 62 coupe, the air outlet tends to chill the backlight (rear window), and on a muggy day, moisture condenses on the outside surface of the glass. Questioned about this on one occasion, I was amused to point out that there are only two '41 Cadillacs on the road today, where this is possible, not to be confused with inside fogging caused by water leaking into the back seat area!

Figure 385. "Cadillac's Air Conditioner," (1941) p. 31, by Doug Houston
The Cadillac-LaSalle Self-Starter Annual, Volume XVII (1991) CLC Terry Wenger, Sr.

Noted in the article, in contrast to the 1941 Packard "Air-Conditioned" emblem on its bonnet, the Cadillac did not adorn the car with a "Hydra-Matic" transmission or an "Air-conditioned" emblem: (Fig. 385)

As on the Hydra-Matic equipped cars, Cadillac had no exterior medallions or notation to indicate that this equipment was installed.

Mr. Houston contacted Mr. Al Haas at the Cadillac Motor Division for the 1941 Cadillac shipping orders. The shipping orders supplied the wholesale pricing for his Series Sixty-Two De Luxe Coupe, officially identified as (Model) Style No. 41-6227D, per the shipping order, and similar wholesale pricing information concerning Mr. Kughn's Series Seventy-Five, Five-Passenger Touring Sedan, officially identified as (Model) Style No. 41-7519. (Fig. 386) Notably, both invoice copies reflected the wholesale pricing for the air-conditioning units at $260.00. At the bottom of Mr. Kughn's invoice, the retail pricing comprised the wholesale price, $260, plus a retail markup of 40%, $104, yielding a list price of $364. No available published literature reflects the 1941 air-conditioning price; however, Mr. Roy A. Schneider's 1999 edition of *Cadillac's of the Forties* quoted the air-conditioning list price as $365 for a Cadillac Series 75, believed to be the only 1942 Cadillac delivered with air-conditioning. (Fig. 413)

Mr. Houston's Cadillac Model 41-6227D originated from Greenlease Moore, Inc., Oklahoma City, Oklahoma. The Cadillac information on Invoice No. 26594 listed the date shipped for car No. 56364 as of June 17, 1941. It listed the wholesale price of Model 41-6227D as $1029.60 and identified five options, including air-conditioning and the color Aberdeen Beige. With taxes, the wholesale price listed as $1,561.92. (Fig. 386)

Mr. Kughn's Cadillac Model 41-7519 originated from Greenlease Ledterman, Inc. of Tulsa, Oklahoma. The Cadillac Invoice No. 19370 listed the date shipped for car No. 39614 as of April 1, 1941. It listed the wholesale price of Model 41-7519 as $1,998.50 plus three options, including the air conditioner. With taxes, the wholesale price listed as $2,475.01. (Fig. 386)

192   Chapter 4: 1941 Cadillac Factory Air-Conditioning

Cadillac Motor Car Division

Shipping order wholesale invoice No. 26594.
Mr. Houston's 1941 Cadillac, Style No. 41-6227D, includes air-conditioning unit, $260.00.
Date shipped: June 17, 1941.

Cadillac Motor Car Division

Shipping order wholesale invoice No. 19370.
Mr. Kughn's 1941 Cadillac, Style No. 41-7519, includes air-conditioning unit, $260.00.
Date shipped: April 1, 1941.

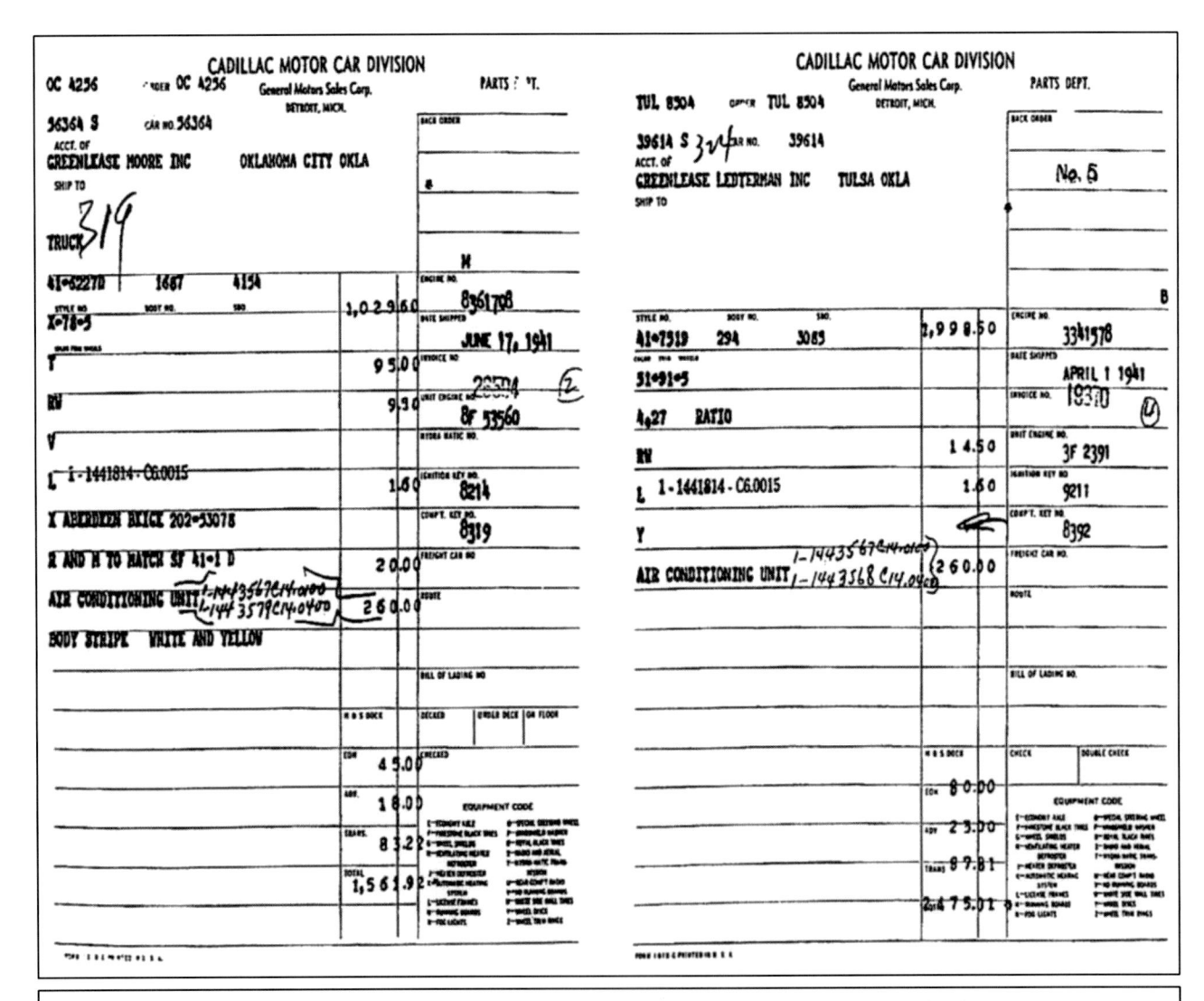

Doug Houston's car was ordered with Air-Conditioning and with the Catalog colors, which required Special Order paint. Prices are wholesale. Date Shipped: June 17, 1941

Richard Kughn's car was ordered with Air-Conditioning. Add 40% to wholesale prices for retail price. Date Shipped: April 1, 1941

SHIPPING ORDERS COURTESY OF AL HAAS, CADILLAC MOTOR DIVISION

Source: 1991 Cadillac Self-Starter Annual, Celebrating 50th Anniversary of 1941 Cadillac
Editor: Bud Juneau   -Terry Wenger, Sr. 6-2-18-   30

1941 Cadillac Factory Air-Conditioning
Retail Price Calculation:
Wholesale Price: $260 + 40% Markup = $364 Retail Price

Figure 386. 1941 Cadillac: Invoice prices for Mr. Doug Houston's (Model) Style No. 41-6227D, Series Sixty-Two De Luxe Coupe, on left.
Invoice prices for Mr. Richard Kughn's (Model) Style No. 41-7519, Fleetwood Series Seventy-Five, on right.
"Cadillac's Air Conditioner," (1941) p. 30 by Doug Houston
The Cadillac-LaSalle Self-Starter Annual, Volume XVII (1991)   CLC   Terry Wenger, Sr.

Factory Air: Cool Cars in Cooler Comfort 193

The only Air-conditioning control is this fan switch, devoid of markings or fanfare.

The engine compartment begins to be crowded by the compressor, bracket, and idler pulley.

The trunk unit puts the cold air in through the package shelf, behind the rear seat.

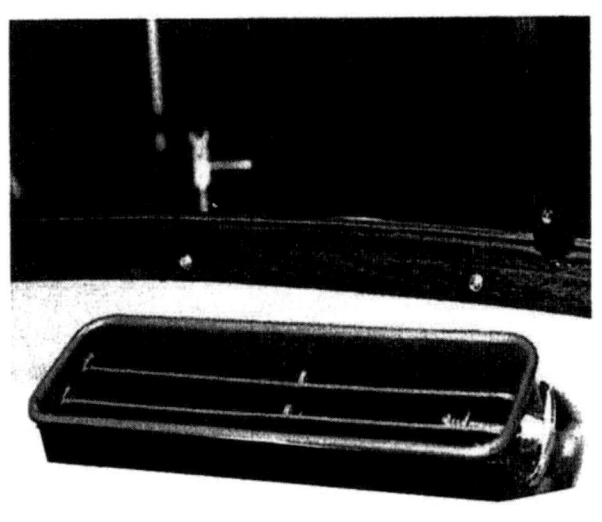

The entire car is cooled by this one outlet, with adjustable vanes.

PHOTOS COURTESY OF DOUG HOUSTON, BUCK VARNON, & RICHARD KUGHN

31

Figure 387. 1941 Cadillac views of Richard Kughn's air-conditioning control knob, Servel air-conditioning compressor, trunk-mounted evaporator, and cooling register with louver control handle "Cadillac's Air Conditioner," p. 31
The Cadillac-La Salle Self-Starter Annual, Volume XVII (1991) by Doug Houston CLC Terry Wenger, Sr.

194   Chapter 4: 1941 Cadillac Factory Air-Conditioning

In the *cadillaclasalleclub.org Technical/Authenticity Forum,* dated January 23, 2006, Mr. Houston described his air-conditioning system more thoroughly. Notably, he stated, "Cooling is reasonably good." (Fig. 388)

forums.cadillaclasalleclub.org/index.php?topic=77445.msg77708#msg77708

Doug Houston

Re: A/C in Caddys - When?
« Reply #4 on: January 23, 2006, 07:51:40 PM »

It was furnished to Cadillac (and Packard) by Bishop & Babcock, of Cleveland. Cadillac built 300 jobs in 41 with air conditioning. Three are known today: Richard Kughn, Detroit (41-75), Richard Zieger(SP?)(41-60S) California, and mine 41-6227D), Ortonville, Mich.

Drive was from the front of the crankshaft with a third pulley, replacing the harmonic balancer. The compressor was mounted on the left front end of the engine on special bracketry. There was no clutch to engage/disengage the compressor, so in winter, it was necessary to remove one of the drive belts.

The radiator was the same as on all other cars. The condenser was mounted in front of the radiator, as on todays cars. Horns were re-located upward to clear the condenser couplings.

The evaporator was in the trunk, with the outlet in the center of the parcel shelf, using the space where the tool compartment normally was. Condensate drained from the evaporator through a tube through the floor pan.

The only control was a 3 speed blower switch beneath the dash, to the left of the steering column, with a knob similar to the fog light knob, but with no markings. The blower motor was immediately below the air outlet in the parcel shelf.

Caillac published a small booklet titled: Cadillac Air Conditioning Manual that year. I had the book before I ever had the car.

Cooling is reasonably good. The system has a tendency to frost over, mainly because the air intake comes from around the back seat, through air filters in the rear seat panel. There should have been a better air flow designed for it. On a muggy day, there is condensation on the OUTER surface of the backlight (the rear window). Ive smugly commented that there are only three 41 Cadillacs where this is possible.

My car has Hydra-Matic on it, and never had a heater before I installed a system. The radio isnt shown on the invoice, so it was installed some time after delivery. It was delivered by Greenlease-Moore Cadillac in Oklahoma City. The car has been on trips at highway speeds in the summer, and has not overheated. Im the sixth owner.

Figure 388. 1941 Cadillac air-conditioning information
cadillaclasalleclub.org Technical/Authenticity Forum, January 23, 2006   Doug Houston   CLC

The "Cadillac & La Salle Club Photo Gallery" featured Mr. Houston's 1941 Cadillac. He presented a brief provenance of his car, including the 1968 purchase date. (Fig. 389)

Cadillac & LaSalle Club Photo Gallery ABS 11-1-12

1941 Cadillac 62 Club coupe   Owner Doug Houston   Ortonville, Michigan

Original owner in Oklahoma City. Car delivered late in the model year. I am the sixth owner. The car is equipped with Hydra-Matic transmission, radio, and factory installed air conditioner.

It was originally ordered in Aberdeen Beige, a special order color. It had black enamel paint when I purchased it in 1968. Stripped and repainted to Cadillac color combination 60: Dusty Gray lower and Rivermist Gray upper.

Figure 389. 1941 Cadillac Series Sixty-Two De Luxe Coupe, Model 6227D, equipped with factory air-conditioning Cadillac & LaSalle Club Photo Gallery, 11-1-12   CLC   Doug Houston photo

Mr. Houston posted more images of his car in the *cadillaclasalleclub.org Technical/Authenticity Forum*, dated February 7, 2014.

He included an engine bay view of the Servel air-conditioning compressor, a left front grille view, the cooling register with adjustable louvers behind the rear seat, the trunk-mounted evaporator, and an image of the "Cadillac Air Conditioner" nameplate. (Figs. 390–394)

Figure 390. 1941 Cadillac Servel air-conditioning compressor mounting. Mr. Houston stated, "Installation is on another car," that of Mr. Kughn's Fleetwood Series Seventy-Five that features a black firewall color." cadillaclasalleclub.org Technical/Authenticity Forum, February 7, 2014   CLC   Doug Houston photo

Figure 391. 1941 Cadillac Series Sixty-Two De Luxe Coupe, equipped with air-conditioning cadillaclasalleclub.org Technical/Authenticity Forum, February 7, 2014   CLC   Doug Houston photo

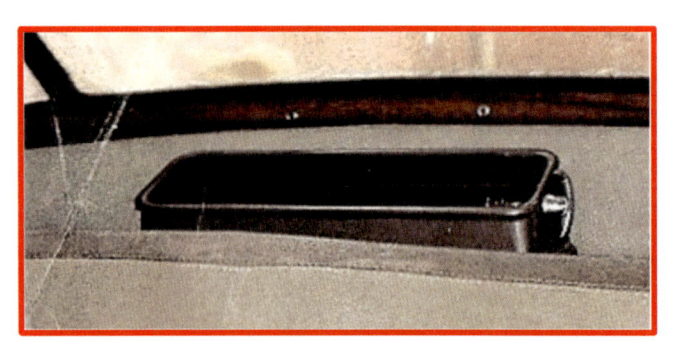

Figure 392. 1941 Cadillac Series Sixty-Two De Luxe Coupe, air-conditioning cooling register with louver control handle cadillaclasalleclub.org Technical/Authenticity Forum, February 7, 2014   CLC   Doug Houston photo

Figure 393. 1941 Cadillac Series Sixty-Two De Luxe Coupe, trunk-mounted air conditioner evaporator with affixed "Cadillac Air Conditioner" nameplate cadillaclasalleclub.org Technical/Authenticity Forum, February 7, 2014   CLC   Doug Houston photo

Figure 394. 1941 Cadillac close-up view of "Cadillac Air Conditioner" nameplate affixed to the trunk-mounted evaporator cadillaclasalleclub.org Technical/Authenticity Forum, February 7, 2014   CLC   Doug Houston photo

Factory Air: Cool Cars in Cooler Comfort   197

After Mr. Houston's passing on February 9, 2016, the Braun & Helmer Auction Service conducted his estate sale on September 20, 2016. Several images reflect their printed advance information. (Figs. 395–397, 402)

Figure 395. 1941 Cadillac Series Sixty-Two De Luxe Coupe, right front view   Former owner Doug Houston   David G. Helmer, Braun & Helmer photo

Figure 396. 1941 Cadillac Series Sixty-Two De Luxe Coupe, equipped with air-conditioning. Right profile view   Former Owner Doug Houston   David G. Helmer, Braun & Helmer photo

Figure 397. 1941 Cadillac Series Sixty-Two De Luxe Coupe, view of unmarked, three-speed air-conditioning control knob under left dash (yellow arrow)   Former owner Doug Houston   David G. Helmer, Braun & Helmer photo

198 Chapter 4: 1941 Cadillac Factory Air-Conditioning

Mr. West Peterson, Editor of *Antique Automobile*, published by the AACA (Antique Automobile Club of America), supplied the following images from the Braun & Helmer Auction on September 20, 2016: (Figs. 398–401)

Figure 398. 1941 Cadillac Series Sixty-Two De Luxe Coupe, Servel air-conditioning compressor profile view   Former owner Doug Houston   West Peterson photo

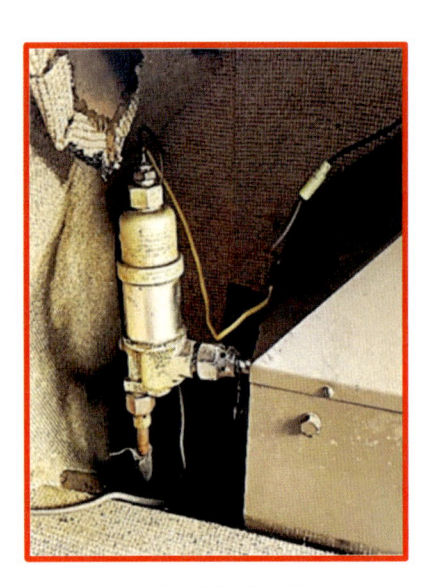

Figure 399. 1941 Cadillac Series Sixty-Two De Luxe Coupe, close-up of air-conditioning expansion valve leading into the evaporator coils   Former Owner Doug Houston   West Peterson photo

Figure 400. 1941 Cadillac Series Sixty-Two De Luxe Coupe trunk-mounted air conditioning expansion valve on left of the evaporator with affixed "Cadillac Air Conditioner" nameplate   Former Owner Doug Houston   West Peterson photo

Factory Air: Cool Cars in Cooler Comfort   199

Figure 401. 1941 Cadillac Series Sixty-Two close-up of air-conditioning cooling register louvers in closed position
Former Owner Doug Houston   West Peterson photo

Mr. Larry Edsall, Editor of the *ClassicCars.com Journal*, provided an image of the Braun & Helmer Auction Service information card placed on the windshield. It listed "Hydra-Matic" and "Rare factory AC." The image captured the cooling register behind the rear seat. (Fig. 402, and yellow arrow)

Figure 402. 1941 Cadillac Series Sixty-Two De Luxe Coupe auction sign by Braun & Helmer Auction Services listed, "Hydra-Matic" and "Rare Factory AC." (cooling register, yellow arrow)   Former owner Doug Houston
Larry Edsall, ClassicCars.com Journal photo

Spotlight on Cadillac, *Featured Car Number 3*
1941 Cadillac Fleetwood Series Seventy-Five, Model 7519, Equipped with Factory Air-Conditioning

Massive in size and strikingly impressive in black, the 1941 Cadillac Fleetwood Series Seventy-Five, Five-Passenger Touring Sedan, Model 7519, previously owned by Mr. Richard Kughn (1929–2019), is one of only three 1941 Cadillacs in existence equipped with air-conditioning.

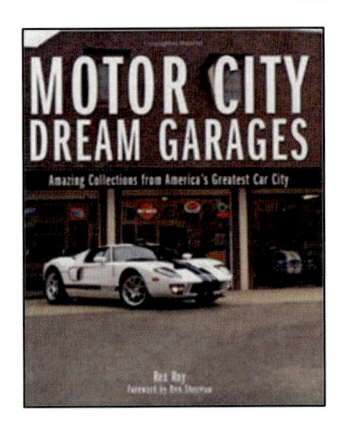

Figure 403. Motor City Dream Garages Cover Rex Roy, author, and photo

Figure 404. 1941 Cadillac Fleetwood Series Seventy-Five air-conditioning cooling register viewed through center-opening rear door (yellow arrow) Former owner Richard Kughn Rex Roy photo

Mr. Rex Roy, the author of *Motor City Dream Garages,* featured a ten-page spread of Mr. Kughn's Detroit garages in his 2007 publication. (Figs. 403–405) Formerly housing about 280 cars in his buildings in the early 1970s, Mr. Kughn reduced the number to 35 remaining cars, including the 1941 Cadillac.

Later, in 2013, he sold it in an RM-Sotheby's-St. John's Auction for $75,900.

Mr. Roy commented:

Another car from General Motors represents an important bit of automotive technology. The 1941 Cadillac Fleetwood Series Seventy-Five Sedan is one of two (air-conditioned) known to remain in the world.

Author note: Three factory air-conditioned 1941 Cadillacs remain, not two. General Motors selected its top luxury car, Cadillac, to introduce factory air-conditioning.

Mr. Kughn notes, "It was truly a modern marvel, and when you turn the blower up, it will freeze you right out." Mr. Ferrand, Mr.

Figure 405. 1941 Cadillac Fleetwood Series Seventy-Five, Five-Passenger Touring Sedan, Model 7519, equipped with factory air-conditioning Former owner Richard Kughn Rex Roy photo

Kughn's mechanic, pointed out that Packard beat GM to the market by one year, but the technology needed more time to mature before the public would accept it *en masse.*

Mr. Gene Dickirson, the author of the 2012 book, *Automotive Climate Control 116 Years of Progress,* visited Mr. Kughn's Detroit collection in 2011. Mr. Dickirson's images focused upon Mr. Kughn's 1941 Cadillac air-conditioning componentry. (Figs. 406–411)

Figure 406. 1941 Cadillac Fleetwood Series Seventy-Five view of air-conditioning compressor under hood (yellow arrow) Former owner Richard Kughn   Gene Dickirson photo

Figure 407. 1941 Cadillac Fleetwood Series Seventy-Five driver side view of Servel air-conditioning compressor Former owner Richard Kughn   Gene Dickirson photo

Figure 408. 1941 Cadillac Fleetwood Series Seventy-Five overhead view of Servel air-conditioning compressor in engine bay. An automatic radiator shutter control rests below top of radiator.   Former owner Richard Kughn Gene Dickirson photo

Figure 409. 1941 Cadillac Fleetwood Series Seventy-Five's unmarked air-conditioning control knob at left of heater controls   Former Owner Richard Kughn   Gene Dickirson photo

Figure 410. 1941 Cadillac Fleetwood Series Seventy-Five's air-conditioning cooling register with adjustable louver handle   Former Owner Richard Kughn   Gene Dickirson photo

Figure 411. 1941 Cadillac Fleetwood Series Seventy-Five displays elevated, trunk-mounted air conditioner evaporator   Former owner Richard Kughn   Gene Dickirson photo

The inviting rear seating welcomes passengers into a posh Cadillac Fleetwood atmosphere. A yellow arrow identifies the air-conditioning register. (Fig. 412)

Figure 412. 1941 Cadillac Fleetwood Series Seventy-Five opulent passenger seating with air-conditioning register louvers behind the seat. (yellow arrow)   Former owner Richard Kughn photo

Refer to the article, "Cadillac's Air Conditioner," by Doug Houston, published in the 1991 *Cadillac-La Salle Self-Starter Annual.* He provided more information about the 1941 air-conditioning system, including the Cadillac-supplied Shipping Order with wholesale pricing for Mr. Kughn's air-conditioned 1941 Cadillac Series Seventy-Five. (Figs. 384–387)

1942 Cadillac Factory Air-Conditioning

While Packard advertised and sold nearly 2,000 units of their innovative air-conditioning system during the 1940–1942 model years, Cadillac offered it only in the 1941 model year. It sold 300 units. No available Cadillac information explains the reasoning behind the discontinuation of its expensive optional air-conditioning after the 1941 model year. However, it produced at least one air-conditioned car in 1942.

Mr. Roy. A. Schneider, the author of *Cadillacs of the Forties,* related an account of one 1942 Cadillac Series Seventy-Five Limousine delivered in December 1941 to movie mogul David O. Selznick. Although Cadillac officially discontinued the accessory option after the 1941 model year, the movie producer of *Gone with the Wind* persuaded the Cadillac Division of General Motors to equip his new 1942 Cadillac Series Seventy-Five, Style 7533, Seven-Passenger Imperial Limousine (includes division window) with the optional $365 air-conditioning system. (Fig. 413) Cadillac priced the limousine, one of 430 built, at $3,613. The disposition of this car is unknown.

Author note: According to the CPI, air-conditioning, listed at $365 in December 1941, would cost the equivalent of $6,159 in January 2021 dollars. The $3,613 Limousine would cost the equivalent of $60,973.

The 1942 Selznick "75" Limousine Movie producer David O. Selznick and his actress wife, Jennifer Jones, took delivery of this limousine in December of 1941. Much of the 44,000 miles accumulated by the mid-1950s were chauffeur driven. The Selznick car is believed to be the only 1942 Cadillac equipped with air conditioning. Cadillac introduced true air conditioning in early 1941. At $365, it was expensive, adding more than 20% to the list price of the average car. These first AC units were well engineered and reliable but the lack of proper insulation limited effectiveness. Nonetheless, air conditioning was a moderate success in sedans and worked well in the "62" coupe. It was perhaps most satisfactory for passengers in the back of limousines. Since the cold air outlet was located in the package tray, and the rear compartment could be isolated, appreciable cooling was achievable. With two additional factory options, a sheepskin (mouton) floormat and rear window venitian blinds, beating the heat was a reality. As might be expected, the majority of these first A/C units were ordered by wealthy clients for installation in Seventy-Five and Sixty-Seven sedan cars. Of the 300 units, only three complete units are known to remain.

Figure 413. 1942 Cadillac Series Seventy-Five, Style 7533, Seven-Passenger Imperial Limousine, equipped with factory air-conditioning. Delivered to David O. Selznick in December 1941   Cadillacs of the Forties   Roy A. Schneider

LOOKING FORWARD... CADILLAC FACTORY AIR-CONDITIONING for 1953

VOLUME 1: THE INNOVATIVE YEARS, 1940 to 1942, represented a bold advance into mechanical refrigeration comfort for Cadillac, Packard, and Chrysler passengers. Factory air-conditioning's initial momentum ended in February 1942, when President Roosevelt and the United States Congress halted all automobile sales in favor of war materiel manufacturing. The post-WWII consumer rushed to purchase all things manufactured, especially newly built, previously war-rationed automobiles. In the rapid conversion process, automakers did not offer such an expensive luxury accessory as air-conditioning. Not until 1949 did automotive production match prewar levels.

Based upon a group of Texas entrepreneurs in the late 1940s, nascent automotive air-conditioning companies produced thousands of units for southwestern car owners, so many that the Big Three Detroit automakers acquiesced and introduced factory-installed air-conditioning in 1953.

Follow the journey in VOLUME 2: 1953–THE MAGICAL YEAR. Detroit automakers introduced factory air-conditioning in the following cars:

1. General Motors launched Frigidaire factory air for Cadillac, Buick, and Oldsmobile.
2. Packard contracted for Frigidaire units from General Motors.
3. Chrysler Corporation introduced Airtemp factory air for Chrysler, Imperial, De Soto, and Dodge.
4. Ford Motor contracted with NOVI, an independent company, for the Lincoln.

Six full-color chapters will examine the 1953 factory air-conditioned cars, accentuated by richly contrasted, legible images from factory brochures, service publications, owner's literature, magazine testing, and advertising. "Spotlight" sections display privately owned air-conditioned cars.

FACTORY AIR:

COOL CARS IN

COOLER COMFORT

AN ILLUSTRATED HISTORY OF AUTOMOTIVE FACTORY AIR-CONDITIONING
VOLUME 2: 1953–THE MAGICAL YEAR.

Continue with VOLUME 3: UP-FRONT FORERUNNERS and OLD-STYLERS,

featuring 1954–1956 factory air-conditioned cars.

Continue with VOLUME 4: UP-FRONT LATE BLOOMERS,

featuring 1957–1960 factory air-conditioned cars.

Chapter 5

1941 Chrysler Factory Air-Conditioning
and one produced in 1942

The 1941 Chrysler brochure announced, "Ride in Refrigerated Air!" (Fig. 414)

Figure 414. 1941 "Chrysler Air Refrigeration System" brochure cover   ABS

The 1940 Packard Weather Conditioner introduced the first complete mechanical refrigeration system option for closed cars. Packard, the forerunner, according to Packard Vice President of Engineering, W. H. Graves, sold nearly 2,000 units before auto manufacturing ceased for the war effort in February 1942. The 1941 Cadillac sold nearly 300 units during its one-year offering, and at least one in 1942. Chrysler, however, never reported its sales figures. Estimates for the 1941–1942 Chrysler, Imperial, and possibly De Soto's air-conditioning production ranged from single digits to 200 units sold. At least one 1942 model exists.

The "Chrysler Air Refrigeration System" offered "an escape from oppressive heat into conditioned coolness—cooled to suit your comfort…" (Fig. 415 and yellow arrow). Their marketing department targeted wealthy clients, the *discriminating car owner who desires additional driving comfort.* (emphasis by author)

It continued:

Even when the mercury threatens to pop out of the top of the thermometer, you can ride in clean, closed-window comfort, *with the temperature inside the car ten to fifteen degrees cooler than outside.* (emphasis by author) (Fig. 415 and orange arrow)

The Chrysler "Air Refrigeration System" offered:

1. A genuine mechanically operated system
2. Air filtration
3. Dehumidification and cooling by refrigerant

It stressed:

On the hottest days of summer, you may revel in coolness inside your Chrysler. A flip of the control button and you bring a breath from the Northwoods.

Mr. Warren H. Erb presents an in-depth article later in the chapter titled "Breath from the Northwoods," written in 1992 for the Chrysler Product Restorers Club (Figs. 423–427).

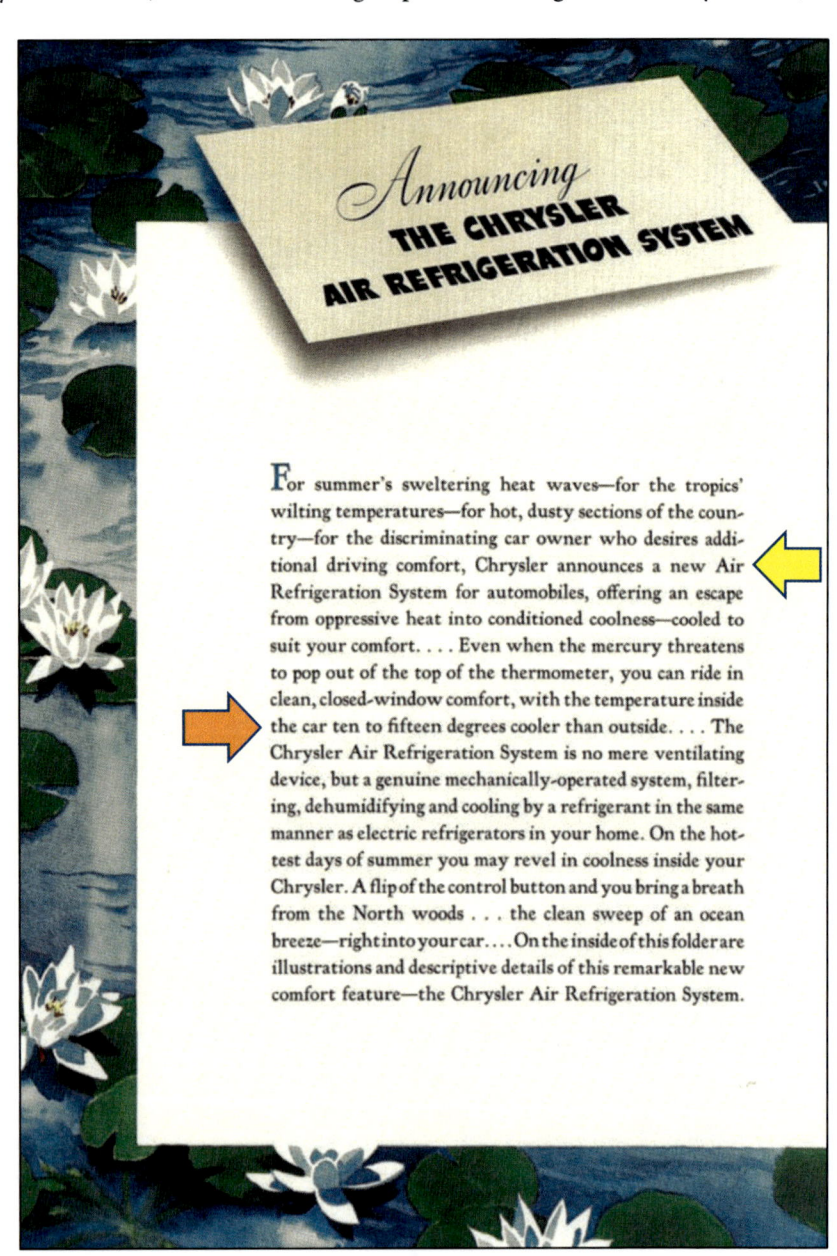

Figure 415. 1941 "Chrysler Air Refrigeration System" brochure information   ABS

The "Chrysler Air Refrigeration System" brochure illustrated the cooled air circulation pattern from the air register behind the rear seat as it blew forward along the headliner, returning under the rear seat through a filter. (Fig. 416)

It presented a 1941 Chrysler phantom view of the air-conditioning components by letter: (Fig. 417)

A. The belt-driven compressor
B. The condenser
C. The liquid reservoir (receiver)
D. The evaporator
E. The fan outlet grille
F. The fan-speed control

Figure 416. 1941 "Chrysler Air Refrigeration System" brochure cooled air circulation pattern   ABS

Figure 417. 1941 "Chrysler Air Refrigeration System" brochure phantom view of system components   ABS

The "Chrysler Air Refrigeration System" process is as follows:

(A) A compressor exerts pressure on the refrigerant, causing it to enter the condenser (B) as a gas. Here, the refrigerant cools, becomes a liquid, travels through tubing to the liquid reservoir (C), and then continues through an expansion valve to the evaporator assembly's cooling coils.

(D) As the refrigerant changes from a liquid to a gas in the cooling coils, it absorbs heat from the
air, thus cooling the car. The gaseous refrigerant returns to the compressor, and the cycle repeats.

(E) Fresh refrigerated air circulates from the cooling fan outlet grille to every part of the car's interior by the fan. The cooled air flowing along the headliner returns to the floor. The fan draws the air under the rear seat through filters as it recirculates and cools the air over the evaporator coils.

(F) The fan control on the instrument panel regulates the circulating airspeed.

"For Your Comfort" described the development of the "Chrysler Air Refrigeration" system with its assurance that passengers could control "the weather inside your car." (Fig. 418)

> **FOR YOUR Comfort**
>
> The Chrysler Air Refrigeration System was developed for your comfort—to assure you of an air cooling system giving you personal control over the weather inside your car.
>
> There is nothing complicated about the Chrysler Air Refrigeration System. It works on the same proved principle as your electric household refrigerator. Turning it on and off is as simple as turning on the car heater. The desired amount of cooling may be regulated to your desires by means of a control switch on the instrument panel, regardless of car speed. Unlike a household refrigerator, no defrosting is required, the system by nature of its construction, handling its own needs.
>
> The cooling effectiveness of the Chrysler Air Refrigeration System is demonstrated by the fact that, rated like household refrigerators, it has a cooling capacity of two and a half tons of ice every 24 hours at a car speed of 60 M.P.H. In other words, it has a cooling capacity at higher car speeds equivalent to the cooling effect of placing two and a half tons of ice inside the car and forcing it to melt inside of 24 hours. Regardless of car speeds—low or high—the system is equally efficient, and cooling starts the moment the engine is started. You feel the effects throughout the car immediately.
>
> COOL—When the thermometer hovers close to the hundred mark—*when it's really hot*—to step inside an air-cooled Chrysler is like diving into a deep pool. The air is refreshing, invigorating, and its pleasant effects are apparent immediately.

Figure 418. 1941 "Chrysler Air Refrigeration System" brochure passenger comfort information   ABS

The brochure compared the air-conditioning system to a household refrigerator and was as easy to use as turning the car heater on and off. It stated that it did not require defrosting like a refrigerator of the era. The fan switch on the dash controlled the amount of cooling. There was no mention of a thermostat to regulate the cool air temperature, just the fan speed.

Chrysler rated its cooling effectiveness similarly to refrigerators, measuring cooling capacity over twenty-four hours. It stated that it produced the equivalent cooling capacity of two-and-a-half tons of melting ice at sixty MPH.

This information probably sounded impressive to a prospective 1941 Chrysler customer. The twenty-four-hour cooling rating listed a maintained speed of sixty MPH. Retrospectively, however, he may have believed that there was more overall cooling power than it produced. In 1941, there were few long-distance freeways for such a sustained speed rating. Most roads and some highways listed the speed limits between thirty-five and fifty-five MPH.

Importantly, Chrysler did not address potential customer inquiries concerning an effective cooldown rate for a car that sat in the sun and the subsequent cabin cooling produced at slow traffic speeds.

The brochure continued with information about the "Chrysler Air Refrigeration System" benefits and highlighted the cooling components installed in the car. (Figs. 419, 420)

CLEAN—When it's so dry the dust hangs like a haze, you can roll up your windows in the air-cooled Chrysler and know you will arrive at your destination as spick and span as when you started. Air is clean because it's filtered.

DEHUMIDIFIES—It may be 90 or 110 in the shade outside, but inside the car it's early Spring, cool, comfortable, and clean, because this marvelous new Air Refrigeration System *dehumidifies* as well as cools—for additional comfort on humid days, rain or shine.

FILTERED AIR—If you can't leave town during the hay fever season, your next best bet is a Chrysler with Air Refrigeration. You can drive all day with the windows closed, breathing clean, filtered air—low in pollen content. A special filter attached to the cooling unit continually cleanses the air inside the car.

The compressor is mounted on the engine block. Small in size, it is large in capacity and operates effectively at all engine speeds.

The fan outlet grille is located behind the rear seat. An electric fan circulates conditioned and cooled air through the car.

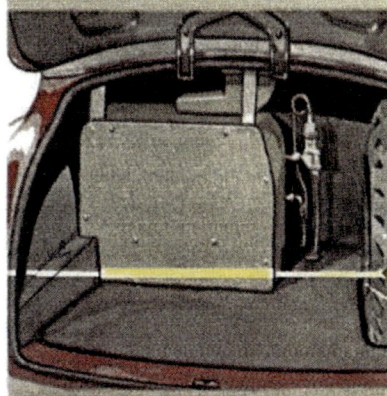

The evaporator assembly, where air is filtered and cooled, is concealed within the trunk. It takes up only 10" of space.

Figure 419. 1941 "Chrysler Air Refrigeration System" brochure   ABS

A transformed interior climate addressed those passengers who suffered as they drove in hot cars in the summer. The system offered clean, dust-free, dehumidified, filtered air. The air-conditioning option added another aspect to motoring—the concept of quietness while driving. Closed windows emphasized the absence of traffic noises and buffeting winds. (Fig. 420)

Artist renderings highlighted the factory-installed components:
1. Overlooked in the publisher's proofing process, the driver-side compressor drawing location (Fig. 419) contrasts with the passenger-side compressor location in the phantom drawing. (Fig. 417)
2. Chrysler identified the cooling register and louvers behind the rear seat as the "fan outlet grille." (Fig. 419)
3. The trunk-mounted evaporator showed a ten-inch footprint depth. (Fig. 419)
4. The air-conditioning control panel showed a paddle-type switch with an "off" and four fan speeds. (Fig. 420) The 1940 Packard also used the identical paddle-type switch.
5. The cutaway drawing revealed the condenser placement behind the grille in front of the radiator. (Fig. 420)

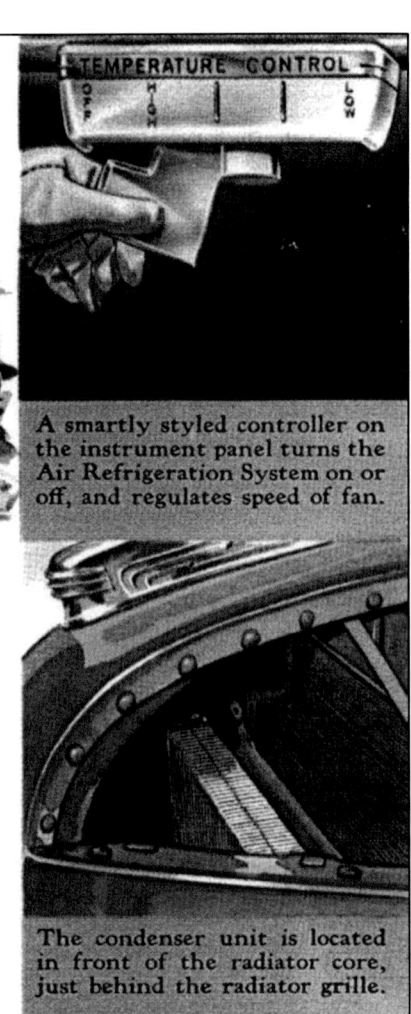

Figure 420. 1941 "Chrysler Air Refrigeration System" brochure   ABS

The brochure depicts a dark blue, top-of-the-line 1941 Chrysler New Yorker in motion (Fig. 421) while it cools its six passengers in one of the few 1941 **FACTORY AIR**-equipped

COOL CARS IN

COOLER COMFORT

Figure 421. 1941 "Chrysler Air Refrigeration System" brochure ABS

A 1941 Chrysler New Yorker sedan presents a contrasting view of the low-slung, sleekly drawn brochure image. (Figs. 421–422)

Figure 422. 1941 Chrysler New Yorker Sedan
Chrysler Historical Collection

212   Chapter 5: 1941 Chrysler Factory Air-Conditioning

Mr. Warren H. Erb of the Chrysler Product Restorers Club presented the history of the 1941–1942 "Chrysler Air Refrigeration System" in the *WPC News, Vol. XXIII No. 9*, dated May 1992. (Figs. 423–427) His account continues with relevant information about the 1953 Chrysler Corporation's Airtemp air-conditioning system discussed in *Volume 2* of *Factory Air: Cool Cars in Cooler Comfort*. Information about the 1954, 1955, and 1956 Chrysler Airtemp air-conditioning systems will continue in *Volume 3* of *Factory Air: Cool Cars in Cooler Comfort*.

Figure 423. 1941 "Chrysler Air Refrigeration System" article, WPC News, Vol. XXIII No. 9, May 1992 Content and Photo(s) courtesy Warren H. Erb, Docent, AACA Museum, Hershey PA

Walter P. Chrysler dedicated his New York City Art Deco-inspired, fully air-conditioned Chrysler Building in 1930. It served as a dual milestone for a significant legacy: opening the first manufactured structure to stand taller than 1,000 feet—at 1,048 feet, seventy-seven stories—and the first use of Chrysler's own high-speed radial-designed air-conditioning compressor in a commercial building. Mr. Erb stated, "With Chrysler's reputation for engineering innovation, it is not surprising that Chrysler cars were among the first to have their interiors artificially cooled."

> ## "BREATH FROM THE NORTHWOODS" CHRYSLER'S AIR CONDITIONED CARS
> by Warren Erb
>
> On May 27, 1930, Walter P. Chrysler spoke at dedication ceremonies in New York City to inaugurate the world's tallest structure to date – the Chrysler Building, now regarded as an Art Deco masterpiece. Among the innovations featured in this new building was air conditioning, which provided cleaned, washed and tempered air throughout its entire interior. The Chrysler building was the first office building in the world ever to be centrally air conditioned.
>
> Walter Chrysler had wanted the building air conditioned from its inception, but could find nothing on the market that would satisfy his specifications. Therefore, he designed a unit that would fulfill his requirements and organized the Temperature Corporation in Detroit to produce the units. This Corporation later became known as Chrysler Corporation's Airtemp Division and, shortly thereafter, moved to Dayton, Ohio. The Airtemp Division grew to become one of the world's leading producers of air conditioning and heating equipment for commercial and home use.
>
> Chrysler's early interest in air conditioning wasn't just restricted to buildings, either. Just ten years after W.P. himself had inaugurated his ultra-modern Chrysler building in New York, Chrysler Corporation began to play a leading, if not pioneering role in air conditioned automobiles. However, ironically, the involvement of Chrysler's Airtemp Division in this field was not to occur until the early 1950's.
>
> In today's world, with homes and offices comfortably air conditioned, and with a high percentage of even the smallest cars equipped with A/C, it is difficult to imagine motoring without artificial climate control in hot weather. Yet, by 1940, air conditioning on hot summer days was still pretty much restricted to a few modern office buildings, movie theaters, hospitals and restaurants, primarily in the southern states, where the luxury could be justified in terms of physical well being or, more importantly, return on investment.
>
> Although many train coaches were air conditioned, the thought of air conditioning an automobile in those days smacked of Buck Rogers or the New York World's Fair's "City of Tomorrow". Indeed, heaters in automobiles were still emerging slowly from the dark ages. The heaters of the day were primitive and awkward devices, usually mounted beneath the dash on the passenger side. They distributed heat poorly, and windshield defrosters were usually an extra option. A few makes, including Chrysler, had just begun to offer heaters which took in outside air. Our story is about cool, not heat. The point is, however, that the attempt to cool car interiors was made before car manufacturers had quite learned how to heat them properly.
>
> With Chrysler's reputation for engineering innovation, it is not surprising that Chrysler cars were among the first to have their interiors artificially cooled. Officially, Packard claims to have introduced the option first, in February, 1940. However, there is unofficial evidence to suggest that Chrysler also had the same system available to its customers in 1940. A photograph accompanying this article clearly shows an air conditioning control mounted beneath the dash of a 1940 Chrysler Crown Imperial, and the January, 1940 Crown Imperial advertisement shown here clearly mentions "air conditioning" as one of the available options.
>
> By 1941, air conditioning, referred to as an "air refrigeration system", was officially available in all Chrysler cars. the new option appeared in the sales literature for the handsome Crown Imperial series, and even rated a nice, four-color folder of its own. Packard continued to offer the system and Cadillac joined the group in 1941. In 1942, DeSoto picked up the option.
>
> Unfortunately, there seems to be no evidence of how many pre-war Chrysler Corporation cars were equipped with the innovative "air refrigeration system". One report states, "Though Chrysler ostensibly offered Airtemp cooling on some 1941-42 models, none are known to have been sold that way". This writer knows, however, that at least one vehicle of the period was so equipped, because he saw and photographed a 1942 Chrysler Crown Imperial limousine with the system at a summer car show in Macungie, Pennsylvania. Its owner claimed that he was told it was one of two 1942 Crown Imperial limousines with factory A/C. Packard is said to have equipped 2,000 cars with the option between 1940 and 1942. Cadillac is said to have outfitted about 300 cars with the same system in 1941. We might guess, therefore, that there were, in fact, several hundred '41 and '42 Chrysler Corporation vehicles delivered with "air refrigeration system".
>
> Neither does any evidence seem to exist as to the extra cost of this early system on the purchase price of a new Chrysler Corporation car. Packard installations were reported to have cost $275 extra,
>
> 7a 7b

(Continued p. 9)
Figure 424. 1941 Chrysler: "Breath from the Northwoods, Chrysler's Air-Conditioned Cars"
WPC News, Vol. XXIII No. 9, p. 7a, p. 7b, May 1992   Warren H. Erb

214 Chapter 5: 1941 Chrysler Factory Air-Conditioning

Figure 425. 1941 Chrysler: "Breath from the Northwoods, Chrysler's Air-Conditioned Cars"
WPC News, Vol. XXIII No. 9, p. 8a, May 1992   Warren H. Erb

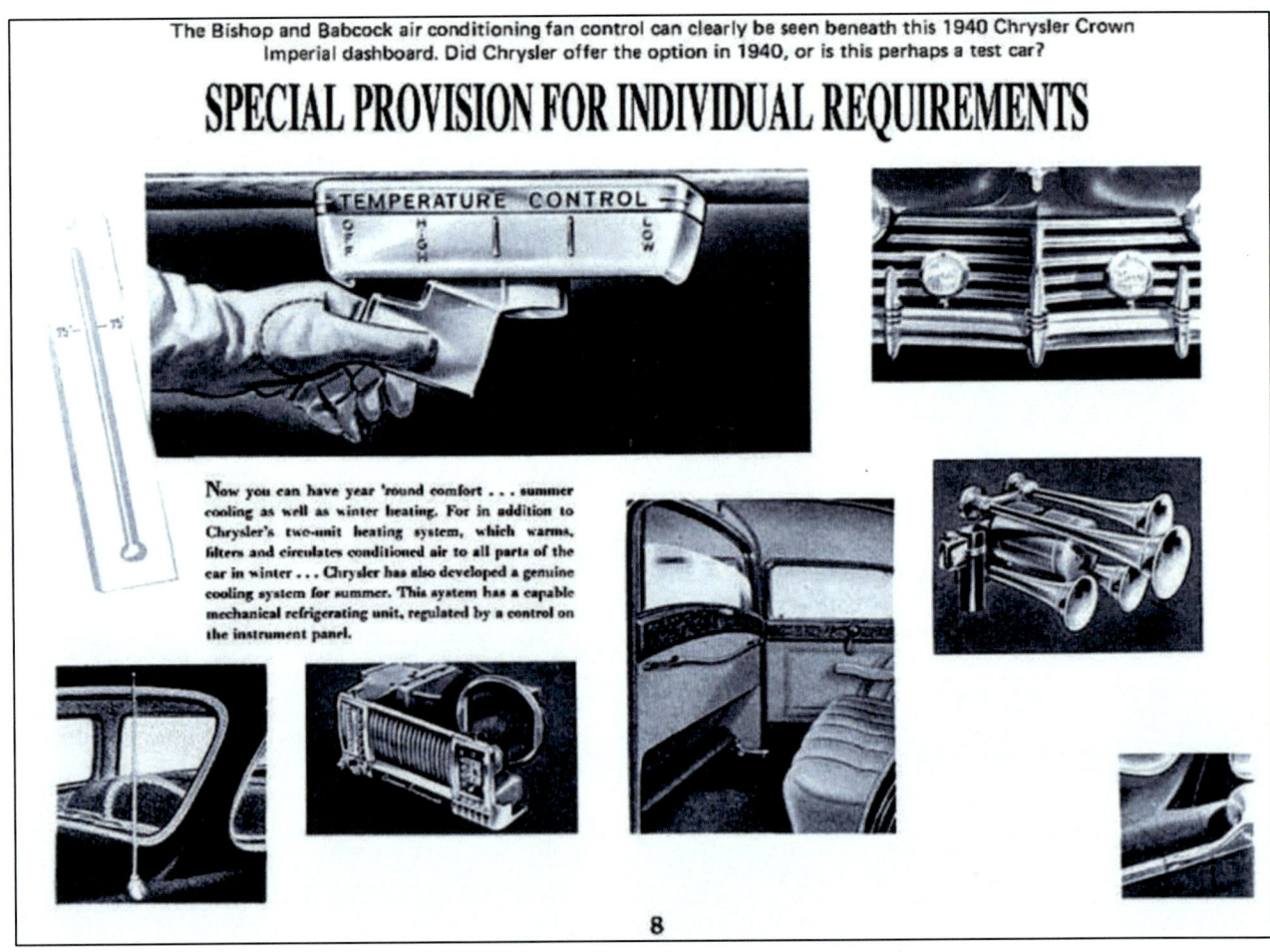

Figure 426. 1941 Chrysler: "Breath from the Northwoods, Chrysler's Air-Conditioned Cars"
WPC News, Vol. XXIII No. 9, p. 8b, May 1992   Warren H. Erb

and we can assume, for lack of better information, that the Chrysler installation would have cost about the same. At first glance, $275 doesn't seem like such a high price to pay for what we now consider to be an almost essential automotive option. We need to remember, however, that, with the price of a Chrysler New Yorker sedan at $1,265 in 1941, A/C at $275 would have cost almost 22% of the base price of the car itself. That would be like adding a $4,000 option to today's New Yorker!

Not Chrysler's own Airtemp Division, but Bishop and Babcock of Cleveland, Ohio, was the company which supplied all the early A/C systems for Chrysler, as well as for Packard and Cadillac. In the case of Packard, cars designated to be outfitted with A/C were specially insulated at the factory in their roofs, doors and floors, then sent to Cleveland to have the units installed at the Bishop and Babcock plant. Chrysler probably handled the installation in a similar manner, although there is no published record of any special insulation in Chrysler's case.

The refrigerant in the Bishop and Babcock system was compressed to 180 lbs/sq.in. by a two-cycle, reciprocating compressor run off the fan-belt pulley at 70% engine speed. The system had a 2-1/2 ton capacity. The cooling coil in the trunk-mounted evaporator operated at a temperature of between 40 and 50 degrees F. Cool air was forced to circulate throughout the car's interior by means of a cage-type fan under the package shelf behind the rear-seat backrest. The cool air followed the contour of the roof toward the front of the car and, because cool air is heavier than warm air, it descended toward the floor near the front. It was then pulled back under the seats and into the cooling system again by the suction action of the blower. A constant flow of cooling air was thus kept in circulation. When the air was drawn under the rear seat by the suction of the blower, it had to pass through filtering elements of oil-coated fiber board, which cleansed it from dirt and dust, and maintained a low pollen count. It was recommended that these filter elements be changed at least once a year, preferably in the spring. To replace the filters, the rear seat had to be removed, the old filters slid out of position, and new ones installed in their place.

Ordinarily, sufficient air leakage from the outside was introduced for adequate ventilation. If a full load of passengers was carried, additional ventilation may have been desired, and could easily have been accomplished by opening one of the front wing windows just enough to create a fresh air intake.

Smoke and body odors would be picked up by the circulating air and carried back to the cooling coil, where they would be deposited. It was, therefore, recommended that this coil be cleaned occasionally with "Diversal" or other bacteria insecticide solution which would not attack the tin-plated interior parts of the coil assembly. The cover plate on the coil had to be removed and sprayed with the solution, which, after passing over the coil, would run out a drain tube at the bottom of the air return duct.

By the time automobile production was halted in February, 1942, to "clear the decks" for war work, it had become very clear that air conditioning as an automotive option, with all its innovation and comfort advantages, had not been a success. There were important reasons, too, for its lack of acceptance:

▪ The required maintenance just mentioned would certainly have turned some potential customers away!

▪ The timing was wrong; in the early 1940's automotive options of any kind, even radios and heaters, were just not as commonly purchased as they are today. It was not common in those days to find very high-priced cars devoid of any of the factory accessories offered. In this environment, no wonder more people didn't opt for air conditioning.

▪ The "air refrigeration system" added about 300 lbs. to the car's weight. Given the 112 to 135 H.P. of the Chrysler Corporation cars of the period, the inherent slippage of Fluid Drive and the power consumption of the A/C compressor, there was, no doubt, an adverse affect on fuel economy and engine performance.

▪ The large evaporator took up considerable space in the trunk compartment, and there were elements of the system located throughout the entire car, with wiring and plumbing to connect them.

▪ There were performance problems, too. For one thing, there was no clutch on the compressor; it ran all the time. The faster the engine ran, the cooler the coils became. Owners were advised to remove the compressor drive belt in the fall and replace it again when the days got warm. In continued hot weather, the lack of a clutch was not a problem. However, the system was constantly producing cold air, and the only means of controlling that air was the three-speed switch on the circulating fan. On cooler summer days, even with the fan turned off, heavy cold air would flow out of the evaporator, downwards and out the under-seat intakes, chilling the ankles and feet of rear-seat passengers.

Bishop and Babcock, the firm which had built the early units, disappeared during World War II and, when the post-war 1946 models appeared, factory air conditioning in cars was no longer available. In those early post-war years, prospective customer names were placed on long waiting lists for almost all makes of new cars. Buyers often had to slip dealers a "tip" to get their names placed higher on those lists. Those buyers were just glad to get a new car - any new car! So, there was little or no pressure for automotive air conditioning at first.

For the discussion of the 1953 Chrysler Airtemp air-conditioning system,

Refer to VOLUME 2 of

Factory Air: COOL CARS IN COOLER COMFORT

Figure 427. 1941 Chrysler: "Breath from the Northwoods, Chrysler's Air-Conditioned Cars"
WPC News, Vol. XXIII No. 9, p. 9, May 1992   Warren H. Erb

The photograph (Fig. 425) shows an under-dash air-conditioning control mounted on a 1940 Chrysler Crown Imperial. The dash photograph of the 1940 Chrysler Crown Imperial may or may not be a production model. It shows a multi-speed, paddle-type air-conditioning control switch, which appears identical to Packard's 1940 air-conditioning control installed by the contractor, Bishop & Babcock Mfg. Co. of Cleveland, Ohio. Concurrently, Chrysler offered "air-conditioning" as an option in a January 1940 ad for the Chrysler Crown Imperial. Unsure of its future accessory designation, the same ad also promoted "All-Weather Aircontrol." (Fig. 428 and yellow arrows)

Figure 428. 1940 Chrysler Crown Imperial promoted "Air-conditioning" and "All-Weather Aircontrol" in the same ad
Online Imperial Club (OIC)

For 1940, Packard introduced automotive air-conditioning and produced nearly 2,000 units through 1942, according to Packard Vice-President of Engineering, W. H. Graves. Cadillac produced 300 units in 1941. Mr. Erb stated, "We might guess, therefore, that there were several hundred '41 and '42 Chrysler Corporation vehicles delivered with the Air Refrigeration System." Unfortunately, no records reflect Chrysler's production. One report stated, "Though Chrysler ostensibly offered Airtemp cooling on some 1941–1942 models, none are known to have been sold that way." Mr. Erb refutes this claim with images and commentary of a 1942 Chrysler Crown Imperial Limousine equipped with factory air-conditioning. (Figs. 430–432, 434–437)

There is no evidence of the air conditioner's price. Based upon the pre-April 30, 1941 purchase price of $275 for the 1941 Packard's air-conditioning system, Mr. Erb estimated a similar price for the Chrysler. He calculated the price to be roughly equivalent to a $4,000 option.

Author note: More correctly, a $325 post-April 30, 1941 price is necessary for correct calculations. According to the CPI, the January 1942 price of $325 is equivalent to $2,883, based on the article's March 1992 date. The $325 option is equivalent to $5,414 in January 2021 dollars.

Although Chrysler owned Airtemp, the Bishop & Babcock Mfg. Co. of Cleveland, Ohio supplied Chrysler, Packard, and Cadillac with the air-conditioning system componentry. Neither Chrysler's, Cadillac's, nor Packard's literature acknowledged the air-conditioning installation location.

At least one writer believes Packard shipped the selected, nearly completely assembled cars to the Bishop & Babcock Mfg. Co. in Cleveland for the air-conditioned componentry installation. Without citing a reference, author L. Morgan Yost stated the following on p. 365 of *Packard: A History of the Motor Car and the Company*:

> Special insulation for cars so equipped was provided at the factory, but, although noted in Packard literature as "factory installed," the refrigerating unit itself was actually fitted into the car by Bishop & Babcock in Cleveland, and in theory, vehicles ordered with it were shipped there, the unit was installed and then the cars were forwarded to the dealers.

Author note: According to James Hollingsworth, author of *Packard 1940: A Pivotal Year*, the Bishop & Babcock Mfg. Co. of Cleveland, Ohio, manufacturer of fans, auto thermostats, radiators, and heaters,

> ...began experimenting with automotive air-conditioning as early as 1937. A test car in 1939 convinced Packard management that this could be a practical accessory in the fine car field. The compressor was designed in 1936 by Servel for use in walk-in coolers and modified slightly for automotive use.

The Bishop & Babcock Air-Conditioning system components included the following:
1. A compressor: Servel, two-cycle reciprocating compressor.
2. It produced 180 PSI by fan-belt pulley run at 70% engine speed.
3. Refrigerant: Freon (R-12), capacity six and one-fourth lbs.
4. Cooling capacity: The rating equaled melting two and one-half-tons of ice in a twenty-four-hour period.
5. Cooling coil temperature in the trunk-mounted evaporator operated at a temperature of 40°F to 50°F.

Mr. Erb stated, "Cool air was forced to circulate throughout the car's interior by means of a cage-type fan under the package shelf behind the rear seat backrest." He discussed the cool air circulation pattern inside the car interior, the air filter location, and changing frequency. The front wing windows allowed fresh air ventilation, especially with a full load of passengers who smoked. Bishop & Babcock's service manual instructions, titled under the Chrysler, Packard, and Cadillac names, recommended deodorizing accumulated smoke and body odors by the periodic evaporator spraying of "Diversal" bacteria insecticide solution onto the cooling coils. The solution drained onto the ground through the condensate tube.

As an automotive option, Mr. Erb believed that air-conditioning had not been a success with all its innovation and comfort advantages. He listed the following reasons:
1. The maintenance required by owners.
2. Automotive accessories, such as radios and heaters, were not customarily purchased, much less expensive auto air conditioners.
3. Chrysler's air refrigeration system weighed 300 pounds, further under-powering the automobiles.
4. The large trunk-mounted evaporator took up much of the small trunk space and required extensive wiring and plumbing throughout the car.
5. The compressor ran continuously and required removing the compressor belt for winter weather.

Author note: The compressor ran continuously, potentially resulting in a mechanical breakdown sooner than the expected design lifespan. Far into the future, a year after the 1953 reintroduction of factory-installed air-conditioning by General Motors, Chrysler Corporation, Ford Motor Company, and Packard, the 1954 models of Cadillac, Buick, Oldsmobile, Pontiac, and Nash introduced an electromagnetic compressor clutch that disconnected the compressor when unused.

Mr. Erb noted that even with the fan turned off during milder weather, the cooler, heavier air flowed from the evaporator cooling coils towards the floor and chilled rear seat passenger's feet and ankles. He also stated that the Bishop & Babcock Mfg. Co. "disappeared during WWII," resulting in no optional automotive air-conditioning availability for the 1946 models.

Author note: Mr. Erb's statement proved partially correct. The Bishop & Babcock Mfg. Co. ceased supplying automotive air-conditioning systems after WWII while it continued to produce commercial air-conditioning blowers. Referencing their *Bulletin 115*, dated September 1947, it listed "blower assemblies, wheels, housings, and accessory parts, designed expressly for manufacturers of warm air furnace and air-conditioning equipment." (Fig. 429) The bulletin cover stated, "Manufacturers of centrifugal blowers for more than 30 years."

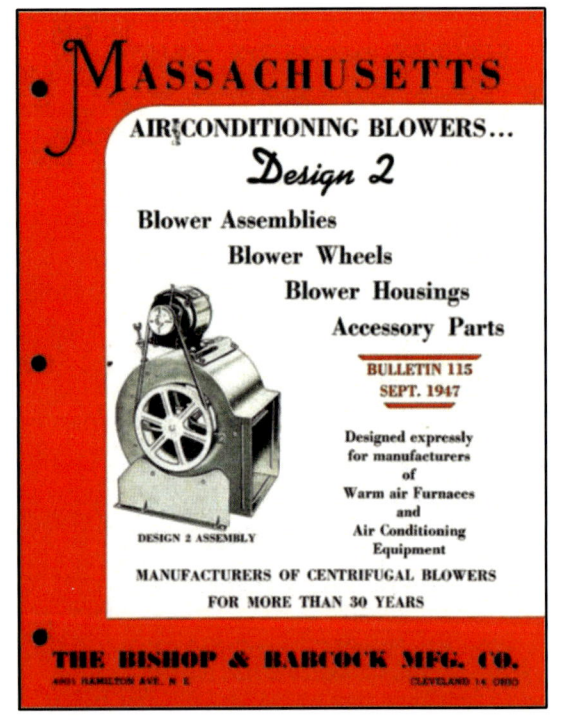

Figure 429. The Bishop & Babcock Mfg. Co. Bulletin 115, September 1947   ABS

Mr. Erb continued by stating that the pent-up demand for new cars far surpassed the supply. Often, buyers offered cash incentives to dealers for placing their new car orders higher on the waiting list. He noted, "Those buyers were just glad to get a new car—any new car! So, there was little or no pressure for automotive air-conditioning at first."

1942 Chrysler Factory Air-Conditioning

Before he wrote the "Breath from the Northwoods" article in 1992, Mr. Erb observed:(Figs. 430–432)

This writer knows, however, that at least one vehicle of the period was so equipped because he saw and photographed a 1942 Chrysler Crown Imperial limousine with the system at a summer car show in Macungie, Pennsylvania. Its owner claimed he was told it was one of the two 1942 Crown Imperial Limousines with factory A/C.

We might guess, therefore, that there were, in fact, several hundred '41 and '42 Chrysler Corporation vehicles delivered with (the) "air refrigeration system."

Figure 430. 1942 Chrysler Crown Imperial Limousine, equipped with the Air Refrigeration System. Left forward trunk displays the evaporator housing the cooling coils. Macungie, PA, circa early 1990s Owner unknown   Warren H. Erb photo

Figure 431. 1942 Chrysler Crown Imperial Limousine, Model C-37, 448 units produced. This Chrysler's equipment included the Air Refrigeration System. Macungie, PA, circa early 1990s   Owner unknown   Warren H. Erb photo

220 Chapter 5: 1941 Chrysler Factory Air-Conditioning

Figure 432. 1942 Chrysler Crown Imperial Limousine, left rear view, equipped with the "Air Refrigeration System."
Macungie, PA, circa early 1990s   Owner unknown   Warren H. Erb photo

A Chrysler factory photo displayed the 1942 Chrysler Crown Imperial Limousine. (Fig. 433)

Figure 433. 1942 Chrysler Crown Imperial Limousine   Chrysler Historical Society

Factory Air: Cool Cars in Cooler Comfort   221

With the same license plate number shown in Mr. Erb's images, the Online Imperial Club site posted four images of the 1942 Chrysler Crown Imperial limousine, equipped with air-conditioning. Unfortunately, their December 2018 update did not identify the owner's name, photo date, or event location. Member Tony (unknown last name) offered insights into the head-turning automobile. (Figs. 434–437)

From the Online Imperial Club website:

Despite some wear, this car is absolutely stunning. The leather is in good shape, the buzzer to "call" the driver still functions, and the owner claims it is the only one made with a/c. He also said that it was used to escort foreign diplomats.

Figure 434. 1942 Chrysler Crown Imperial Limousine, equipped with the "Air Refrigeration System"
OIC (Online Imperial Club) website update for December 2018   Owner unknown   Member Tony photo

The owner listed the accessories and special features of the 1942 Chrysler Crown Imperial Limousine, the "only one known to exist with factory air." (Fig. 435 and yellow arrow)

Figure 435. 1942 Chrysler Crown Imperial Limousine, equipped with the Air Refrigeration System
OIC   Owner unknown   Member Tony photo

Flag holder brackets are in use in the Online Imperial Club images (Figs. 434, 436); the right bracket shows in Mr. Erb's similar right front image view. (Fig. 431)

Figure 436. 1942 Chrysler Crown Imperial Limousine, equipped with "Air Refrigeration System"
OIC   Owner unknown   Member Tony photo

The painted pot metal license plate holder reflects the WWII rationing efforts of automotive chromium and nickel products. (Figs. 432, 437)

Automakers installed exterior metal trim as long as their supplies existed. Chrysler's 1942 Crown Imperial models distinguished themselves from the New Yorker series with added stainless streamlining trim carried to the rear fenders.

Figure 437. 1942 Chrysler Crown Imperial Limousine, equipped with the "Air Refrigeration System." Rear view displays WWII-era blackout license holder
OIC   Owner unknown   Member Tony photo

The 1941 Chrysler Crown Imperial prestige brochure (Fig. 438, 439) described its optional air-conditioning accessory under the heading "Special Provision for Individual Requirements."

Rather than identifying and promoting the air-conditioning system by the name from its marketing department, the "Air Refrigeration System" from the concurrent 1941 Chrysler full-color brochure, "Ride in Refrigerated Air!" shown initially in this chapter (Fig. 414), it merely noted that it developed a "genuine cooling system for summer." (Fig. 439 and yellow arrow)

Author note: This inconsistency possibly stems from various Chrysler divisional inputs, timing deadlines, or lack of coordination and communication with the marketing/advertising department.

Figure 438. 1941 Chrysler Crown Imperial prestige brochure cover
oldcarbrochures.com

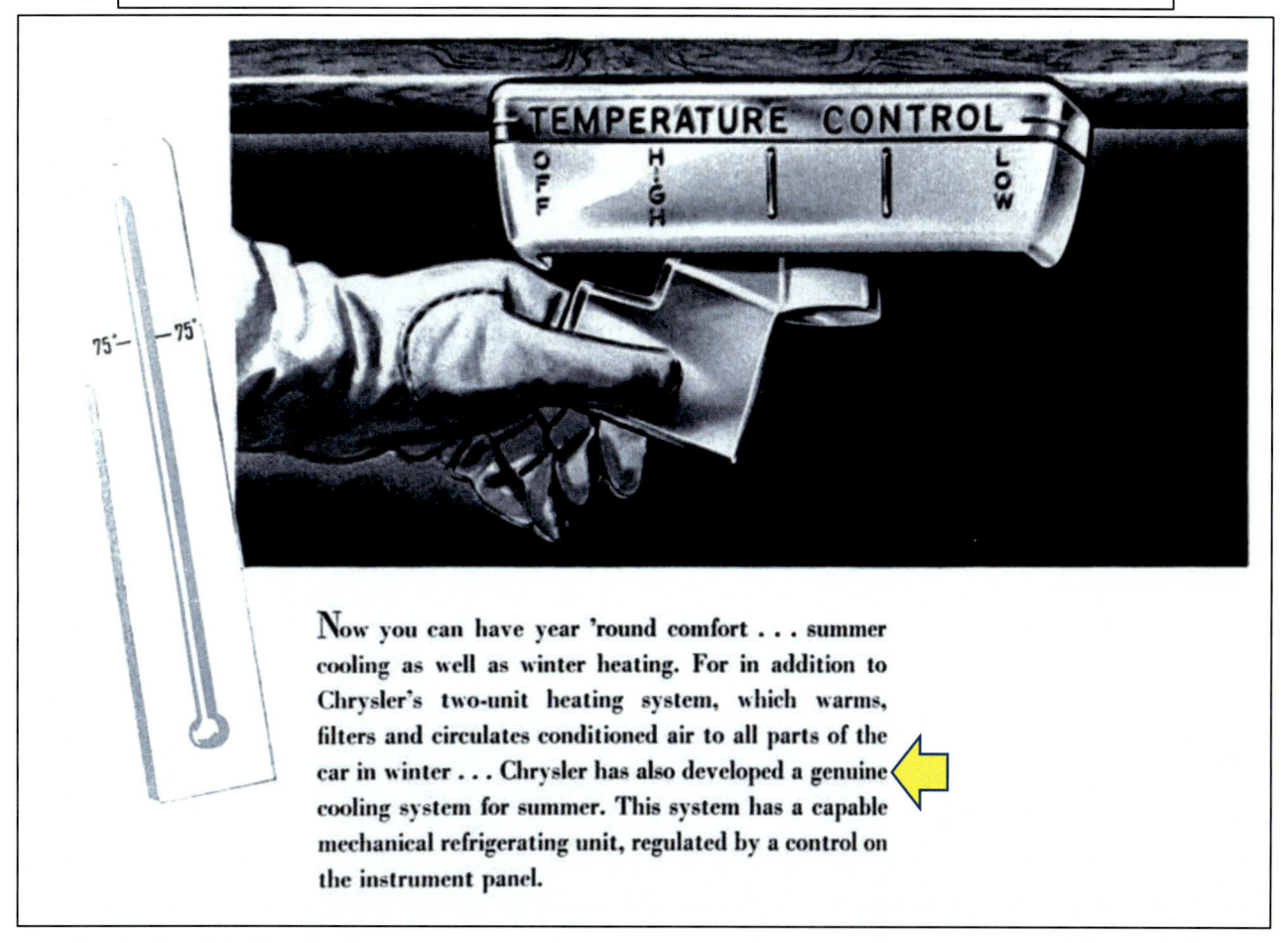

Figure 439. 1941 Chrysler Crown Imperial prestige brochure listed the air-conditioning accessory   Warren H. Erb

224   Chapter 5: 1941 Chrysler Factory Air-Conditioning

An ad for a 1941 Chrysler Crown Imperial offered "Fluid Drive" or "Vacamatic Transmission." It also offered "All-Weather Aircontrol" (with refrigeration, if desired) air-conditioning, rather than using their "Air Refrigeration System" nomenclature described in their separate full-color brochure. (Figs. 414, 440, and yellow arrow)

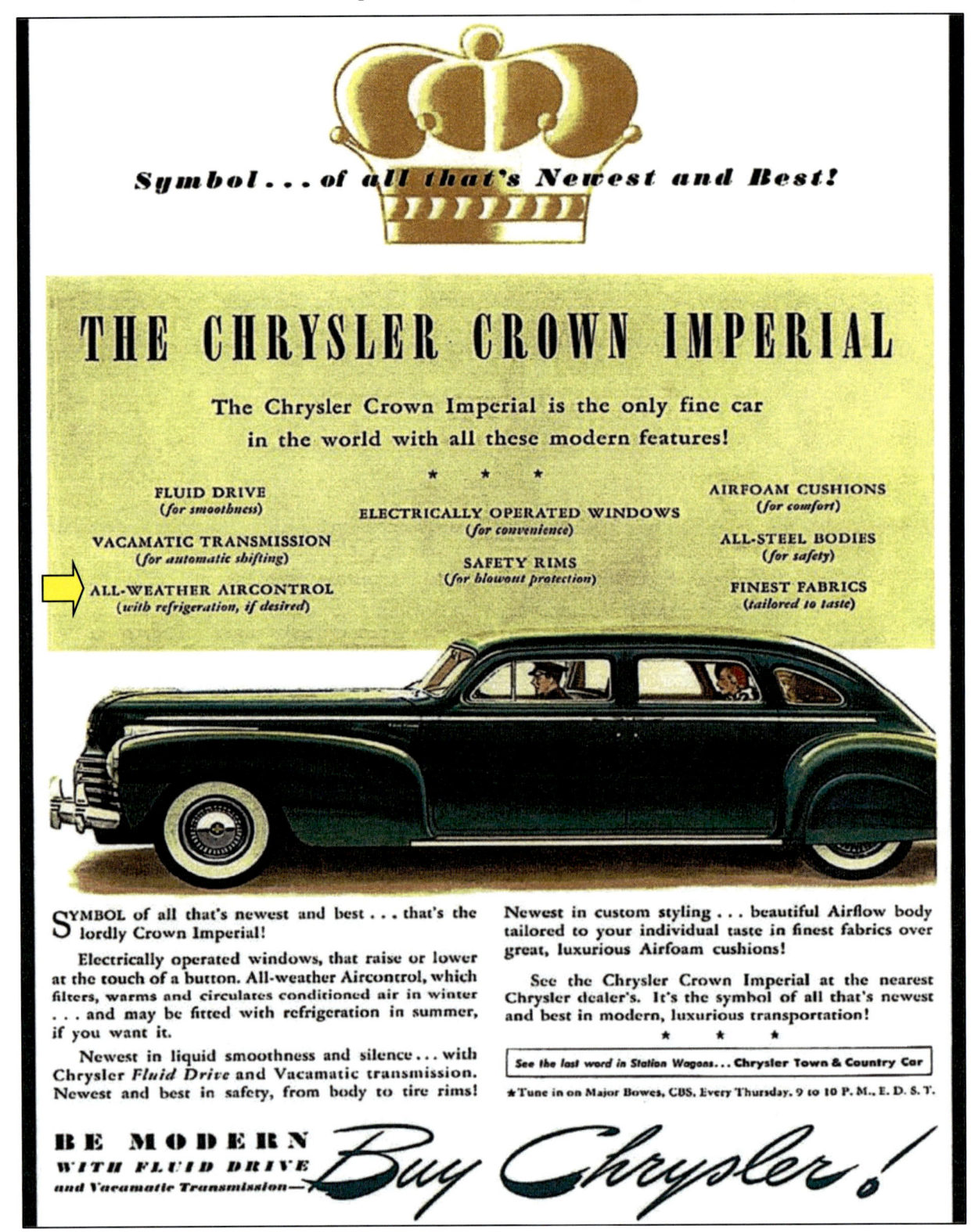

Figure 440. 1941 Chrysler Crown Imperial ad promoted "All-Weather Aircontrol"   OIC

It is fortunate that all three factory air-conditioning service manuals exist, representing the 1941 Chrysler, Packard, and Cadillac air-conditioning systems. The manufacturer of the air-conditioning componentry, Bishop & Babcock Mfg. Co. wrote the original air-conditioning service manual. Compared side to side, the Detroit automakers duplicated Bishop & Babcock's phrasing, headings, subtitles, wording, images, and entire paragraphs in their separately produced knockoffs, titled under the auspices of the automakers.

The question arises again: Why did Packard delay its service manual publication until 1941 when it exclusively introduced its Weather Conditioner system for the 1940 model year? The respective publications' timing suggests that Bishop & Babcock suffered delays of its master publication and completed it during the 1941 model year. Upon receipt Packard, Cadillac, and Chrysler published their manuals for 1941. The 16-page copy of the 1941 *Chrysler Air Refrigeration System Operation—Service* manual follows. (Figs. 441–456)

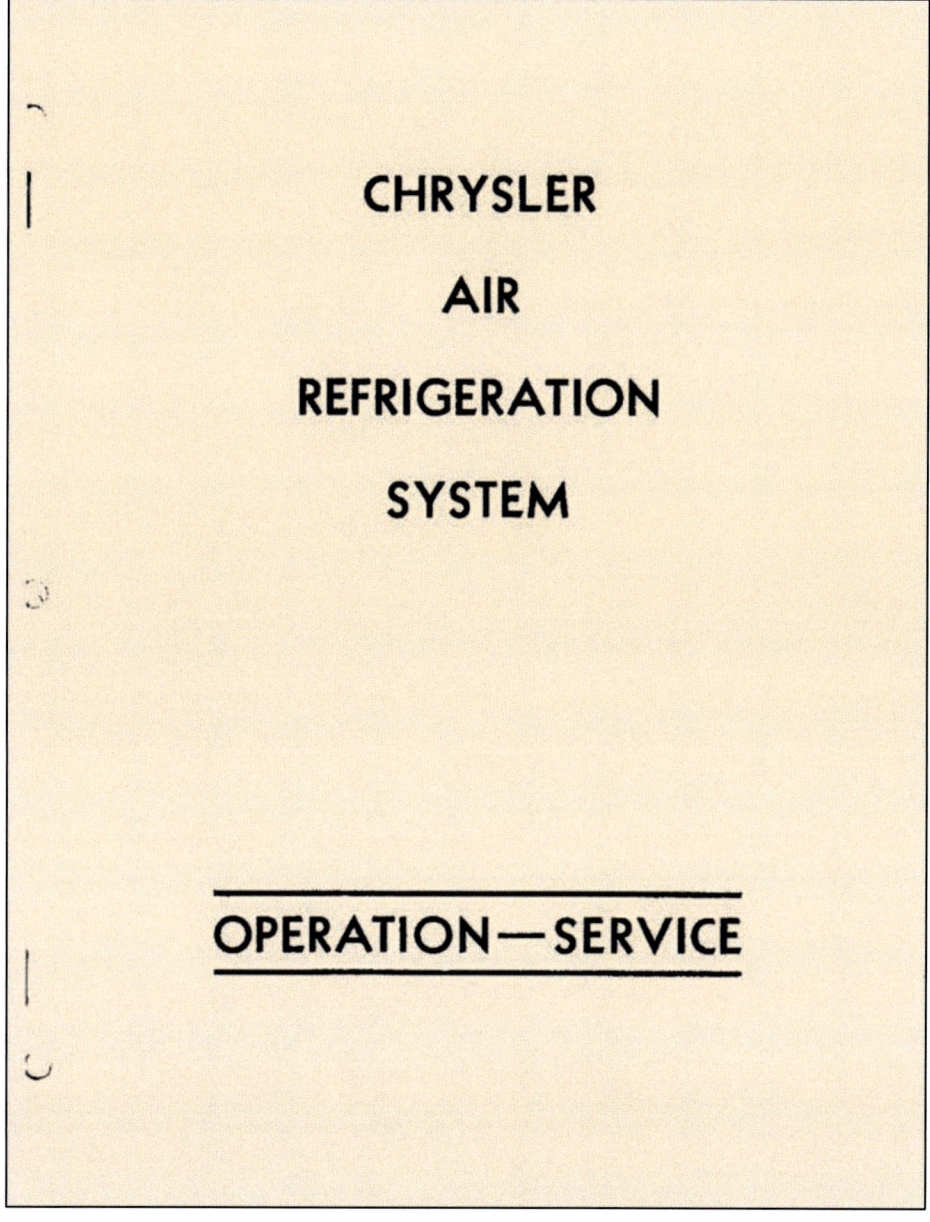

*Figure 441. 1941 Chrysler Air Refrigeration System Operation—Service Manual cover
Content and Photo(s) courtesy Warren H. Erb, Docent, AACA Museum, Hershey PA*

OPERATION

Chrysler Air Refrigeration System which is supplied on all 1941 Chrysler closed body models as factory-installed special equipment at extra cost, is a mechanical refrigeration system which provides cool, filtered, dehumidified air for passenger comfort.

With the compressor belt installed and the engine running, Air Cooling is always available. It can be turned on by merely regulating a switch on the instrument panel.

THE INSTRUMENT PANEL SWITCH is conveniently located on the instrument panel, to the left of the steering column. When the switch is turned on the cooling fan is in operation. Similar to a heater switch, there are four positions on the switch, so that the amount of cool air delivered into the car can be controlled.

If Air Refrigeration is not desired, simply turn the switch to the "off" position.

THE ADJUSTABLE LOUVRES, located in the rear seat shelf panel, direct the flow of air into the car. One of the louvres is fixed, the others may be adjusted to any desired position by simply turning a small knob. Under certain conditions, when only a slight amount of cooling is desired, the adjustable louvres may be closed and the cool air is then discharged into the car through the opening controlled by the fixed louvre.

Some atmospheric conditions will cause condensation to form on the outside of the rear window if the blast of cold air strikes the window. To overcome this condensation, adjust the louvres so that the air is directed forward, away from the window.

VENTILATION—For maximum cooling, the cowl ventilator and all windows should be closed. Ordinarily, ventilation is adequate under these conditions. However, if additional ventilation is required, it may be obtained by partially opening the cowl ventilator.

If the car is parked for a long period of time, high temperature may develop inside the body. Before turning on the cooling switch, drive the car with the windows open for a few minutes, to allow the air to circulate. Close the windows and turn on the switch.

REFRIGERATION CYCLE — Fundamentally the refrigerant, which is circulated through the system by the compressor, picks up heat at the evaporator coil, (Fig. 1), carries it to the condenser, and there discharges it to the outside air.

The refrigerant, Freon (F-12), a non-toxic, non-inflammable and practically odorless gas, is stored in the receiver in a liquid state under relatively high pressure. (Fig. 1.)

From the receiver, the liquid passes through the expansion valve and into the evaporator coil, where it expands into a gas at relatively low pressure. This expansion or evaporation from a liquid to a gas, absorbs heat from the metal coil and the air drawn over it by the fan, thus cooling the air. (Fig. 1.)

The compressor, mounted on the engine block, draws the refrigerant gas from the evaporator at relatively low pressure, and discharges it at high pressure into the condenser. In the condenser, the gaseous refrigerant is cooled sufficiently to condense into a liquid. From the condenser the liquid refrigerant flows into the receiver, and the cycle starts again. (Fig. 1.).

AIR CIRCULATION—When the cooling fan is turned on, by means of the 4-position switch located on the instrument panel, air in the car is drawn under the rear seat, through the air filter and evaporator coil, (Fig. 1). It is then discharged into the car, through the adjustable louvres, so that it follows the contour of the roof.

If the car is driven at high speed for a period of time with the cooling fan not in operation, a coating of frost builds up on the evaporator coil. If the fan is then turned on, some fine frost particles may be blown into the car through the discharge grille. This will last for only a short time and is entirely normal.

AIR FILTERING is accomplished by passing all the air discharged into the car by the cooling fan through an oil coated fibre board filter, which removes dust and other impurities from the air. (Fig. 1.).

AIR COOLING takes place when the air drawn from the car passes over the evaporator coil. The air gives up its heat to the coil, where the temperature is normally from 40 to 50 degrees, and is discharged into the car at a temperature a few degrees higher than the coil temperature.

WINTER OPERATION—During cold weather, when air cooling is not required, remove the compressor drive belt. Nothing else need be done. (Fig. 1.).

When Air cooling is again desired, simply install the belt, and the unit is ready for operation.

Figure 442. 1941 Chrysler Air Refrigeration System Operation—Service, p. 2   Warren H. Erb

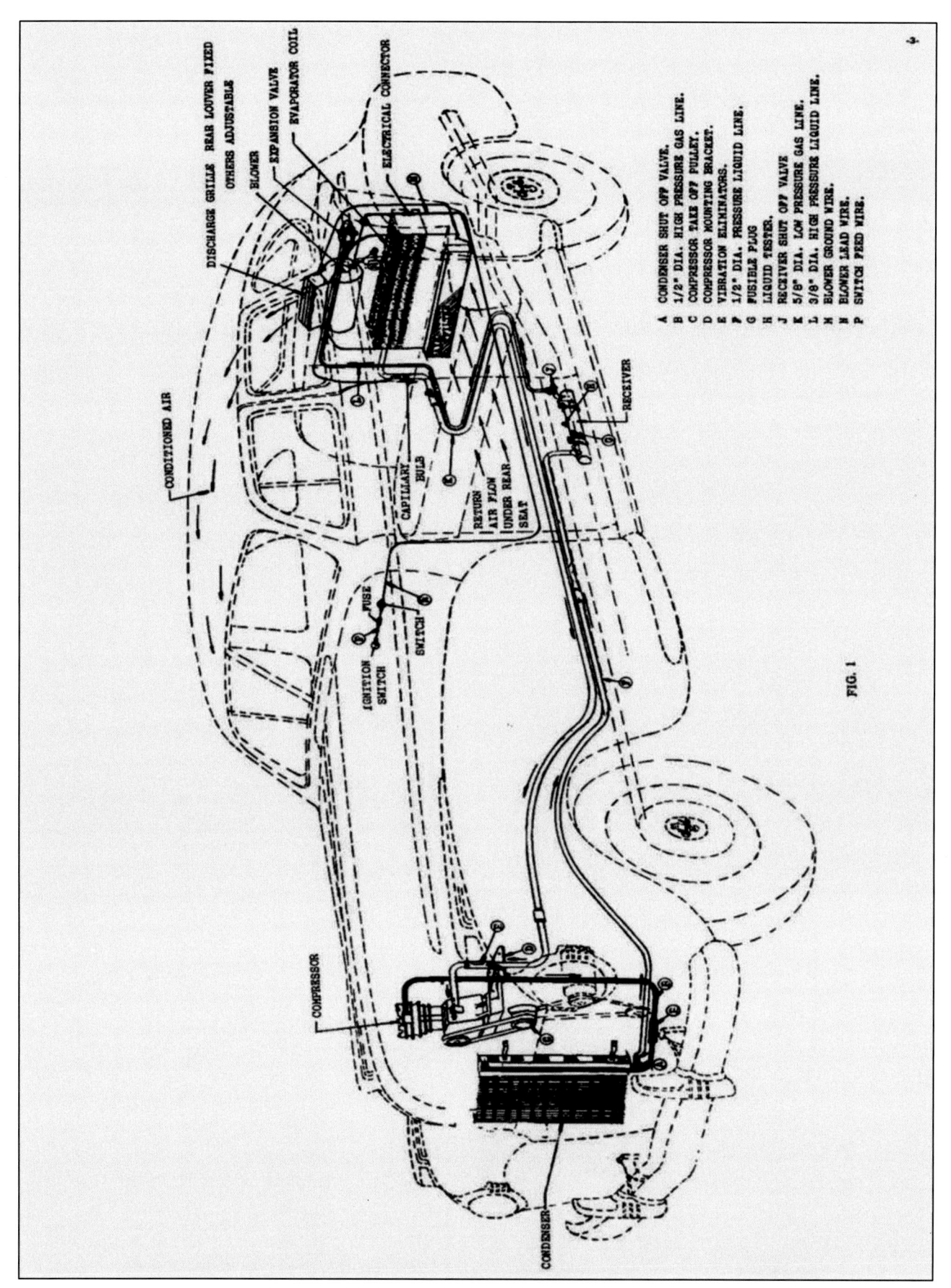

Figure 443. 1941 Chrysler Air Refrigeration System Operation—Service, p. 3   Warren H. Erb

During the winter months, the engine cooling system should be protected against freezing by the use of an anti-freeze solution of the ethylene glycol type. Anti-freeze solution having an alcohol base boils at too low a temperature for satisfactory operation of the refrigeration system.

When removing the filler cap from a hot radiator, rotate it towards the left until the stop is reached. This is the vented position which allows the pressure to escape. Keep in this position until the pressure in the cooling system has been relieved, then turn more forcibly to the left to remove. Turn cap all the way to the right when installing.

SERVICE

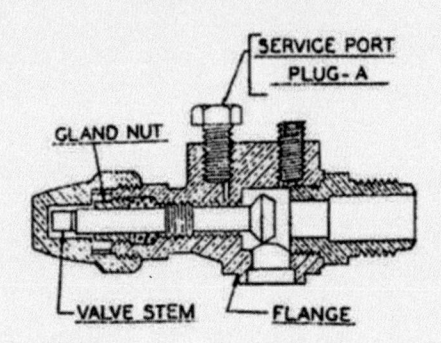

FIG. 2

THE COMPRESSOR SUCTION SERVICE VALVE is located at the compressor inlet connection. It is of the double-seating type which seats or closes when it is screwed in all the way and also when it is screwed out all the way. When the valve is in the "in" position, the suction line is shut off from the Compressor. When the valve is in the "out" position or "back-seated" it is in the operating position. The Service Plug may be removed to permit attachment of the low-pressure or compound gauge or the charging line.

THE COMPRESSOR DISCHARGE SERVICE VALVE is located at the Compressor outlet connection. This valve is identical to the Compressor Suction Service Valve, except that it is for ½" tubing instead of ⅝" tubing. The Discharge Service Valve controls the outlet for refrigerant gas from the Compressor to the Condenser, and provides a connection for the attachment of the high pressure gauge. NOTE: Always back seat valve before removing plug "A" for attaching gauges to Compressor Service Valves.

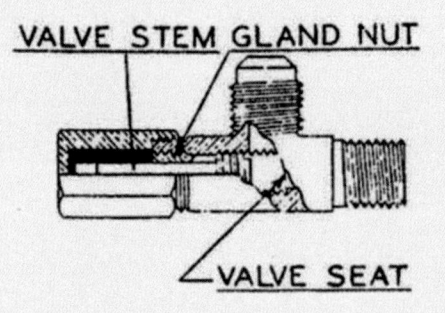

FIG. 3

THE RECEIVER SHUT-OFF VALVE "J," Fig. 1, is of the single-seating type as shown in Fig. 3. It is located at the outlet of the Receiver and is used when it is desired to pump all the refrigerant into the Receiver to permit removal of some part of the refrigeration apparatus.

THE CONDENSER SHUT-OFF VALVE is of the same type as shown in Fig. 3. It is located on the inlet of the Receiver and is used to retain the gas in the Receiver, after pumping back, so that the Condenser may be removed without loss of charge.

F—Temperature Bellows
G—Thermostatic Tube
K—Pressure Bellows
M—Thermostatic Feeler Bulb
P—Inlet Connection
S—Needle Valve
T—Needle Valve Seat
U—Outlet Connection

FIG. 4

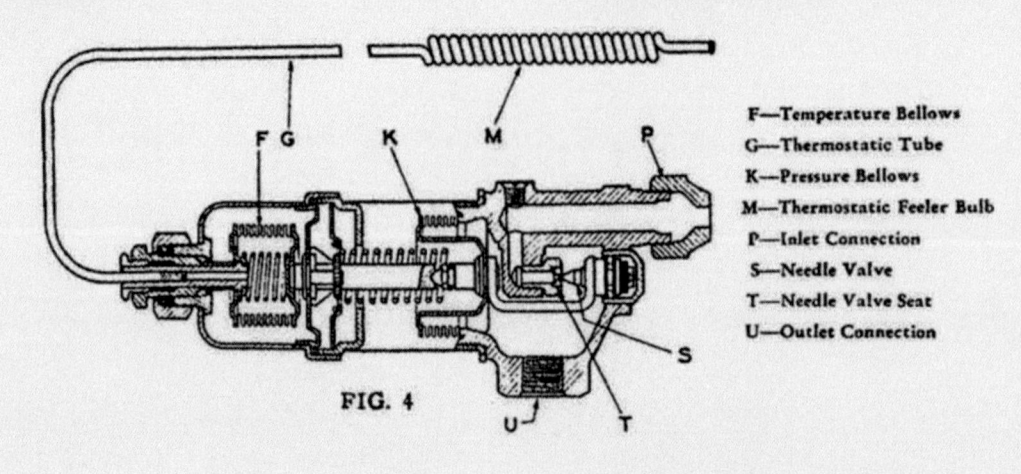

Figure 444. 1941 Chrysler Air Refrigeration System Operation—Service, p. 4 Warren H. Erb

THE EXPANSION VALVE is mounted in the ⅜" high pressure liquid line at the inlet of the evaporator coil, and its thermostatic feeler bulb is clamped to the ⅝" low pressure gas line at the outlet of the Evaporator Coil, Fig. 1. The purpose of the Expansion Valve is to meter the amount of liquid refrigerant passing into the Evaporator Coil. The flow of refrigerant through the valve is controlled thermostatically by the feeler bulb. If too much refrigerant passes into the coil it will not evaporate completely and some liquid will flow into the low pressure gas line. This liquid will cool the bulb, which causes the valve to close partially, thus reducing the amount of liquid entering the coil. If not enough liquid is entering the coil, the bulb will warm up, opening the valve wider to let more liquid enter the coil.

Do not attempt to adjust the Expansion Valve.

The Expansion Valve is of the Thermostatic type, ⅜" S.A.E. Inlet, ½" S.A.E. Outlet, 5/32" Orifice, 55 pound Freon 12, Gas charged power element, with 60" capillary.

FUSIBLE PLUG

The Liquid Receiver is equipped with a fusible plug, Fig. 1-G, set to discharge at 190°F. This is a safety device to prevent excessive pressure in the event the system is overcharged with Freon gas. When an excessive pressure is reached the Fusible plug will melt, allowing all of the Freon to escape. It is then necessary to replace this plug with a new one having the same temperature setting (190°F.) before recharging. In replacing the plug make sure it is screwed in tightly. NOTE: If the fusible plug melts, be sure to check the condenser to see if it is stopped up with bugs and dirt. If it is, clean it before recharging the system.

AIR FILTERS

The Air Filters used in the Chrysler Air Refrigeration System are a replacement type of filter, constructed of oil coated, corrugated fibre board.

The filters should be replaced with new clean filters at least once a year, in the Spring. This insures that the air supplied by the Air Refrigeration System is properly cleaned at all times. Also, it insures maximum operating efficiency of the Air Refrigeration System.

To replace the filters, remove the rear seat, slide the old filters out of position and replace.

TO CLEAN AND DEODORIZE THE EVAPORATOR OIL

In any air Refrigeration System, smoke and body odors will be picked up by the circulating air and carried back to the evaporator coil, where they will be deposited. As a result, unless the evaporator coil is cleaned occasionally, the odors will accumulate and be carried back into the air cooled space.

To prevent the possibility of objectionable odors thus created being carried into the passenger compartment, it is recommended that the evaporator coil be cleaned occasionally with a solution of Diversol or other bacteria insecticide which will not attack the tin-plated interior parts of the cooling coil assembly. Use a solution strength of 4 ounces of cleanser to a gallon of water.

To clean and deodorize the coil, remove the cover plate on the coil housing in the trunk compartment and spray the coil, using about a gallon of the cleaning solution. After passing over the coil, the solution will drain from the return duct through the drain tube located in the bottom of the duct.

TO TEST FOR LEAKS

HALIDE TORCH METHOD

Connect the Halide Leak Detector to a presto-Lite Tank or to the acetylene tank on the welding equipment.

Light the torch on the Halide Gas Leak Detector.

Pass the end of the Leak Detector searching tube around the joint or connection to be tested. (After service work, all joints and connections in the system should be checked.) If there is a leak in the joint the color of the flame in the torch will turn to a brilliant green. This is a positive indication of a leak.

If a leak is detected in a flare connection, draw up the flare nut tightly. If the leak still exists the flare on the tubing is probably defective. The gas will then have to be pumped into the Receiver, as described in the section "Pumping Down Entire System," and the tubing disconnected and another, new flare made in the tubing.

If a leak is detected in a soldered joint, relieve the pressure in that part of the system down to zero on the gauge, either by pumping the refrigerant from the part of the system to another part, or, in the case of the Receiver, purging the gas out of the system to the air or into a small Refrigerant cylinder (Service Drum). Do not attempt to solder the joint while there is pressure in that part of the line.

If the system has lost its charge of refrigerant, there must be a leak, and it must be found rather than simply to add more refrigerant which will in turn be lost unless the leak is found and re-

Figure 445. 1941 Chrysler Air Refrigeration System Operation—Service, p. 5   Warren H. Erb

paired. Do not give up until the leak is found. NOTE: If the system has completely lost its charge, it will be necessary to add some refrigerant before the leak can be found.

To determine whether there is any refrigerant in the system, open the Liquid Tester "H," Fig. 1, on the Receiver, using liquid receiver key. If gas escapes from the tester, there is refrigerant in the system. If no gas escapes, the refrigerant has been lost and it is quite probable that air has entered the system. In this case, proceed as described in the first four steps in the section "To Completely Recharge the System from a Freon Drum." As soon as some refrigerant has been pumped into the system, turn off the engine and close the valve on the Freon drum. It will then be possible to detect the leak.

SOAP AND WATER SOLUTION METHOD

Make a solution of soap (yellow laundry soap) and water. Prepare it at least an hour or so before using so that all bubbles have disappeared and the solution is of a thick "ropey" consistency about the same as heavy oil. Spread this solution on all joints or connections with a soft brush. Examine closely under a strong light. Leaks will show up by the presence of bubbles under or bursting through the film of the solution. Use a small mirror in examining the rear sides of joints otherwise not directly visible.

ATTACHING GAUGES TO READ SUCTION AND DISCHARGE PRESSURES

1. Remove cap from Compressor Suction Service Valve.

2. Back seat valve by turning valve stem counter clockwise as far as possible.

3. Remove plug from service port (Fig. 2.).

4. Attach hose from compound gauge of Charging and Testing Gauge Unit to service port.

5. Purge gas line by opening valve on compound gauge side of gauge manifold and then opening compressor suction service valve ¼ turn. This will allow the refrigerant to blow the air out of the low pressure line through the center charging line. After a few seconds close valve in gauge block.

6. Repeat the above operations with the Compressor Discharge Service Valve, attaching the high pressure gauge line to service port in this valve and purging the line as before by opening the valve on the pressure gauge side of the gauge block.

7. Start engine and adjust suction and discharge service valves until gauges show pressure with minimum fluctuation of the needles.

8. Check all connections for leaks.

PUMPING DOWN LOW PRESSURE SIDE

1. Attach gauges to Compressor as described under section "Attaching Gauges to Read Suction and Discharge Pressures."

2. Remove cap from ⅜" Receiver Shut-off valve, "J," Fig. 1, and close valve by turning valve stem clockwise as far as it will go.

3. Start engine and run at slow speed. The compressor will now pump all the gas from the ⅜" and the ⅝" Low Pressure Gas Line (Suction Line) and store it in the Receiver.

Run the engine until the Compound Gauge reads 2" vacuum. Then stop the engine, place the suction service valve in the "in" or closed position. It is desirable not to open the refrigerant lines while the Compound Gauge registers a vacuum. If necessary, "crack" the Receiver Shut-off Valve "J" until the Compound Gauge reads one pound pressure, when the lines may be opened.

4. It is now possible to remove any piece of equipment on the low side of the system without losing the Freon charge.

Watch both gauges while pumping down the system. If the high pressure gauge goes above 250 lbs., stop the engine, as either the Compressor has been run at too high speed or the installation has received too much Freon gas charge when initially charged. If the pressure is due to an excessive compressor speed the pressure can be reduced by allowing the compressor to remain idle for a few minutes until the pressure drops to normal, when the engine can be restarted. However, if there has been too much Freon charged initially the pressure will not drop when the engine is stopped and it will be necessary to vent the excess gas by loosening the Flare nuts on the inlet and outlet of the Expansion Valve (See Fig. 4.) and allowing the gas to escape slowly to the atmosphere until the lines are empty. IMPORTANT: Do not run Compressor while low pressure side is pumped down.

PUMPING DOWN ENTIRE SYSTEM

1. Follow directions given in first three steps under section "Pumping Down Low Pressure Side."

2. When all the Freon has been pumped from the low pressure side into the Receiver close the Condenser Shut-off Valve, Fig. 1, located at the front of the Receiver.

Figure 446. 1941 Chrysler Air Refrigeration System Operation—Service, p. 6 Warren H. Erb

3. It is now possible to remove any piece of equipment from the system without losing the Freon charge.

NOTE: It is not possible to pump all the gas in the Condenser into the Receiver and therefore a small amount of gas will be lost when a connection is broken on the high pressure side. In some cases it may be necessary to add Freon to the system to make up for this loss.

IMPORTANT: Do not run Compressor while system is pumped down.

AIR AND MOISTURE IN THE SYSTEM DEHYDRATORS

All outside air has some moisture in it in vapor form. Just how much moisture it has, varies considerably from the dry air of Arizona to the humid area of any location on a hot, sultry or rainy day. If moist air is allowed to get into the refrigeration system, both the air and water are harmful. If the air itself were perfectly dry it would do no more harm than to increase the discharge pressure (Thus decreasing the capacity of the compressor), cause excessive heating of the compressor and in general take up room in the system that could be advantageously used for the Freon which performs the cooling.

The introduction of moisture is more serious as it may freeze up in the expansion valve and evaporator coil and stop or retard the cooling action and also corrode many of the finely finished metal surfaces such as the discharge and suction valves and other parts of the compressor and the seat and needle of the expansion valve.

Air and moisture can be accidentally sucked into a system if a leak is present on the evaporator or other portions of the low pressure part of the system when for any reason the low pressure part of the system is on a vacuum. Or they may be introduced by allowing lines or apparatus to stand open for long periods when the system is being opened for repairs either to the system or to some part of the car that is inaccessible except by removal of a part of the conditioning system.

AIR has a tendency to collect in the condenser as it does not condense or turn into a liquid and pass on to the receiver. Air can be purged by removing the plug from the service port of the compressor discharge service, closing the compressor suction service valve, and "cracking" the discharge service valve. This can best be done after the engine has been stopped a few minutes as the air tends to collect in the upper part of the condenser. It may be necessary to run and stop the engine a few times to remove all the air from the system, and thus reduce the excessive discharge pressure.

MOISTURE—There is only one effective correction to moisture in the system—remove it! Special non-freezing liquids may be put in the system that combine with the moisture and, diluting it, prevent freeze-ups at the expansion valve needle and seat. However, (disregarding freeze-ups), the moisture is still in the system to cause corrosion and some of these liquids are corrosive in themselves. Do not use this method.

The dehydrator, is a cylinder with inlet and outlet connections in which is a material known as a desiccant. The approved desiccant is "Drierite," but activated alumina or silica gel may also be used. (DO NOT USE CALCIUM-CHLORIDE.) These are in granular form that allow Freon to pass but absorb the moisture in the Freon.

If it is suspected or known that there is moisture in the system proceed according to the instructions under the section "Pumping Down Low Pressure Side," and then as follows:

1. Loosen nut and remove 3/8" line from inlet connection to Strainer, Fig. 1.

2. Connect the dehydrator to the 3/8" line and, by means of another short piece of 3/8" tubing, to the inlet connection to the Strainer, leaving 3/8" nut on the Strainer connection loose.

3. "Crack" the Receiver shut-off valve "J," Fig. 1, and purge the air from the 3/8" tubing and dehydrator.

4. Tighten the 3/8" nut at the Strainer inlet connection.

5. Open the receiver shut-off valve and start engine.

6. Allow the dehydrator to remain in the line while the cooling system is in operation for at least an hour, then remove by repeating operations for pumping the refrigerant out of this line.

7. Remove dehydrator and replace original 3/8" connection to inlet on Strainer, pull up tightly, open receiver shut-off valve and test for leaks.

8. System should now be free of moisture and ready for use.

NOTE: Always be sure that both ends of the dehydrator are sealed when not in use. If this is not done the desiccant will absorb moisture from the air and may become saturated with moisture. If it were used in this condition it would do

Figure 447. 1941 Chrysler Air Refrigeration System Operation—Service, p. 7   Warren H. Erb

more harm than good. The safest procedure is to always put a fresh charge of desiccant in the dehydrator each time it is to be used.

TO REPLACE EXPANSION VALVE

1. Pump all the Freon from the low pressure side of the system as described in section "Pumping Down Low Pressure Side."

2. Remove the Capillary Bulb from its clamp on the 5/8" line, Fig. 1, leaving the evaporator coil. NOTE:—WHEN THE EXPANSION VALVE IS REMOVED OR AT ANY TIME WHEN THE LINES ARE OPEN, CARE MUST BE TAKEN NOT TO START THE ENGINE UNLESS THE COMPRESSOR HAS BEEN UNBELTED.

3. Remove Expansion Valve by loosening inlet and outlet flare nuts. Some gas may escape, but the loss should be negligible if the Receiver Shut-off Valve is closed tightly.

4. Install new valve and pull flare nuts tight.

5. Clamp Capillary Bulb to 5/8" line as before. Fig. 1.

6. "Crack" Receiver Shut-off Valve until Compound Gauge reads 30 to 40 pounds, then close.

7. Test around Expansion Valve nuts for leaks.

8. Open Receiver Shut-off valve by turning valve stem counter-clockwise. Tighten gland nut around valve stem. Replace cap tightly.

9. Start engine and watch gauges. The Compound Gauge should read 20 to 40 pounds, and the high pressure gauge 140 to 190 pounds, after the engine runs at slow speed for a few minutes (with the blower running).

10. "Back seat" the Compressor Discharge and Suction Service Valves by turning the valve stems counter-clockwise as far as they will go. Remove gauges, replace plugs, tighten gland nuts and replace caps tightly.

TO REMOVE AND REPLACE CONDENSER

1. With the Blower running, start the engine and run at slow speed until the Condenser is warm.

2. Stop engine, remove caps from the Compressor Discharge Service Valve and the Condenser Shut-off Valve "A," Fig. 1, and close both valves by turning stems clockwise as far as they will go.

3. Loosen flare nut on the Condenser Inlet Connection and allow the gas in the Condenser to escape slowly.

4. Unscrew this flare nut after all the gas has escaped and also unscrew the flare nut on the Condenser outlet connection.

5. Remove Condenser, cover inlet and outlet to keep damp air out of Condenser.

6. When it is desired to replace the Condenser, put it back in place.

7. Connect flare nut to Condenser Outlet Connection and tighten.

8. Connect flare nut to Condenser Inlet Connection and tighten.

9. Loosen service port plug "A" in Compressor discharge service valve (Fig. 2).

10. "Crack" the Condenser shut-off valve "A," Fig. 1, allowing gas from the 1/2" line to the Receiver to pass up through the Condenser and out at the loose plug on the Compressor discharge service valve, thus purging the air from the Condenser. It is necessary to purge only a few seconds.

11. Tighten the plug in the Compressor discharge service valve.

12. Open Compressor discharge service valve and Condenser shut-off valve by turning counter-clockwise. Replace and tighten valve caps.

13. Check all joints for leaks.

TO REPLACE COMPRESSOR

1. Run engine at slow speed for a few minutes until Compressor is warm.

2. Stop engine, remove caps from Compressor Discharge and Suction Service Valves and close both valves by turning stems clockwise as far as they will go.

3. Loosen Plug, (Fig. 2) in Service Port of Suction Service Valve and allow gas in the Compressor to slowly escape. Also remove Plug in service port of Discharge Service Valve and allow gas in Compressor head to escape.

4. Remove cap screws securing the Compressor Service Valves to the Compressor and carefully lift the valves away from the Compressor.

Figure 448. 1941 Chrysler Air Refrigeration System Operation—Service, p. 8   Warren H. Erb

5. Loosen bolts securing the Compressor to the Compressor base, and remove the drive belt.

6. Remove bolts and Compressor from base.

7. Take off nut holding Compressor pulley to shaft and remove pulley. See that Woodruff key is in place on replacement Compressor and transfer pulley to replacement compressor. Slide it in place and tighten shaft nut.

8. Put replacement Compressor on base, put in bolts loosely, put on belt and pull up Compressor until belt is tight—not too tight, just so it does not slip. Also see that belt and pulleys are lined up. Tighten bolts.

9. Put in new copper flange gaskets between the Service Valves and Compressor. Put valve flanges down against the copper gasket, run in cap screws finger tight. Tighten cap screws evenly so that flange fits flat against the gasket.

10. Replace plug in the service port of the Suction Service Valve and crack this valve by opening ¼ turn to blow gas from the suction line up through the Compressor and out the service port of the Discharge Valve. Turn Compressor pulley over by hand to assist in purging the air from the Compressor.

11. Tighten plug in Service port of Suction Service Valve, open valve stem to back seat, tighten gland nut, put on and tighten valve cap.

12. Put in and tighten plug in service port of Discharge Service Valve, back seat valve stem, tighten gland nut, replace and tighten cap.

13. Test for leaks around Service Valve flanges, service ports, and caps. If there are no leaks, the Compressor is ready for use.

CHECKING OIL LEVEL IN COMPRESSOR

If there has been a loss of liquid Freon from the system some of the oil (which mixes readily with Freon and is present in all parts of the system) may also have been lost. The level in the Compressor may be checked as follows:

1. With the blower running, start the engine and run at slow speed for a few minutes until the Compressor crankcase is warm. Then stop engine.

2. Remove cap and close (clockwise) the Compressor Suction Service Valve as far as it will go. Loosen the plug in the Service port and allow the gas in the Compressor to slowly escape.

3. When there is no further escape of gas, remove the Oil Filler Plug, a Hex Head Plug on the side of the Compressor Crankcase.

4. Insert a clean rod to the bottom of the Compressor crankcase, and measure the height of the oil level. It should be up to, or within ⅜" below the centerline of the Compressor shaft.

5. If the oil level is low, add oil as necessary. USE SPECIAL CHRYSLER COMPRESSOR OIL ONLY.

6. Replace Oil Filler Plug, tighten plug in Service Port on Suction Service Valve, open valve to back seat, tighten gland nut and replace, and tighten cap. Test for leaks. No purging of the compressor is necessary for this operation.

CHARGING REFRIGERANT INTO THE SYSTEM

It is very simple to determine if there is a sufficient amount of Freon in the system for normal operation. With the Compressor running, open (slightly) the Liquid Tester "H", Fig. 1, which is a small test cock on the Receiver. If there is liquid refrigerant up to the level of the tester, liquid Freon will come out in a milky white flow. If gas only blows out, this indicates that the system is short of refrigerant and some should be added, as described below.

If neither liquid nor gas escapes from the tester, the system must be completely recharged, either from a Freon drum or from a charged Receiver. Both of these methods are described below. A complete charge is 6¼ pounds of Freon.

TO PUMP AIR FROM SYSTEM (See Fig. 5)

When the entire charge of Freon has been lost, it is quite probable that air has been introduced into the system at the point where the leak occurred. It is necessary to remove this air before adding a new charge of Freon. By using the compressor as a suction pump, the air can be removed as follows:

1. Install gauges and adjust valves as shown in Fig. 5.

2. Be sure both valves in receiver tank are open.

Figure 449. 1941 Chrysler Air Refrigeration System Operation—Service, p. 9 Warren H. Erb

3. Start engine and run at slow speed until compound gauge reads 20-28 inches vacuum. NOTE: If oil is discharged through the charging line along with the air during this operation, stop the engine for a few minutes, then start up again and proceed until the proper vacuum is reached.

4. When the system has been pumped down to 20-28 inches vacuum, stop the engine. If the vacuum holds for several minutes, this is an indication that any leak in the system is comparatively small and it will be safe to recharge the system before finding the leak.

NOTE: If the vacuum does not hold when the engine is stopped, there must be a bad leak in the system. Be sure to find this leak before recharging the system. To find the leak, attach the Freon drum to the end of the charging line and open the valve in the drum, allowing Freon to enter the lines until both gauges register 60-70 pounds. Now check the entire system for leaks as described under the section "To Test for Leaks."

When the leak has been found and repaired, proceed as described below.

TO COMPLETELY RECHARGE THE SYSTEM FROM A FREON DRUM

1. Remove all the air from the system, as described in the section "To Pump Air from System."

2. Install gauges and adjust valves as shown in Fig. 6.

   NOTE: Be sure to purge the air from both gauge lines, also from the charging line.

3. Open valve on Freon drum.

4. Start the engine and run at slow speed. The Compressor is now drawing gaseous Freon out of the drum.

   NOTE: Stand the Freon drum upright so that only gaseous Freon is drawn into the Compressor. Do not invert the drum or lay it on its side as this will allow liquid Freon to enter the Compressor, perhaps damaging the Compressor valves and causing oil pumping and slugging. If the drum gets cold, set it in pail of warm water to hasten the vaporizing of the liquid Freon.

5. Keep trying the liquid tester on the Receiver. When a milky white spray comes from it, shut the valve on the Freon drum and stop the engine.

6. Find out where the Freon leaked out and repair the leak.

7. Start engine and check liquid level in receiver tank. Charge in more refrigerant if necessary.

8. Remove gauges from Compressor.

TO COMPLETELY RECHARGE THE SYSTEM BY THE CHARGED RECEIVER METHOD

1. Remove the old Receiver from car.

2. Install the new Receiver, which is charged with Freon.

   NOTE: Receiver should contain 6¼ lbs. of Freon.

3. Loosen flare nut on ½" line at Compressor discharge service valve.

4. Purge ½" line from Receiver to Compressor by cracking Condenser Shut-off Valve "A", Fig. 1, and allowing gas to pass through the line forcing the air out ahead of it through the loose connection at the Compressor. When all the air is purged and Freon begins to escape (use Halide torch to detect the Freon), tighten the ½" nut at Compressor.

5. Loosen flare nut on ⅝" line at Compressor suction service valve.

6. Purge the ⅜" and ⅝" lines from Receiver to Compressor in same manner as ½" line was purged, by cracking Receiver Shut-off Valve "J," Fig. 1, and allowing air to escape through loose connection on ⅝" line at compressor. When the air is purged, tighten the ⅝" nut.

7. Check entire system for leaks.

Figure 450. 1941 Chrysler Air Refrigeration System Operation--Service, p. 10   Warren H. Erb

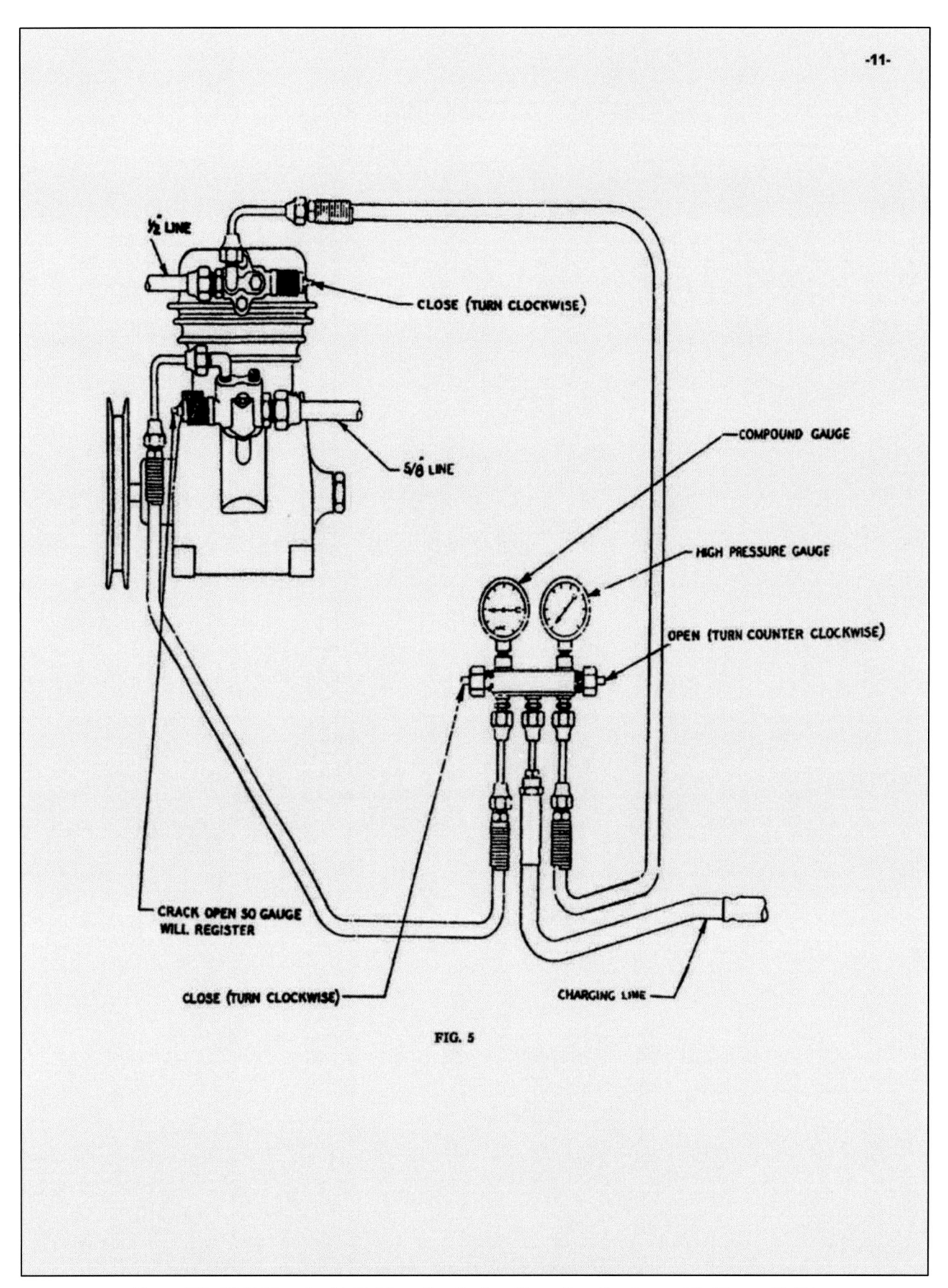

Figure 451. 1941 Chrysler Air Refrigeration System Operation--Service, p. 11   Warren H. Erb

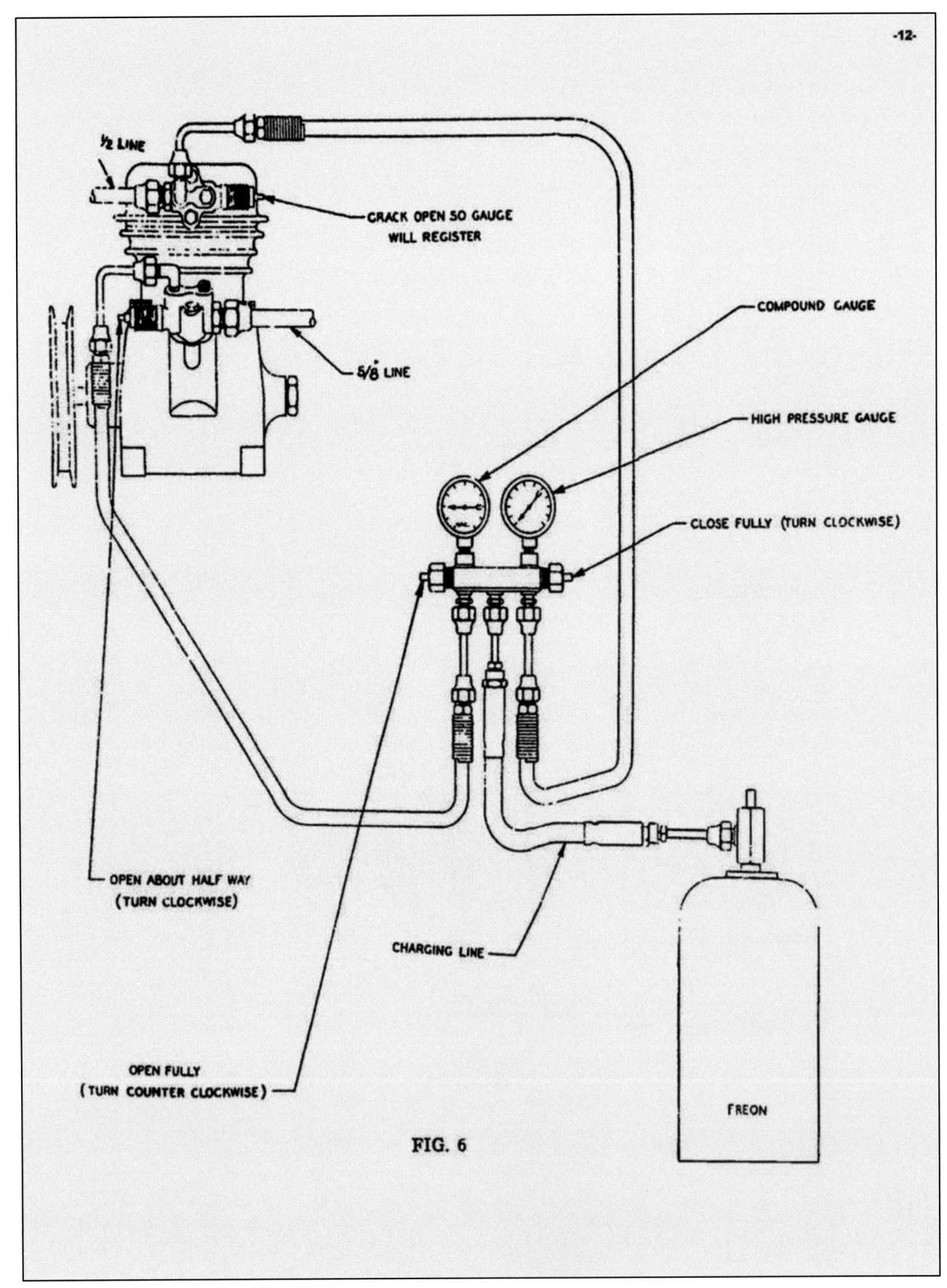

236 Chapter 5: 1941 Chrysler Factory Air-Conditioning

Figure 452. 1941 Chrysler Air Refrigeration System Operation--Service, p. 12   Warren H. Erb

SERVICE CHART

CAUSE	CHECK	CORRECTION
INTERIOR OF CAR NOT COOL, BUT NORMAL AMOUNT OF COLD AIR FROM BLOWER.		
1. Windows or cowl ventilator open. Doors opened too much of time.	1. Obvious. NOTE: Ventilation requirements for crowded car, especially if occupants are smoking, may impose abnormal load, particularly in hot, moist weather.	1. Use care in regulating ventilation for smoke and in leaving doors open.
2. Car engine idling or running very slowly large portion of time.	2. High suction pressure, high discharge pressure.	2. Set engine idling speed somewhat higher. Pull out handle throttle when standing with engine running. Run blower full speed.
3. Extra moist weather—perhaps raining.	3. A large part of the cooling capacity is used to remove excess moisture from the air. The inside of the car will be more comfortable than an unconditioned car. However the temperature as read on an ordinary thermometer may not read any lower than a shaded thermometer outside the car.	3. Use very minimum of ventilation and operate blower at maximum speed.
NOT ENOUGH AIR FROM BLOWER.		
1. Air filters stopped up.	1. Low suction pressure. Suction line cold.	1. Remove air filters and replace with new ones.
2. Blower running under speed—Loose or corroded connections, switch broken, battery charge low.	2. Insufficient air circulation.	2. Trace circuits for bad connections. Check switch and replace if necessary, check battery and recharge if low.
3. Evaporator coil stopped with dirt or lint.	3. Low suction pressure, suction line cold.	3. Cleanse evaporator coil as described in section "To Clean and Deodorize the Evaporator Coil."

Figure 453. 1941 Chrysler Air Refrigeration System Operation—Service, p. 13 Warren H. Erb

CAUSE	CHECK	CORRECTION
AIR FROM EVAPORATOR NOT COLD		
1. Compressor not running, or running slowly. Belt broken or loose and slipping.	1. Obvious upon inspection.	1. Loosen bolts from compressor to base and shift compressor to tighten belt. Be sure to line pulleys up properly.
2. Loss of Refrigerant.	2. Open liquid tester on receiver. If gas only comes out, some Freon has been lost.	2. Test entire system for leaks thoroughly. Find and repair leaking joint, then add refrigerant as described under "Adding Refrigerant to the System." If Fusible Plug in Receiver has melted, check for dirt or bugs in Condenser.
3. Expansion Valve Strainer stopped up with foreign matter.	3. Low suction pressure. Suction tube out of evaporator coil warm.	3. Stoppage will probably be found in Strainer at inlet of valve. Remove and wash in clean naptha. Follow instructions under "Pumping Down Low Pressure Side" to remove Strainer.
4. Condenser stopped with dirt, bugs or lint.	4. High discharge pressure. Discharge line from Compressor extra hot.	4. Clean Condenser thoroughly with hose.
5. Moisture in Expansion Valve.	5. Same as 3 above, except symptoms may not appear every time unit is operated. Moisture sometimes passes Valve and does not appear for several hours.	5. Same as 3 above and also install Dehydrator, see section "Air and Moisture in the System."
6. Loose or improperly insulated Expansion Valve feeler bulb.	6. Obvious upon inspection.	6. Tighten feeler bulb clamp and insulate bulb from outside air.
COMPRESSOR UNIT NOISY		
1. Loose drive pulley or Compressor pulley.	1. Inspect if nut on Compressor shaft is tight, also key from Compressor shaft to pulley may be loose in keyway.	1. Tighten shaft nut and replace key if loose.
2. Squeaky drive belt, loose or greasy.	2. Belt should not have more than ⅜" slack. Inspect for oil on belt.	2. Tighten belt if loose. If greasy, wipe clean with naptha and coat with powdered talc.

Figure 454. 1941 Chrysler Air Refrigeration System Operation—Service, p. 14   Warren H. Erb

CAUSE	CHECK	CORRECTION
COMPRESSOR UNIT NOISY (Cont'd)		
3. Expansion Valve Strainer stopped up with foreign matter.	3. Warm suction line. Vacuum reading low on compound gauge. (See Fig. 5). This causes oil pumping by Compressor—results in knocking sound in Compressor.	3. Remove and wash Strainer in clean naptha. Follow instructions under "Pumping Down Low Pressure Side" to remove strainer.
4. Moisture in Expansion Valve.	4. Same as 3 above except noise does not appear every time unit is operated. Moisture sometimes passes valve and does not appear for several hours.	4. Install dehydrator as described in section "Air and Moisture in the System."
5. Liquid Freon instead of gas being returned to Compressor through suction line. This causes oil pumping and inadequate lubrication of the Compressor. Usually caused by defective Expansion Valve or by capillary bulb loose on suction line.	5. Wet and cold suction line-even the suction service valve and the Compressor cylinder housing and crankcase may be cold, although the engine and Compressor are running — causes a knocking sound similar to loose bearings.	5. If capillary bulb is loose from suction line, tighten it and also be sure it is properly insulated. If this does not correct the trouble, replace the Expansion Valve.
6. Compressor loose on base.	6. Obvious upon inspection.	6. Tighten four Compressor hold-down bolts.
7. Air in system. Stoppage in valves or lines. Overcharge of Freon.	7. High discharge pressure. Causes pounding, laboring sound.	7. Correction depends on cause of the high discharge pressure. For air in system, purge air or discharge system and entirely recharge. Stoppage: remove stoppage. Overcharge of refrigerant: purge Freon to tester level.
8. Too much oil in Compressor crankcase.	8. Evidenced by a dull, thumping sound.	8. Check at Compressor oil filler plug as under "Checking Oil Level in Compressor."
9. Not enough oil in Compressor crankcase.	9. Usually not evidenced until bearings and pistons are worn and knocking or seal leaking. Denoted by very hot crankcase. Seal may be squeaking due to insufficient oil in Compressor or oil passages to seal stopped with foreign matter.	9. This condition frequently the result of leakage of oil from the system and if noticed and caught soon enough, may be corrected by adding oil to the Compressor, but if damage has already resulted. it will be necessary to change the Compressor.

Figure 455. 1941 Chrysler Air Refrigeration System Operation—Service, p. 15 Warren H. Erb

CAUSE	CHECK	CORRECTION
COMPRESSOR UNIT NOISY (Cont'd)		
10. Broken parts in Compressor.	10. Evidenced by a knocking sound similar to loose bearings.	10. Replace entire Compressor.
11. Compressor valve noise telephoned to dash.	11. Tubing fastened to dash may be improperly insulated from dash so that noise carried from Compressor is amplified by dash.	11. Insulate tubing from dash with soft rubber.
NOISY OPERATION OTHER THAN COMPRESSOR.		
1. Loose tubing, straps, brackets or sheet metal parts.	1. See correction.	1. Trace sound for location and cause and repair as may be required.
2. Noisy blower motor.	2. See correction.	2. Remove blower motor and fan and replace motor.
3. Low Freon charge.	3. Hissing sound at Expansion Valve. May be no more than normal, but if excessive may indicate low Freon charge.	3. Check liquid tester on Receiver. If necessary, add Freon till proper level is reached.
4. Freon passing through evaporator coil.	4. Gurgling sound in evaporator coil. Not audible except under very quiet conditions.	4. Entirely normal. No correction required.

-16-

Figure 456. 1941 Chrysler Air Refrigeration System Operation—Service, p. 16   Warren H. Erb

The brevity of the 1941 manual left many questions and issues unanswered, such as:
1. The specifications, tolerances, and manufacturer's information of the Servel compressor.
2. On-car photos of the car's componentry, such as the air-conditioning controls, compressor, condenser, bracketing, hoses, tubing, receiver, louvered cooling register, evaporator with/without the inspection panel to show the cooling coils, expansion valve, and electrical blower assembly.
3. Photos of a technician operating the pressure gauges during Freon measurement.
4. Illustrations/photos for air filter replacement.
5. Instructions for removal and installation of the compressor belt.
6. Outside temperature/inside cabin temperature cooling charts under varying loads and conditions.
7. The CFM, cubic feet per minute, airflow rate from the cooling register.
8. Schematics for the system wiring.

The lack of photos and other information is surprising, considering the delayed publication date until 1941, especially since Bishop & Babcock equipped the air-conditioning system in the 1940 Packard.

1942 De Soto Air-Conditioning

De Soto's styling for 1942 emphasized a unique frontal appearance. Above the waterfall grille, the headlights disappeared by using a mechanism to operate the driver-controlled metal doors.

"Your Next Car, De Soto" headlined the sleekly formed 1942 De Soto on its brochure cover. The cover emphasized its *avant-garde* appearance with a subtitle, "Chrysler Corporation's *Style Leader*." (Fig. 457)

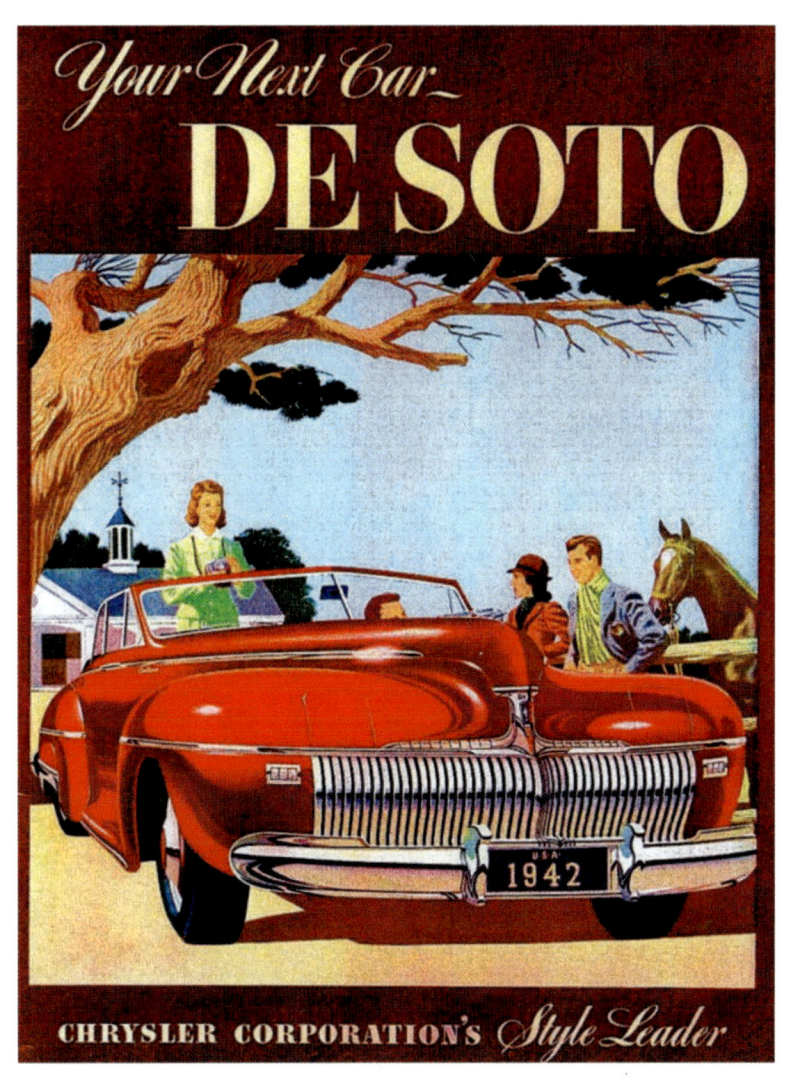

Figure 457. 1942 De Soto brochure cover   ABS

De Soto offered "Tomorrow's Style Today," "New Airfoil Lights Out of Sight Except at Night." (Fig. 458 and blue arrow)

De Soto's heavily stylized image displayed a grille featuring narrowly spaced, waterfall-style vertical bars. Above the grille, a horizontal chrome bar left a clean area for disappearing, owner-operated headlight doors.

Figure 458. 1942 De Soto ad featured "New Airfoil Lights" that remained "Out of Sight Except at Night." (blue arrow)   ABS

An image of a 1942 De Soto sedan contrasts significantly with De Soto's streamlined renditions. (Fig. 459)

Figure 459. 1942 De Soto   Classiccars.com id# 1144796

Expecting an even better sales year than the record 1941 De Soto sales of 91,004, the 1942 De Soto lineup planned to follow Chrysler and offer its customers the "Air Refrigeration System." (Fig. 460) The headline announced, *De Soto Refrigerated Air-Conditioning!* as an optional accessory in the brochure. It offered clean, cool, dehumidified air that *"is 10 to 15 degrees cooler than outside air!"* (emphasis by author and indicated with yellow arrow) It compared itself to an all-mechanical system of cooling found in movie theaters.

In a 1999 article referencing the interior cooling capability, Mohinder S. Bhatti, Ph.D., wrote in the *ASHRAE* (American Society of Heating, Refrigeration and Air-conditioning Engineers) *Journal*:

> At that time, it was believed that if the difference between the outside air and the conditioned air exceeded 10°F, the occupant of the conditioned space could experience a *thermal shock* (emphasis by author) upon emerging into the outside air!

Author note: Reflecting on the air-conditioning theme and related potential illness, in 1940, Houston's wealthy oil wildcatter, Glenn McCarthy, built an air-conditioned mansion. Although a rarity in homes, his son, Glenn, Jr., recounted that his mother refused to operate it, fearing the family could contract polio!

Figure 460. 1942 De Soto brochure offered "De Soto Refrigerated Air-Conditioning"
Dave Duricy

Unfortunately, the condensed, one-page 1942 De Soto air-conditioning information, based upon the 1941 Chrysler air-conditioning brochure, ends there. There are no records of any De Soto sold with the "Chrysler Air Refrigeration System." No promotional advertising presents itself, nor does any automotive article discuss a De Soto that features air-conditioning. No owners acknowledge factory air-conditioning in their 1942 De Soto. The possibility that De Soto produced three air-conditioned cars rates only as a legendary tale, according to author and De Soto expert Dave Duricy. He recalls:

> To my knowledge, no one knows how many 1942 De Soto models were equipped with air-conditioning. The number "three" comes from an article I wrote a few years ago for *Collectible Automobile* (February 2007). I stated that after an extensive search of used car ads from the '40s, I found three 1942 De Soto's with air-conditioning advertised. A reader of the article subsequently understood that to mean "three" had been built and said as much in, I think, an AACA forum.

Dave Duricy's research adds credibility to an indeterminate production and sale of 1942 De Soto models equipped with air-conditioning. The 1940-style term used in classified ads for air-conditioning meant merely a heater, defroster, and ventilating system.

The pre-WWII Chrysler Corporation factory air-conditioning story presents limited information. Chrysler records either were not kept or were not retained or both. It is evident, however, that at least one 1942 Chrysler Corporation air-conditioned vehicle exists. Mr. Warren H. Erb photographed a 1942 Chrysler Crown Imperial Limousine equipped with the "Air Refrigeration System" in the early 1990s at a car show in Macungie, PA. (Figs. 430–432) The Online Imperial Club presented the same car in a December 2018 update. (Figs. 434–437)

An S. A. E., Society of Automotive Engineers, meeting in 1952 hosted Chrysler Corporation's Mr. T. C. Gleason, Head-Fluid Dynamics Laboratory, who presented "A Survey of Heating, Ventilating, and Air-Conditioning of Car Bodies." Mr. Gleason accompanied his lecture with Chrysler air-conditioning-related images:

1. 1942 Chrysler (?) "Conventional cool air discharge and grille" (*Chrysler Fig. 6*, Fig. 461)
2. Chrysler (?) "1942 production blower/evaporator air-cooling unit" (*Chrysler Fig. 8*, Fig. 462)
3. 1942 Chrysler (?) "Conventional under hood installation of compressor and condenser" (*Chrysler Fig. 9*, Fig. 463)
4. 1942 Chrysler (?) "Belt-driven reciprocating compressor" (*Chrysler Fig. 10*, Fig. 464)

Figure 461. (Chrysler Fig. 6) 1942 Chrysler (?) "Conventional cool air discharge and grille" S.A.E. (Society of Automotive Engineers) T. C. Gleason

Factory Air: Cool Cars in Cooler Comfort 245

Figure 462. (Chrysler Fig. 8) Chrysler (?) "1942 production blower/evaporator air cooling unit"   SAE   T. C. Gleason

Figure 463. (Chrysler Fig. 9) Chrysler 1942 (?) "Conventional under hood installation of compressor and condenser"   SAE   T. C. Gleason

Figure 464. (Chrysler Fig. 10) Chrysler 1942 (?) "Belt driven reciprocating compressor   SAE   T. C. Gleason

246   Chapter 5: 1941 Chrysler Factory Air-Conditioning

Figure 465. 1940 Packard air-conditioning evaporator Unknown owner

Figure 466. 1941 Packard Model 1442 air-conditioning evaporator   Owners M/M Sal Saiya   Dave Czirr photo

Figure 467. 1942 Packard One-Eighty air-conditioning evaporator Former owner and photo Hyman Ltd. Classic Cars

While Mr. Gleason's 1952 presentation primarily offered information covering post-WWII automotive fresh air heating and ventilating systems, he commented, *"Summer cooling by refrigeration of air in car bodies is in a very elementary stage of progress."* (emphasis by author) He supplied low-resolution images of Chrysler Corporation's air-conditioning componentry from 1942: one image of the rear louvered cooling register behind the rear seat (*Chrysler Fig. 6*, Fig. 461) and the (Chrysler) *"1942 production blower/evaporator air cooling unit."* (emphasis by author) (*Chrysler Fig. 8*, Fig. 462)

"Mr. Gleason's caption (*Chrysler Fig. 8*, Fig. 462) identified a 1942 production car evaporator. The smaller evaporator size in that image does not appear similar to the phantom renderings in the 1941 Chrysler and 1942 De Soto "Air Refrigeration System" brochures and *Operation Manual* images. (Fig. 417, 443, 460) Mr. Gleason's image (*Chrysler Fig. 8*, Fig. 462), however, shows a marked similarity to other smaller brochure renderings of the trunk-mounted evaporator. (Figs. 419, 460)

By far, the best evaporator comparisons are those photographed in the trunks of the 1940–1942 Packard and the 1941 Cadillac (Figs. 465–468) when compared with the only evaporator photograph known to exist of the 1942 Chrysler Crown Imperial. (Fig. 469)"

Figure 468. 1941 Cadillac Model 6227D air-conditioning evaporator Former owner and photo Doug Houston

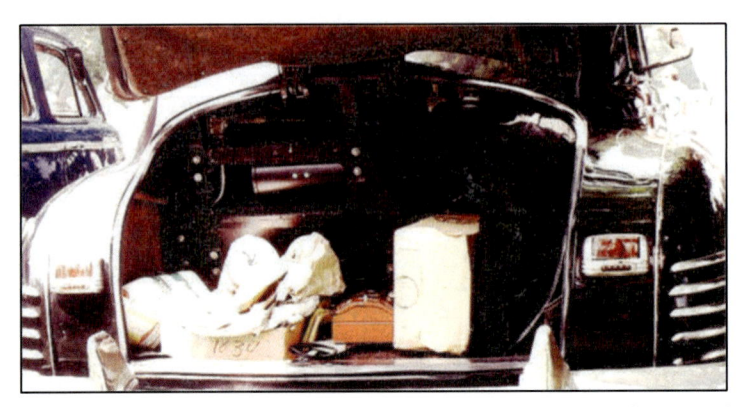

Figure 469. 1942 Chrysler Crown Imperial limousine air-conditioning evaporator   Owner unknown   Warren H. Erb

A question arises as to the source of Chrysler's smaller-sized evaporator. The Bishop & Babcock Mfg. Co. contracted with Packard, Cadillac, and likely Chrysler for their air-conditioning systems. The company remained obscure; none of the three makers ever acknowledged the Bishop & Babcock supplier name in any of their makes' air-conditioning literature. The obfuscation continued as it supplied the service and maintenance literature published under the Chrysler, Packard, and Cadillac names.

As stated previously, neither Packard's, Cadillac's, nor Chrysler's literature acknowledged the air-conditioning installation location. At least one writer believes Packard shipped the selected, nearly completely assembled cars to the Bishop & Babcock Mfg. Co. in Cleveland for the air-conditioned componentry installation. Without citing a reference, author L. Morgan Yost stated the following on p. 365 of *Packard: A History of the Motor Car and the Company:*

> Special insulation for cars so equipped was provided at the factory, but, although noted in Packard literature as "factory installed," the refrigerating unit itself was actually fitted into the car by Bishop & Babcock in Cleveland, and in theory, vehicles ordered with it were shipped there, the unit was installed and then the cars were forwarded to the dealers.

While Packard had no relationship with a refrigeration supplier, Cadillac and Chrysler did through Frigidaire and Airtemp, respectively. Cadillac's research with Dupont introduced Freon-12. In 1939, it installed a trunk-shaped air-conditioning system prototype on the rear of a 1938 Cadillac. In 1941, Cadillac followed Packard and contracted with Bishop & Babcock, not Frigidaire, to supply about 300 air-conditioning units for their cars.

The Chrysler Corporation owned the commercial refrigeration and cooling company, Airtemp Corporation. In 1930, Airtemp installed the first commercial air-conditioning in New York City's Chrysler Building. Is it possible that Chrysler collaborated with Bishop & Babcock to modify their 1941 and 1942 "Chrysler Air Refrigeration System" evaporator construction? Unfortunately, there is no answer.

The evaporator image shows that the service plate is missing, revealing the blower's housing above the evaporator's cooling coils, and does not appear to display a name badge on the outer housing. (Figs. 430, 469) Would the service plate carry the Chrysler Airtemp name and the Bishop & Babcock manufacturer name with two name badges, like the 1942 Packard evaporator (Fig. 470) or the Chrysler name badge *sans* the Bishop & Babcock name badge as displayed upon the 1941 Cadillac evaporator? (Fig. 471) Unfortunately, there is no answer.

Figure 470. 1942 Packard One-Eighty name badges on the air-conditioning evaporator
Former owner and photo Hyman Ltd. Classic Cars

Figure 471. 1941 Cadillac-Fleetwood Series 60 Special "Cadillac Air Conditioner" name badge on the air-conditioning evaporator
Owner Dr. Richard Zeiger   Ron Verschoor photo

With one exception, Mr. Gleason's caption list showed only one specific date: 1942. He referred to "Experimental" and "Conventional" in the other listing descriptions. The compressor images (*Chrysler Fig. 9,* Fig. 463, *Chrysler Fig. 10,* Fig. 464) lack resolution and size to identify Servel-based compressor equipment shown in the 1940–1942 Packard and 1941 Cadillac air-conditioning systems. (Figs. 472–475)

Notably, Mr. Gleason's presentation featured experimental testing with cool air distribution images along the headliner (Fig. 476 and yellow arrows), an air scoop for fresh air introduction into the passenger cabin via the evaporator (Fig. 481), and a smaller-sized evaporator. (Fig. 483)

Chrysler's experimental headliner-mounted cool air outlets and grilles appeared in a modified version of Oldsmobile's initial 1953 design for air-conditioning cool air distribution. (Figs. 477, 479, and yellow arrows)

Figure 472. 1940 Packard One Eighty Servel compressor
Owner and photo   West Peterson

Figure 473. 1941 Packard Model 1442 Servel compressor
Owner M/M Sal Saiya   Dave Czirr photo

Figure 474. 1941 Cadillac-Fleetwood Series 60 Special Servel compressor
Owner Dr. Richard Zeiger
Ronald Verschoor photo

Figure 475. 1942 Packard Clipper compressor Servel
Former owner General D. Macarthur
James Hollingsworth photo

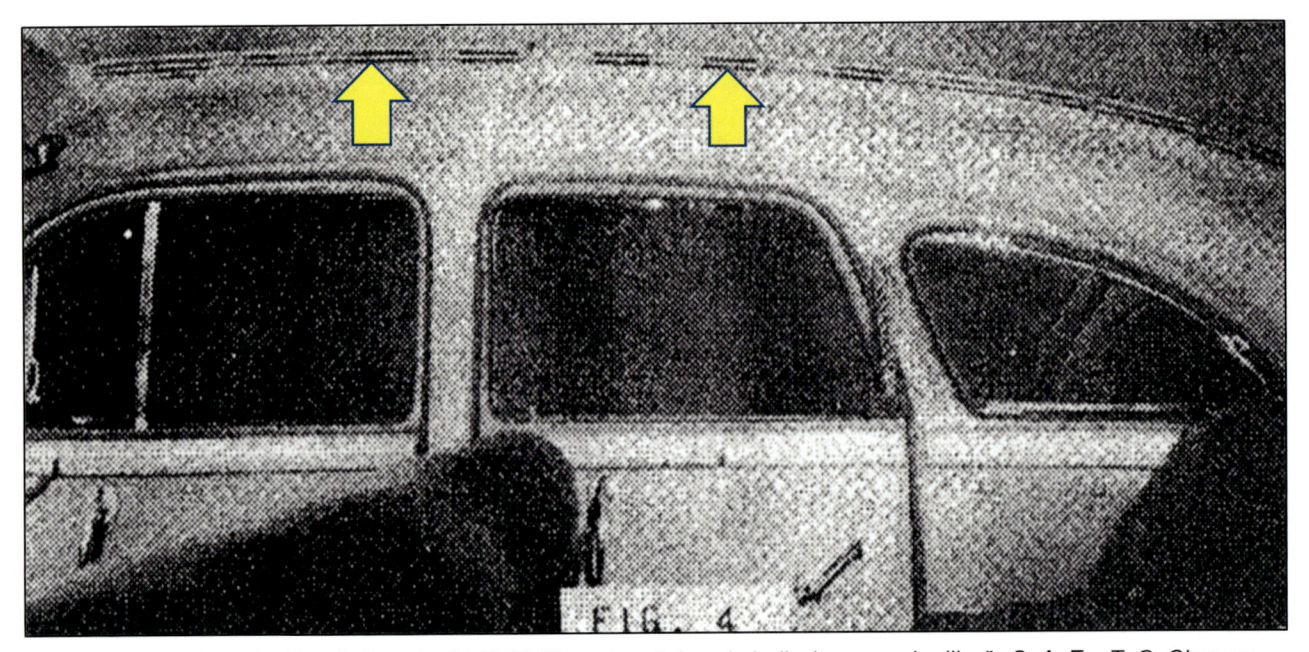

Figure 476. (Chrysler Fig. 4) Chrysler 1942 (?) "Experimental cool air discharge and grilles"   S. A. E.   T. C. Gleason

Factory Air: Cool Cars in Cooler Comfort 249

The initial 1953 Oldsmobile Frigidaire Car Conditioner blew cooled air into sleeves through approximately 1,500 holes (yellow arrows) and directional nozzles above passenger windows. (Figs. 477–479)

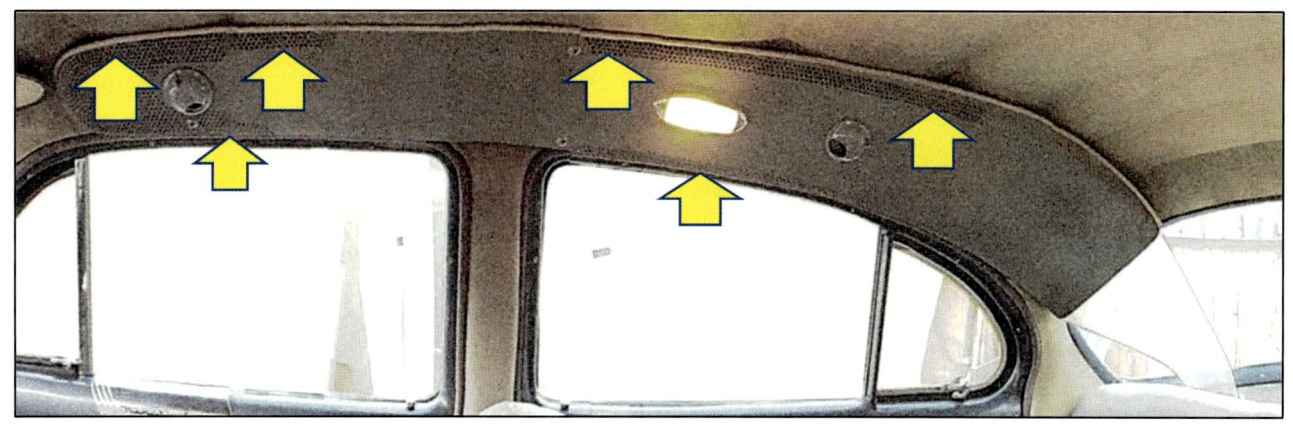

Figure 477. 1953 Oldsmobile headliner-mounted air-conditioning sleeve that contained multi-hole venting along the upper and lower border (yellow arrows), combined with directional passenger nozzles   Former owner Paul Wnuk

TINY HOLES—1,500 of them—in the side ducts that distribute cooled air through the car prevent drafts. Airplane-type vents are for those who prefer the air to hit their faces.

Figure 478. 1953 Oldsmobile air-conditioning offered "Tiny Holes" that delivered cooled air to passengers   GM MEDIA ARCHIVE

Figure 479. 1953 Oldsmobile close-up view of multi-holes for cool airflow distribution, coupled with a passenger nozzle   Former owner Paul Wnuk

Mid-year, Oldsmobile abruptly redesigned the cool air delivery method by installing louvered air vents that incorporated a nozzle for each window passenger. (Fig. 480)

Figure 480. 1953 Oldsmobile features redesigned, louvered air-conditioning vents with passenger nozzles   Owner and photo Phil Gaffney

Chrysler's body-mounted, fresh air scoop closely resembles the later design of the 1953 Cadillac equipped with factory air-conditioning. (Figs. 481, 482)

Figure 481. (Chrysler Fig. 5) Chrysler 1942 (?) "Experimental fresh air intake for cooling unit," similar to the later design of the 1953 Cadillac fresh air scoop (Fig. 482)   S. A. E.   T. C. Gleason

Mr. Gleason presented another version of an "experimental blower/evaporator air cooling unit." (Fig. 483, *Chrysler Fig. 7*)

*Figure 482. 1953 Cadillac Fleetwood Series 60 Special fresh air intake scoop, a similar design first shown on "Experimental" Chrysler designs
Owner Robert Ober   ABS photo*

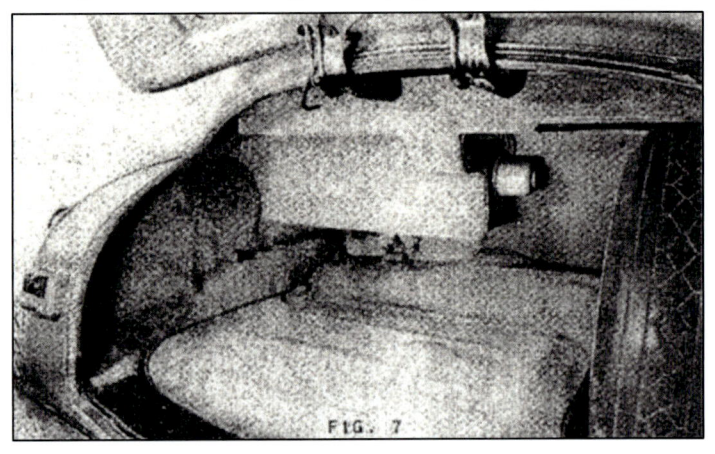

Fig. 483. (Chrysler Fig. 7) Chrysler 1942 (?) "Experimental blower/evaporator air cooling unit"   S. A. E.   T. C. Gleason

Overall, the Chrysler Air Refrigeration System sales appeared minimal, compared with Packard's and Cadillac's sales. During the prewar years, automotive air-conditioning availability initiated a spark of enthusiasm for more comfortable passenger travel, though limited to wealthy owners.

For the few owners of **FACTORY AIR**-equipped Chryslers, each believed they owned one of the

COOL CARS IN

COOLER COMFORT

LOOKING FORWARD: CHRYSLER AIRTEMP AIR-CONDITIONING for 1953

VOLUME 1: THE INNOVATIVE YEARS, 1940 to 1942, reflected the power of refrigeration engineering adapted to passenger car comfort. Packard, Cadillac, then Chrysler offered this luxury option that cooled and dehumidified the passenger cabin. The initial momentum ended in February 1942, as President Roosevelt and the United States Congress halted all automobile sales in favor of war materiel manufacturing. Automotive air-conditioning remained a low priority until the late 1940s when the population and the economy rebounded from the war.

In 1948–1949, a few wealthy Texas customers requested auto air-conditioning from a Ft. Worth company and Lone Star Cadillac of Dallas. The companies installed engine-mounted compressors, tubing, and trunk-mounted evaporator cooling coils similar to the prewar units. The auto air conditioners sold. The auto air conditioner business boomed. Several independent brand air conditioner manufacturers entered the competition to cool southwestern autos.

Follow the journey in VOLUME 2: 1953–THE MAGICAL YEAR.

Detroit automakers introduced factory air-conditioning in the following cars:
1. Chrysler Corporation introduced Airtemp factory air for the Chrysler, Imperial, De Soto, and Dodge.
2. General Motors introduced Frigidaire factory air for the Cadillac, Buick, and Oldsmobile.
3. Packard contracted with the Frigidaire Division of General Motors.
4. Ford Motor contracted with NOVI, an independent company, for the Lincoln.

Six full-color chapters will present the 1953 factory air-conditioned cars, supplemented with full-color images from factory literature, service publications, owners' literature, magazine testing, and advertising while spotlighting privately owned 1953 air-conditioned cars.

FACTORY AIR:

COOL CARS IN

COOLER COMFORT

AN ILLUSTRATED HISTORY OF AUTOMOTIVE FACTORY AIR-CONDITIONING
VOLUME 2: 1953–THE MAGICAL YEAR

Continue with VOLUME 3: UP-FRONT FORERUNNERS and OLD-STYLERS,

featuring 1954–1956 factory air-conditioned cars.

Continue with VOLUME 4: UP-FRONT LATE BLOOMERS,

featuring 1957–1960 factory air-conditioned cars.

Index

"Spotlight" Series and other factory air-conditioned automobiles

1940 Packard Custom Super-8 One Eighty, owned by Peterson, West, 8, 34–39, 41–47, 49, 94, 131, 137

1941 Cadillac Fleetwood Series Seventy-Five, previously owned by Kughn, Richard, 151, 175, 192, 200–202

1941 Cadillac Fleetwood Series Sixty Special, owned by Zeiger, Dr. Richard, 147, 151, 175–188, 247–248

1941 Cadillac Series Sixty-Two De Luxe Coupe, previously owned by Houston, Doug, 150, 152, 175, 181, 189–199, 246

1941 Packard Clipper Custom, owned by Weiss, Terry, 78, 89–95, 128, 137

1942 Cadillac Fleetwood Series Seventy-Five, previously owned by Selznick, David O., 203

1942 Chrysler Crown Imperial Limousine, owner unknown, 213, 217, 219–222, 244, 246

1942 Packard Clipper Eight, previously owned by MacArthur, General Douglas, 121–130

1942 Packard One Eighty Touring Sedan, previously owned by Hyman Ltd. Classic Cars, 130–135, 246–247

Dates

1930 Nash 8, air-conditioning by Kelvinator unit on rear, Hamman, John Jr., 5

1934-1940 Packard lower-priced models percent of sales, 23

1934-1940 Packard sales, 23

1939 Cadillac prototype air-conditioner, 6, 153

1940 Chicago 40th National Automobile Show, AKA Chicago Auto Show, 11

1940 Chrysler Crown Imperial "Air-Conditioning," and "All Weather Aircontrol," ad, 216
"Temperature Control" panel, 214

1940 Packard Accessory Listing, brochure, heater prices, 32

1940 Packard "Air-Conditioned" emblem accessory, 78

1940 Packard Custom Super 8 One Eighty, hardbound brochure, 35

1940 Packard Custom Super-8 One Eighty, air-conditioned, owned and photos by Peterson, West, 34–39, 41–47, 248
  air-conditioner air filters, 44, 47
  auxiliary seating, 35, 38–39
  blower grille, 30, 42–43
  Bonnet Louver Nosepiece Medallion, 36
  Cormorant hood ornament, 34, 36
  CPI (Consumer Price Index), Touring Sedan price; 2021 equivalent, 51
  discharge grille, 24
  evaporator, inspection plate removed to reveal cooling coils, 42
  evaporator compared to Henney-Packard Ambulance, 8

Index 253

expansion valve, 42

grille comparisons, 35–36

insulation installed in the air-conditioned car: front floor, kick panel, underneath seat frame, rear seating area, 44–45

insulation close-up views, 46

insulation, cowl, firewall, 45

lap robe for passengers, 34, 39

mouton carpeting, optional, in the car, 34, 38–39, 43–44

    brochure image, 39

"Packard Weather Conditioner," "The Bishop & Babcock Mfg. Co.," nameplates on the evaporator, 42

paddle switch, four-speed, on "Temperature Control" panel, 3, 30, 37

Peterson, West, photo, 44

profile view, twilight setting, 38

rear compartment floor-mounted heater grille, 43

rear seat cushion removal with labels, 44

rear seating with air-conditioning return airflow space underneath, 24, 30, 43–44

rear seating, auxiliary seating, air-conditioning blower grille, lap robe, 34, 39

rear seating, auxiliary seating, floor-mounted heater grille, air-conditioning blower grille, 38

receiver-reservoir tank, 42

return airflow space under the rear seat, 43

road test of similarly equipped air-conditioned Packard, 49–50

seating, auxiliary, 34–35, 38–39

Servel air-conditioning compressor, 41, 45, 131, 248

U-shaped support bolt, 41

"Tenite" dash with views of air-conditioning "Temperature Control" panel, 37

1940 Packard One Eighty Club Sedan, air-conditioned, previous owner, Hollingsworth, James; photos by Temple, David W., 25–26

ads, 20, 33

blower grille and louvers, 26

CPI (Consumer Price Index), Club Sedan price; 2021 equivalent, 51

heater, 20

deleted heater core redesign, 26

left front view in Miami Sand color, 25

"Packard Weather Conditioner," "The Bishop & Babcock Mfg. Co.," nameplates on the evaporator, 26

    "Summer Cooling," and "Winter Heating," instructions on evaporator nameplate, 26

right rear view in Miami Sand color, 26

1940 Packard One Sixty Weather Conditioner demonstrator, 17

1940 Packard One Twenty Deluxe Touring Sedan, owner Bates, Nelson; photos by Simons, Allen B., 18

CPI (Consumer Price Index), Deluxe Touring Sedan price; 2021 equivalent, 51

1940 Packard phantom view of the compressor, condenser, receiver, evaporator coil, expansion valve, conditioned air, 24
    discharge grille, blower, heater, air filter, return airflow under the rear seat, damper, 24
1940 Packard "Weather Conditioned" cars, 40, 108
1940 Packard "Weather Conditioner," 8-9, 17, 19, 25–26, 28–32, 49, 51, 108, 206
1940 Senior Packards brochure cover, 32
1940–1942 Cadillac, Chrysler, and Packard evaporators, 246
1940–1942 Packard air conditioning units sold, 10, 90, 206
1940–1942 Packard grilles, 137
1940–1942 Packard retail prices of "Weather Conditioner" and "Air Conditioner":
    price of $274, 14–15, 51, 61
    price of $275, 27, 61
    price of $325, 61, 114
1940–1942 Packard sales, 52

1941 Cadillac Fleetwood Series Seventy-Five, air-conditioned, previously owned by Kughn, Richard, 151; 175, photo by Roy, Rex; 191–192; 196, photo by Houston, Doug; 200, photos by Roy, Rex; 201–202, photos by Dickirson, Gene and Kughn, Richard
    1941 Cadillac brochure, Fleetwood Series Seventy-Five, 151
    air conditioning control knob, 202
    air-conditioning cooling register, 200, 202
    air-conditioning cooling register with louver control handle, 202
    evaporator, 193, 202
    left front view, 175
    rear seating and cooling register, 202
    right front view, 200
    Servel air-conditioning compressor with engine bay views, 196, 201

1941 Cadillac Fleetwood Series Sixty Special, air-conditioned, owned by Zeiger, Dr. Richard; photos by Verschoor, Ronald, 147, 175–188, 247–248
    1941 Cadillac brochure, Fleetwood Series Sixty Special, 151
    air-conditioning control knob, 180
    air-conditioning cooling register with louver control handle, 181
    "Cadillac Air Conditioner" nameplate on evaporator, 182, 247
    Cadillac crest, block letters, and "Flying Lady" hood ornament, 177
    Cadillac crest on the trunk lid, 183–184
    Cadillac wheel disc with inset red medallion, 184
    expansion valve, 182
    "Fleetwood" logo on the door sill, 184
    "Fleetwood" script on the front fender, 184
    "Flying Lady" hood ornament, 147, 176–177
    gas filler cap, concealed, open and closed, 185
    left front view, 175

left front view, with "Sunshine Turret Top" sunroof open, 176

left rear view, 183

passenger side dash view, 185

rear seating; view through "Sunshine Turret Top" sunroof; and convenience features, 186–187

Servel air-conditioning compressor with engine bay views, 178–179, 248

"Sunshine Turret Top" views, accessory price listing, and *Cadillac Data Book* description, 151, 175–176, 180, 187–188

three-piece backlite, 183

Venetian blinds, 180–181

ventilated hood louvers, 177

Zeiger, Verschoor photo, 188

1941 Cadillac Fleetwood Series Sixty Special designer, Mitchell, William L. (Bill), 147

1941 Cadillac phantom views of the compressor, condenser, cooling unit, dehydrator-reservoir, expansion valve, 154–155, 164

    Compressor image, 165

1941 Cadillac Series Sixty-Two De Luxe Coupe, air-conditioned, previously owned by Houston, Doug, 152; 175, photo by Houston, Doug; 181; 189, photo by Houston, Doug; 190–192; 193, photos by Houston, Doug, Kughn, Richard and Varnon, Buck; 194; 195–196, photos by Houston, Doug; 197, photos by Braun & Helmer Auction Service; 198–199, photos by Edsall, Larry and Peterson, West; 246, photo by Houston, Doug

    1941 Cadillac brochure, Series Sixty-Two Coupe, referral, image not shown, 150

    air-conditioning cooling register with closed louvers, 199

    evaporator, expansion valve, and "Cadillac Air Conditioner" nameplate, 198, 246

    lower left dash view of the unmarked air-conditioning control knob, 197

    "Rare Factory AC" and "Hydra-Matic" sign, 199

    right front view, 197

    right profile, 197

    Servel air-conditioning compressor, 193, 198

Cadillac & LaSalle Club Photo Gallery image, 189, 195, photos by Houston, Doug

cadillaclasalleclub.org Technical Authenticity Forum, 2006, January 23, 194

    "A/C in Caddys-When?" email to Houston, Doug, 194

        "Bishop & Babcock, Cleveland," 194

        "Cadillac built 300 jobs in '41 with air-conditioning," 194

        "Cooling is reasonably good." 194

        "There is condensation on the outer surface of the backlite." 194

        "The system has a tendency to frost over." 194

        "Three (1941 air-conditioned Cadillacs) are known today," 194

cadillaclasalleclub.org Technical Authenticity Forum, 2014, February 7, 196, photos by Houston, Doug

    air-conditioning cooling register showing louver control handle, 196

    "Cadillac Air Conditioner" nameplate close-up view, 196

    evaporator with "Cadillac Air Conditioner" nameplate, 196

    left front view, 196

Servel air-conditioning compressor with a photo of Kughn's, Richard, compressor, 196

"Cadillac's Air Conditioner" (1941 Cadillac), *Cadillac-LaSalle Self-Starter Annual, Vol. XVII (1991)*, written by Houston, Doug, 190, edited by Juneau, Bud, contributed by Wenger, Terry, Sr.; 191-192; 193, photos by Houston, Doug, Kughn, Richard, and Varnon, Buck

    300 air-conditioned Cadillacs produced in 1941, 191

    "air outlet tends to chill the backlite," 191

    "Air-Conditioned" emblem not displayed on Cadillac, 191

    air-conditioning control, 191; 193

    air-conditioning cooling register with louver control handle, 193

    "Bishop & Babcock," air-conditioning designer, 190

    "clutching pulley not available," 190

    condenser, 191

    evaporator, 1941 Cadillac Fleetwood Series Seventy-Five, 193, previously owned by Kughn, Richard

    expansion valve, evaporator with "Cadillac Air Conditioner" nameplate, 193; 196, photo by Houston, Doug; 198, photo by Peterson, West; 246, photo by Houston, Doug

    Servel air-conditioning compressor, 193; 198, photo by Peterson, West

    "Shipping Orders courtesy of Haas, Al, Cadillac Motor Division,"

        1941 Cadillac, Style No. 41-6227D, air-conditioned, Houston, Doug, 192

        1941 Cadillac, Style No. 41-7519, air-conditioned, Kughn, Richard, 192

car-nection.com, *Original Cadillac Database*, Sanders, Yann, 189

1941 Chrysler Crown Imperial air-conditioning brochure, 223

    ad, 224

1941 Chrysler receiver, reservoir tank, 207

1941 New York National Auto Show, 70

1941 Packard Air-Conditioning Operation, Service, and Owner's Instruction, 40, 97–108

1941 Packard Clipper, 61, 86–88, 96–97

    ad, "First streamlined car planned for 'lookers' and riders," 87

    "A normally tall man can easily see over it," 88

    Darrin, Howard "Dutch," Clipper styling, 93

    FadeAway front fenders, 86

    grille and rear views, 87

    interior dimensions, 88

    Packard Clipper Data Book, April 1941, 88

    profile view, 86–87

    Servel air-conditioning compressor, 92

1941 Packard Clipper Custom, air-conditioned, owned by Weiss, Terry, 89–95; photos by Simons, Allen B.

    "Air-Conditioned," bonnet emblem, 78, 89, 93, 95

    cool air outlet, 89–90, 93, 95

    "Cooling," two-speed push-pull switch on dash, 91

        switch drawn in error for 1946 literature, 91

    Darrin, Howard "Dutch," Clipper styling, 93

    evaporator, expansion valve, 94

comparison to 1940 Packard evaporator, 94
FadeAway front fenders, and tapered trunk line, 93
grille views, 89, 92–93, 95, 137
hood opens from either side, 92
interior front and rear seating, 90–91
profile views, 89, 93
Sanden air-conditioning compressor, engine bay, 92
Servel air-conditioning compressor, 92
"That Packard Look," 93
Weiss, Terry, owner photo, 93

1941 Packard Clipper sales, 86

1941 Packard One Eighty, Body 1442, air-conditioned, owned by Saiya, Sal, and JoAnn, 78–80; photos by Czirr, Dave, Heinmuller, Dwight, and Swaney, Jack; 137, photo by Swaney, Jack; 246, 248, photos by Czirr, Dave
"Air-Conditioned," emblem on the side panel, 78
air-conditioning control knob, 79
air conditioning cool air outlet, 80
evaporator and nameplates for "Packard Air Conditioner" and "The Bishop & Babcock Mfg. Co.," 80, 246
right front views, 78–79
Servel air-conditioning compressor views, 79, 248

1941 Packard One Eighty, Body 1442, air-conditioned, with Cellarette bar, previously owned by McGahey, David (Dave), E.; passengers, Simons Story, Annette, and Story, Jim, 76
1941 Packard One Eighty, Body 1442, owned by Baier, Royce, and Sheila, 77; photos by Kietzman, Charlie
left profile, left front views, 77
"Robe Compartment," storage bin views, 77

1941 Packard One Ten Deluxe Touring Sedan, ads, 62, 66, 81

1941 Packard One Twenty Club Coupe, air-conditioned, previously owned by Hollingsworth, James, 85; photos by Temple, David W.
air-conditioning control knob, 85
cool air outlet with louver control handle, 85
evaporator and nameplates for "Packard Air Conditioner" and "The Bishop & Babcock Mfg. Co.," 85
right rear view, 85

1941 Packard phantom view of the adjustable louvers, air filter, blower fan, compressor, condenser, cooling coil, expansion valve, receiver, strainer, thermostatic bulb, 98
1941–1942 Cadillac air-conditioning units sold, 191, 203, 206
1941–1942 Chrysler air-conditioning units sold, 206, 217, 219–222

Index 258

1942, February 9, automobile production ceased for WWII, 10, 52, 136, 145, 175

1942 Cadillac Series Seventy-Five Limousine, air-conditioned, previously owned by Selznick, David O., 203

1942 Cadillac Series Sixty-Two pontoon fenders, 118

1942 Chrysler Crown Imperial Limousine, air-conditioned, owner unknown, 213, 219–222
    evaporator, left rear view, right front view, 219–220; photos by Erb, Warren H.
    left front view, rear view, right front view, standard, and accessory equipment list, 221–222; photos by OIC (Online Imperial Club) member Tony

1942 De Soto brochure cover, 241

1942 "De Soto Refrigerated Air-Conditioning!" 243
    cooling air that "is 10°F to 15°F cooler than outside air!" 243
    "thermal shock," Bhatti, M, S., Ph.D., 6, 159, 243

1942 Packard Clipper ad, "Brand new Clipper styling in every Packard price range," 112

1942 Packard Clipper Eight, air-conditioned, previously assigned to Eisenhower, General Dwight, 130

1942 Packard Clipper Eight, air-conditioned, previously owned by MacArthur, General Douglas, 121-129,
    "Destined to Survive" (1942 Packard), *Cormorant, The,* Spring 1975, article and photos by Hollingsworth, Jim
    air-conditioner control, "Cooling" switch, 128; 128, "Cooling" switch, owned by Weiss, Terry; photo by Simons, Allen B.
    canvas backlite cover, 124
    cool air outlet with louver control handle, 127
    Cormorant painted hood ornament; driving lights, auxiliary; headlamp shutters; siren; Packard delivery plate; 126
    Cormorant, The, Spring 1975, cover, 121
    corn-cob pipe, MacArthur, General Douglas, 128
    evaporator, trunk-mounted, with illegible nameplates, 127
    five-star flag and plate for General of the Army, MacArthur, General Douglas, 122, 124
    grille, painted, displays optional accessories, 122, 124, 126
    MacClellan, Don, owner of the car from 1968, per 1975 article, "Destined to Survive," 129
  Packard images, 123, 129
    Packard Motor Car Company gift letter of car, 123
    Servel air-conditioning compressor, 127
    U. S. Government motor vehicle release, 125
    weaponry mounted between front seat and dash, canteen, first aid kit, fire extinguisher, 128

1942 Packard One Eighty Limousine, air-conditioned, previously owned and photos by Fluckiger, Paul, 139
    air-conditioning control knob label, "See warning notice on unit in trunk," 138–139
    Methyl Chloride refrigerant required in "Warning notice on unit in trunk," *Packard Service Letter,* 138

1942 Packard One Eighty Touring Sedan, air-conditioned, previously owned, and photos by Hyman Ltd. Classic Cars, 113, 130–135, 137, 246–247
    air-conditioning compressor on car, compared with 1940 Servel air conditioning compressor, photo by Peterson, West, 131
    air-conditioning control knob, 131
    automatic window regulators (power windows), 134
    Bonnet Louver Nosepiece Medallion, 135
    cool air outlet with louver control handle, 131–134
    Cormorant hood ornament, 135
    driver seating, 131, 134
    evaporator and nameplates for "Packard Air Conditioner" and "The Bishop & Babcock Mfg. Co.," 133
    left front view, 113
    rear seating, cool air outlet with louver control handle, 132
    right front view, 130
    "Speedline Stripes," 135

1942 Packard One Sixty for U.S. Army use, 130
1942 Packard sales, 136
1953 Cadillac Fleetwood Series Sixty Special, air-conditioned, owned by Ober, Robert, 250; photo by Simons, Allen B.
1953 Oldsmobile Ninety-Eight, air-conditioned, owned and photo by Gaffney, Phil, 43, 249
1953 Oldsmobile Ninety-Eight, air-conditioned, previously owned and photo by Wnuk, Paul, 249
1954 Buick, air-conditioning compressor with electromagnetic clutch, 57
1954 Cadillac, air-conditioning compressor with electromagnetic clutch, 57
1954 Nash, air-conditioning compressor with electromagnetic clutch, 57
1954 Oldsmobile, air-conditioning compressor with electromagnetic clutch, 57
1954 Pontiac, air-conditioning compressor with electromagnetic clutch, 57
1955 Packard, air-conditioning compressor with electromagnetic clutch, 57
1964 Cadillac *Climate Control* air conditioning, 58

A

AACA and Library (Antique Automobile Club of America, and Library & Research Center, Hershey, PA), 32, 35, 46–47, 52–60, 109–111, 114–115, 198, 212, 225, 244
ABS (Simons, Allen B., Collection of,), 7, 11, 13, 18, 20, 24–25, 48, 75, 78, 81, 87, 89–96, 120, 128, 136–137, 145, 148–152, 205–211, 218, 241–242, 250
Adelson, Kathy, 158–161
Air conditioner, Air-conditioning:
    1940-1942 compressors, 248
    air filtered, air filters, disposable, filtering, filters, filtration, 2–4, 7, 9, 11, 21, 24, 30–31, 39, 43–44, 47–48, 50, 53–56, 58, 83, 85, 98, 100, 105, 109, 114–115, 117, 120, 143, 154, 156, 159–160, 162, 166, 206, 210, 218, 229, 241
    disposable cabin filter, 48
    airflow, 3–4, 26, 34, 43, 48, 68, 85, 109, 116–117, 155–156, 162, 181, 241, 249

"conditioned air," circulation, 30

"return airflow under rear seat," 24, 30, 43–44, 117

"blower grille," or "cool air outlet," or "cooling register," or "fan outlet grille," 3–4, 26, 29, 34, 38–39, 42–44, 54–56, 67, 80, 83, 85, 91, 109, 116, 133, 156, 159–160, 180–181, 200, 218, 241, 244–247, 250

circulates, 53, 83, 120

componentry phantom views, 3, 12, 14, 24, 26, 28, 30, 55, 64, 116, 120, 207, 210, 246

compressor, 3–4, 7, 12, 14, 19, 21, 24, 29–31, 40–42, 45, 49, 55–57, 64, 67, 79, 84, 92, 98–109, 115–117, 120, 127, 131, 140–142, 155, 158, 161–174, 178–179, 190, 193–194, 196, 198, 201, 207, 209–210, 213, 215, 217–218, 226–228, 230–236, 238–241, 244–245, 248, 251

    Servel air-conditioning compressor, 4, 7, 14, 41, 45, 79, 92, 109, 127, 131, 158, 162, 178–179, 196, 198, 201, 217, 241, 248

compressor electromagnetic clutch, 57, 117, 178

condensate, 48, 218

condenser, 3, 14, 24, 29–30, 36, 42, 55–56, 84, 98–99, 101–102, 105, 109, 116, 120, 142, 155, 161–165, 169–170, 178–179, 191, 207, 210, 226–228, 231–232, 238, 241, 244–245

control knob, 79, 85, 131, 180, 197, 202

    blower switch, 159

    control switch, 164

control panel, four-speed paddle switch, "Temperature Control" panel, 3–4, 37, 44, 55, 57, 109, 117, 178, 180, 207, 210

"cool air outlet," name changed in 1941 Packard from 1940 "blower grille," 56

cooler damper inside evaporator, 30

cooling

    capacity, 4, 6, 11–14, 21, 24, 29, 31, 50, 54, 57, 67–68, 82, 117, 157, 208, 217

    coils, 3–4, 9, 14, 24, 30, 42, 44, 48, 56, 83, 90, 94, 109, 116–117, 120, 133, 155–156, 160–162, 180, 182, 198, 207, 218–219, 241, 247, 251

        "Cooling" switch, 91, 128

        dehumidification, dehumidifies, 31, 48, 53, 56, 83, 117, 120, 160, 206

        effectiveness, 31, 57, 117, 208

        performance, 6

        principle, 31, 41, 50, 115

        Review: "during milder weather, (it) chilled rear seat passenger's feet and ankles" (1941 Chrysler), 218

        evaporator nameplate(s), 26, 42, 80, 85, 131, 133, 182, 196, 198

evaporator, trunk-mounted, 3–4, 8–9, 14, 21, 24, 26, 42–44, 48, 54, 56–57, 63–64, 67, 80, 84–85, 91–92, 94, 97, 109, 116–117, 120, 127, 131, 133, 153, 155, 161–162, 178, 180, 182, 189, 196, 198, 202, 207, 210, 217–219, 241, 244–248, 250–251

"Exclusive," air-conditioning system (1940–1941 Packard), 2, 16, 21, 31–32, 34, 54

expansion valve, 3, 14, 24, 29, 42, 56–57, 94, 98, 102, 105, 109, 116, 133, 155, 161–163, 166, 168, 182, 198, 209, 241

"factory air," 1, 24, 43, 46, 48, 50–52, 55, 58, 89, 109–110, 121, 146–147, 150, 152, 176, 188–189, 195, 200, 203–205, 209, 211–212, 217, 221, 225, 244, 250–251

Factory Air: Cool Cars in Cooler Comfort, 24, 51, 58, 109, 146, 204, 211–212, 250–251 factory installation, installed, 2–4, 6–8, 11–14, 31, 35–37, 41, 43, 45, 48, 50, 57–58, 75, 82, 85, 91–92, 109, 117, 131, 146, 151, 153–154, 158, 162, 196, 204, 209–210, 217, 241, 244–245, 247

factory installation contractor location of air-conditioning componentry, 57, 117, 158, 217, 247

Freon, Freon-12, 3, 6, 14, 42, 56, 58, 72, 103–104, 106, 109, 116, 138–139, 153, 155, 157–158, 160–162, 164–165, 169–170, 217, 241, 247

"genuine Air-Conditioning" (1940–1942 Packard), 1-2, 4, 8, 51, 53, 67, 119

"Genuine Air-Conditioning" dealer album (1941 Packard), 66–67

"genuine cooling system for summer" (1941 Chrysler), 223

"genuine mechanically operated system" (1941 Chrysler), 206

handle, for louver control on the cool air outlet, cooling register, or fan outlet grille, 56, 85, 127, 132, 181, 193, 196, 202

    1941 Packard factory photo of the new louver control handle, 85

heater unit, removal from the evaporator, 7–9, 11–14, 20, 24–32, 35, 37–38, 42–43, 50, 61, 91, 114, 121, 154, 160, 180, 202, 208, 217–218, 244

    heater and floor grille, rear compartment, 38, 43

"Packard Weather Conditioner," name changed to "Air Conditioner," 32

receiver, dehydrator, 3, 14, 29, 42, 56, 84, 98, 101, 104, 109, 120, 155, 162–163, 170, 174, 207, 241

refrigerates, 53, 67, 83

retail prices of "Weather Conditioner" and "Air-Conditioner" price of $274, 14–15, 51, 61

    price of $275, 27, 61

    price of $325, 61, 114

"This Is What I Call Cool Comfort" (1941 Packard) brochure, 81–84

"Simple adjustment," to change heating to cooling (1940 Packard Weather Conditioner), 31

"Air-Conditioned," emblem (1940–1941 Packard, Clipper), 78, 89, 93, 95

"Air-Conditioned," emblem not displayed on 1941 Cadillac, Houston, Doug, 191

Air-Conditioning & Refrigeration News, 12

"Air-Cool-ditioned" ad (1940 Packard), 12, 33, 39

Allen, A. H., 70

"All-Weather Aircontrol," and "Air-Conditioning" (1940 Chrysler Crown Imperial) ad, 216

American Society of Heating, Refrigeration and Air-conditioning Engineers (ASHRAE), Journal, 5–6, 153, 159, 243

Anthony, Earle C., 66

Antique Automobile, 198

Antique Automobile Club of America, and Library & Research Center, Hershey, PA, (AACA and Library), Hocker, Matt, Reilly, Mike, and Ritter, Chris, 32, 35, 46–47, 52–60, 109–111, 114–115, 198, 212, 225, 244

ARA auto air conditioners, 175

Ark-La-Tex Packard Club of Texas, 18, 85

"Ask the Man Who Owns One" (1940–1942 Packard), 20, 22, 31, 33, 62–63, 69, 87, 114

Austin American, Texas, 13

Automotive Air-conditioning Association, factory air-conditioning milestone sales years: 1953, 50

Automotive Climate Control 116 Years of Progress, Dickirson, Gene, 201–202

Automotive Industries, 13, 63–64, 87

Automotive News, 10, 21–22

Weather Conditioner: "lower the temperature inside a car 19 degrees below that of the outside air," 21

auxiliary seating, 34–35, 38–39

Ayres, Paul, 109, 162–174

B

Baier, Royce, and Sheila, 77
    Robe Compartment storage bin, 77

Barclay, Rod, *Boy! That Air Feels Good!* 175

Bates, Nelson, 18

Benson Ford Research Center, The Henry Ford, 53, 111–114

Bhatti, Mohinder S., Ph.D., 5–6, 159, 243
    "thermal shock," 6, 159, 243

Bishop & Babcock Mfg. Co., 6–7, 19, 22, 41, 45, 48, 57, 79–80, 85, 90, 92, 108–109, 117, 131, 133, 153, 158, 162, 175, 178, 181, 216–218, 225, 241, 247
    "Bishop & Babcock Weather Conditioner," 7
    evaporator nameplates, 26, 42, 80, 85, 131, 133, 182, 196, 198

Block, Joe, 55, 71–72, 74

Boeing 314 Clipper, Pan American World Airways, 96, 111

Bonnet Louver Nosepiece Medallion, 18, 36, 135

Boyer, Wesley, 63–64, 87

Bradley, General Omar, 130

Bradley, James J., 10

Braun & Helmer Auction Service, 197–199

"Breath from the Northwoods, Chrysler's Air-Conditioned Cars," Erb, Warren H., 213–215

"Brimming with Beauty, Bursting with News!" ad, 62

British Thermal Units (BTU), 6

C

Cadillac Air-Conditioning Manual (1941), 109, 158, 161–174

Cadillac & LaSalle Club (CLC), 154–157, 189–196

Cadillac & LaSalle Club Museum and Research Center (CLCMRC), Ayres, Paul, 109, 162–174
    Cadillac Data Book (1941), 154–157, 181,187
    cover, GM Heritage Center, 154
    "Sunshine Turret Top sunroof," GM Heritage Center, 187
    Supplement pages, Rauch, Warren, 154–157
        Cadillac Air-Conditioning, 154
        Comfortable Interior Temperature, 157
            adequate insulation, 157
            immediate operation, 157
            large capacity, 157
        Controlled Air Circulation, 156
            compressor, 155

Index 263

        condenser, 155

        expansion valve and cooling coils, 155

        reservoir, 155

    Simple in Operation, 155

cadillaclasalleclub.org Technical/Authenticity Forum, 2006, January 23, 194

cadillaclasalleclub.org Technical/Authenticity Forum, 2014, February 7, 196

Cadillac-La Salle Self-Starter Annual, (1991), Juneau, Bud, Editor, 190–193, 202

Cadillac Milestones 1902-1942, 149

Cadillac prototype air-conditioner, 1939, 6, 153, 247

"Cadillac's Air Conditioner" (1941 Cadillac), written by Houston, Doug; contributed by Wenger, Terry, Sr., 190–193, 202

Cadillacs of the Forties, Schneider, Roy, 152, 191, 203

Carpeting, mouton, 34–35, 38–39, 43

CCCA (Classic Car Club of America), 18, 188

Cellarette, 1941 Packard, Body 1442, 69–77, 81, 84

    cabinet, 71, 76–77, 84

    cabinet, lower right door open, 71

    Chicago Auto Show flyer for the Cellarette, "See the Motor Car That Makes Ice Cubes," 69

    "Conditioner" (air) or "Cellarette" switch, 72

    demonstration by a woman mixing a cocktail, 73–74

    ice cube tray, glassware, liquor flasks, mixers for six passengers, 72–74

    "ice cubes may be frozen in both trays in 15 to 20 minutes," 72

    "Make Ice Cubes as You Drive with a Packard Cellarette," brochure section, 84

    McGahey, David (Dave) E., the previous owner of a 1941 Packard with Cellarette, 76

        passengers: Simons Story, Annette, and Story, Jim, 76

    Packard Data Book for 1941, "Cellarette Supplement to page 80," 70, 72

        a woman mixing cocktails, 73

        a woman prepares cocktails, 75

CFC, 6, 153

Chicago Auto Show, 1940, AKA, "40th National Automobile Show," 11

Chicago Auto Show (1941) flyer, "See the Motor Car That Makes Ice Cubes," Frumkin, Mitch, 69

Christopher, George, 136

Chromed horizontal speedlines, 1941 Cadillac styling, 152

"Chrysler Air Refrigeration System" (1941), 109, 205–212, 225–240, 244, 246–247, 250

Chrysler Air Refrigeration System, Operation-Service (1941), AACA Library, 109, 225–240

"Chrysler Air Refrigeration System, The" (1941), brochure, 205–211

Chrysler, Walter P., 213

"Class of '41, the," 52–54, 58, 62, 66

Classic Car Club of America (CCCA), 18, 188

ClassicCars.com Journal, Edsall, Larry, 199, 242

Collection of Mauck, Fred, and Carol, *PackardInfo.com,* Waltman, Kevin, 33, 39

Collection of McPherson, Thomas A., Hanson, Howard, 10

Collection of Simons, Allen B. (ABS), 7, 11, 13, 18, 20, 24–25, 48, 75, 78, 81, 87, 89–96, 120, 128, 136–137, 145, 148–152, 205–211, 218, 241–242, 250

"Conditioned air" (1940 Packard), 3-4, 6, 17, 30, 43, 48

Cool air circulation (1941–1942 Packard), 54, 56, 116, 218

Cool air outlet louver control handle, new for 1941, 56, 85, 127, 132, 181, 193, 196, 202

"Cooled by Mechanical Refrigeration in Summer" (1940 Packard), 20-21

cooling cycle, the (1941 Packard), 56

Cormorant hood ornament (1940–1942 Packard), 18, 34, 36, 79, 135

    painted, during WWII (1942 Packard Clipper), 121, 124, 126

Cormorant, The, "Destined to Survive" (1942 Packard Clipper), written by Hollingsworth, Jim, 121–129

Cowl, firewall insulation, (1940 Packard), 45

CPI (Consumer Price Index Inflation Calculator), 51, 61, 114, 123, 152–153, 188, 203, 217

cubic feet per minute (CFM), 109, 162, 241

Czirr, Dave, 79–80, 246, 248

D

Darrin, Howard "Dutch," 87, 93

Defense Media Network, Zimmerman, Dwight Jon, 136

Dehumidification, dehumidified, 2–4, 6, 9–11, 13–14, 20–21, 24, 31–33, 39, 48, 50–51, 56, 67, 83, 117, 159–160, 206, 210, 243, 251

Detroit Public Library National Automotive History Collection (DPL NAHC), Minor, Romie, and Reczek, Carla, 14, 17, 36, 41, 49–50, 78, 85, 92, 96, 136

Dickirson, Gene, *Automotive Climate Control 116 Years of Progress,* 201–202

DuPont, 6, 153, 158, 247

Duricy, Dave, 243–244

E

Edsall, Larry, 199

80 Years of Cadillac LaSalle, McCall, Walter, 118, 148

Eisenhower, General Dwight, 130

Electromagnetic clutch, compressor, 57, 117, 178, 218

Erb, Warren H., 206, 212–215, 217–223, 225–240, 244, 246

Evaporator

    cooling coils, 24, 30, 42

    expansion valve, 3, 14, 24, 29, 42, 56–57, 94, 109, 116, 133, 155–156, 161–162, 182, 198, 209, 241

    nameplate, 26, 42, 80, 85, 133, 182, 196, 198, 247

    trunk-mounted, 3, 6, 9, 14, 42, 44, 56, 80, 84, 92, 116–117, 120, 127, 131, 133, 153, 175, 178, 180, 182, 196, 198, 202, 210, 217–219, 243, 245–246, 250–251

F

Factory air, Foreword, 1, 24, 43, 46, 48, 50–52, 55, 58, 89, 109–110, 121, 146–147, 150, 152, 176, 188–189, 195, 200, 203–205, 209, 211–212, 217, 221, 225, 244, 250–251

Factory Air: Cool Cars in Cooler Comfort, Preface, 24, 51, 58, 109, 146, 204, 211–212, 250–251

FadeAway fenders, 86–87, 89, 93, 112, 118

Fender-wheel shields (1942 Packard), 112, 183

"1st Air-Conditioned Automobile" (1940 Packard), *Air-Conditioning & Refrigeration News*, 12

"First Completely Weather Conditioned Car, The" (1940 Packard), brochure, 28–31

    phantom view of componentry, 3

Flack, David, 61

Floors (1940 Packard), 45, 59–60

Fluckiger, Paul, 79, 139

Fluid Drive (1941 Chrysler), 224

"Flying Lady" hood ornament, 147, 177

Fortune, 1940 Packard One Eighty Club Sedan ad, 20–21

"4-Year Plan," 22

Frigidaire, 6, 146, 153, 158, 204, 247, 249, 251

Frumkin, Mitch, 69

G

Gaffney, Phil, 1953 Oldsmobile Ninety-Eight, air-conditioned, 43, 249

Gas filler cap, concealed views (1941 Cadillac), 152, 185

General Motors, 6, 146, 148, 153, 158, 200, 203–204, 218, 251

"Genuine Air-Conditioning," dealer album (1941 Packard), 67–68

Gilman, M. M., 22, 61, 123

GM Heritage Center, 154, 187

GM Media Archive, Adelson, Kathy, 158, 161

Graves, W. H., Packard Vice President of Engineering, air-conditioning, 10, 90, 115, 206, 217

Grille comparisons, 1940 Packard, 35–36

Guscha, 13, 70

H

Hamlin, George, 23

Hamman, John, Jr., 5

Handle added in 1941 for louver control on the cool air outlet, cooling register, or fan outlet grille, 56, 85, 127, 132, 181, 193, 196, 202

    1941 Packard factory photo of the new louver control handle, 85

Hanson, Howard, 10, 91

Headlamps, faired-in, 53, 79, 81, 86

Heater, hot water, 9, 12, 14, 20, 25–29, 31–32, 35, 37–38, 42–43, 61, 91, 114, 121, 154, 158, 160, 180, 208, 218, 243–244

    heater and floor grille, rear compartment (1940 Packard), 38, 43

    heater damper, 30

    prices, 32

Heinmuller, Dwight, 8, 23, 28–31, 61, 71–73, 76, 78, 81–86, 97–103, 105–107, 115–117

Henney "Program of Progress" brochure (1940), 8

  Packard Weather Conditioner, 8–9

Henney-Packard Ambulance (1940), 7–9

    "genuine, mechanically refrigerated air-conditioning," 8

    Henney-Trane Ambulance (1938), 10

"Here's What an Air-Conditioned Packard Can Mean to You" (1940), 2

History Channel, The, 11

Hollingsworth, James, 9–10, 25–27, 39, 42, 85, 90, 121–129, 217, 248

    1941 Packard One Twenty Club Coupe, air-conditioned, previously owned by Hollingsworth, James, 85

    air-conditioning control knob, 85

    handle for cool air outlet louver control, 85

    nameplates on the evaporator, 85

Hollingsworth, James, *Packard 1940: A Pivotal Year*, 9–10, 25–27, 39, 46, 90, 217

Hollingsworth, Jim, "Destined to Survive" (1942 Packard), *Cormorant, The,* Spring 1975, 122–129

Hook, Andrew, via Kelley, Steven, 55, 70

Houston, city of, "World's Most Air-Conditioned City," *Preface,* 5–7, 243

Houston, Doug, 150, 152, 175, 181, 189–199, 202, 246

    1941 Cadillac Series Sixty-Two Deluxe Coupe, air-conditioned, previously owned, and photos by Houston, Doug, 189–199

    "Cadillac's Air Conditioner" (1941 Cadillac), 190-193, 202

"How Packard Air-Conditioning Works" (1940 Packard), 4

Hydra-Matic automatic transmission, 152, 191, 199

Hyman Ltd. Classic Cars, 1942 Packard One Eighty, air-conditioned, previously owned and photos, 113, 130–135, 137, 246–247

I

Ice cubes, Cellarette (1941 Packard), 69–76, 84

Inflation Calculator, Consumer Price Index (CPI), 51, 61, 114, 123, 152–153, 188, 203, 217

Information on 1942 Packard Cars, 114

Information on the 1941 Packard, 61

    Accessory Price List, Air-Conditioning (price) without heater, $275, 61

"Innovative Years," 51, 109, 146, 204, 251

Insulation, 31, 44–46, 57, 59–60, 117, 131, 157–158, 181, 217, 247

    foil-backed flooring insulation and close-up photos, 45–46

J

Juneau, Bud, Editor, *Cadillac-LaSalle Self-Starter Annual*, (1991)
   "Cadillac's Air Conditioner" (1941 Cadillac), written by Houston, Doug, 190–193, 202

Juneau, Bud, Editor, *Cormorant, The*, Spring 1975,
   "Destined to Survive" (1942 Packard), written by Hollingsworth, Jim, 121–129

K

Kelvinator, 5
Kettering, Charles, 6, 153
Kinetic Chemicals, 6, 153
Kughn, Richard, 1941 Cadillac Fleetwood Series Seventy-Five, air-conditioned, previous owner, 151, 175, 191–192, 196, 200–202

L

Lawrence, John, 21–22
Leland, Henry, 148
Lone Star Cadillac of Dallas, 175, 251
Lowering inside temperatures as much as 10°F to 15°F (1941 Cadillac air-conditioning), 160

M

MacArthur, General Douglas, 121–129
Macauley, Alvan, 22, 87
MacClellan, Don, 129
Martin, Terry, 10
Mauck, Fred, and Carol, Collection of, The, 33, 39
Mayo, Edward L., 7
McCarthy, Glenn, Glenn, Jr., 6, 243
McGahey, David (Dave) E., 1941 Packard One Eighty, Body 1442, air-conditioned, equipped with Cellarette, previous owner, 76
     passengers: Simons Story, Annette; and Story, Jim, 76
McPherson, Thomas A., Collection of, The, 10
Methyl Chloride, "See Warning Notice," 138-139, 144, 153
Midgley, Thomas, Jr., 6, 153
Mitchell, William (Bill) L., designer of 1938-1941 Cadillac Fleetwood Series Sixty Special, 147, 149
Motor, "Air-Conditioning for Packards–$274" (1940 Packard), 14-15
Motor City Dream Garages, Roy, Rex, 200
Mouton carpeting, 34–35, 38–39, 43

Index 268

N

Nash, 1930, 5, 21, 57
Nash, 1954, 57, 117, 178, 218
"Never Was Quality So Important," *Packard Senior Cars 1942* brochure, 119, 136
"New! Air-Conditioning-A Packard First!" (1941), 63
New York Times, 25

O

Ober, Robert, 1953 Cadillac Fleetwood air-conditioning fresh air intake scoop, 250
Oldsmobile, 1953, cool air distribution by louvered air vents with nozzles, Gaffney, Phil, 43, 249
Oldsmobile, 1953, cool air distribution sleeves, with nozzles, Wnuk, Paul, 249
    "Tiny holes, 1500 of them," air distribution, GM Media Archive, 248–249
Oldsmobile, 1954, air-conditioning compressor with electromagnetic clutch, 57
olive-drab.com, 130

P-Q

Packard 1940: A Pivotal Year, written by Hollingsworth, James, 9–10, 25–27, 39, 46, 90, 217
Packard 1940 Data Book, 40, 46, 55
"Packard Announces the First Completely Weather Conditioned Car," 1940 brochure, 28–31
Packard Clipper Data Book, April 1941, 88
Packard Clippers brochure (1942), 110–114
    "Brand new Clipper styling in every Packard price range," 112
    models, 112–113
Packard Dallas, 17, 36
    dealer demonstration of air-conditioned 1940 Packard One Sixty in showroom, 17
Packard Data Book (1942), 115–117
    air-conditioning, 115
        air filtration, 117
        cool air circulation, 116
        cooling cycle, the, 116
        cooling effectiveness, 117
        dehumidification, 117
        diagram of components, 116
        factory installation, 117
        insulation, 117
        quick cooling, 117
Packard Data Book for 1941, 24, 40, 46, 54–60, 70–72, 82, 157
    Air-conditioning, 55
        air filtration, 56
        compressor speed and relation to crankshaft speed, 57
        cool air circulation, 56

"cool air outlet," description by 1941 Packard that replaced "blower grille" from 1940 description, 56
cooling capacity, 54
cooling cycle, the, 56
cooling effectiveness, 57
dehumidification, 56
factory installation, 57
insulation, 57
proved by Packard standards, 57
quick cooling, 57
system in operation, the, 55
Comprehensive body insulation: materials, phantom view, placement, 60
Insulation: cowl, doors, floors, panels, roof, trunk, 59
Supplement to p. 80, Packard Cellarette, Hook, Andrew, via Kelley, Steven, 70

Packard for 1940: The One Ten-The One Twenty, brochure cover, 18
Packard News Service, 65, 75–76
Packard One Ten and One Twenty for 1941, brochure cover, 18, 52
"Packard Presents a New Miracle, the Air-Conditioned Motor Car" (1940 Packard), brochure, 1
Packard Promotional Pointers, 15–16, 118
　　$274 Weather Conditioner Price, 15
Packard Senior Cars 1942, brochure, 119
　　"Genuine Air-Conditioning, It Cools-It Filters-It Dehumidifies," 119
　　"Never Was Quality So Important," 119
　　P. T. (Patrol Torpedo) boat Packard Supermarine engines, 119, 136
　　Rolls-Royce aircraft engines redesigned, manufactured by Packard, 119, 136
Packard Service Letter, 40, 108, 138, 140–143
　　"Weather Conditioned Cars," 40
Packard Weather Conditioner, 7–10, 12, 15–17, 19–21, 25–32, 40–42, 49–51, 55, 108, 153, 158, 162, 206, 225
　　"Bishop & Babcock Weather Conditioner," 7
　　evaporator nameplate, "Summer Cooling," and "Winter Heating," 26
　　"lower the temperature inside a car 19 degrees below that of the outside air" (1940 Packard), 21
　　"Packard Weather (or Air) Conditioner," "Bishop & Babcock Mfg. Co.," nameplates on evaporator (1940–1942 Packard), 26, 42, 80, 85, 133, 247
　　price of $274, 14–15, 51, 61
　　price of $275, 27, 61
　　price of $325, 61, 114
　　"Weather Conditioner" name changed to "Air Conditioner," 32
"Packard: A Clean Sweep in Every State!" (1941) ad, 66
Packard: A History of the Motor Car and the Company, written by Bradley, James J., Hamlin, George, Heinmuller, Dwight, and Yost, L. Morgan, 10, 19, 23, 117, 136, 158, 217, 247
PackardInfo.com, 12–13, 15–16, 33, 39–40, 61, 65, 70, 79–80, 104, 108, 118, 138, 140–144
　　Czirr, Dave, 79–80
　　Flack, David, 61

Index 270

    Guscha, 13

    Waltman, Kevin, 12, 15–16, 33, 39–40, 65, 104, 108, 118, 138, 140–144

Packard, James Ward, 22

Packer, W. M., 11, 27

Pan American World Airways, Pan Am, Clipper, 96, 111

Patrol Torpedo (P. T.) boat Packard Supermarine engines, 119, 136

"Penalty of Leadership, The" (1915 Cadillac), written by MacManus, Theodore F., 148

Peterson, West, 8, 34–39, 41–47, 49, 94, 131, 137, 198–199, 248

    Peterson photo, 44

"Pioneered by Packard" (1941 Packard), 68

Polio, the threat of, McCarthy, Glenn, Glenn, Jr., 6, 243

Popular Mechanics, 24

Popular Science, 7, 120

prices for the heater (1940 Packard), 32

"Quick Cooling," phrases (1940, 1941 Packard), 28, 57, 117

R

Rauch, Warren, *Cadillac Data Book (1941),* "Supplementary pages," 154-157

    Cadillac Air-Conditioning, 154

        Comfortable Interior Temperature, 157

            adequate insulation, 157

            immediate operation, 157

            large capacity, 157

        Controlled Air Circulation, 156

        Simple in Operation, 155

            compressor, 155

            condenser, 155

            expansion valve and cooling coils, 155

            reservoir, 155

"Real Air-Conditioning A Packard First" (1941–1942 Packard) ads, 66, 82, 114

"Regarding the Air-Conditioning System on Your Cadillac" (1941), GM Media Archive, Adelson, Kathy, 158–161

    Action of System, 159

    Controls, 159

        lowering inside temperatures as much as 10°F to 15°F, 160

    Maintenance, 161

    Operation of System, 160

    Refrigeration cycle, 160

    Touring, 161

"Riding in Comfort II," Bhatti, Mohinder S., Ph.D., 5

Rogers, N. C., 17

Rolls-Royce aircraft engines redesigned, manufactured by Packard, 119, 136

Roosevelt, Franklin D., 136
Rose, W. G., 41
Roy, Rex, *Motor City Dream Garages,* 200

S

Saiya, Sal, M/M, 78–80, 137
Sanden compressor, 92
Sanders, Yann, *car-nection.com, Original Cadillac Database,* 189
San Francisco Chronicle, 66
San Francisco Examiner, 75
Santa Ana Register, 75
Saturday Evening Post, The, 33, 62–63, 81, 148
Schneider, Roy A., *Cadillacs of the Forties,* 152, 191, 203
"See the Motor Car That Makes Ice Cubes," Chicago Auto Show (1941 Packard), Frumkin, Mitch, 69
Selznick, David O., 1942 Cadillac Limousine, air-conditioned, 203
Servel air-conditioning compressor, 4, 7, 14, 41, 45, 79, 92, 109, 127, 131, 158, 162, 178–179, 196, 198, 201, 217, 241, 248
  u-shaped support bolt, 41
Simons, Allen B., Collection of, (ABS), 7, 11, 13, 18, 20, 24–25, 48, 75, 78, 81, 87, 89–96, 120, 128, 136–137, 145, 148–152, 205–211, 218, 241–242, 250
Simons Story, Annette; and Story, Jim, passengers of McGahey, David (Dave) E., in 1941 Packard with Cellarette bar, 76
"Skipper the Clipper," slogan (1942 Packard Clipper), 111
"Socially-America's FIRST Motor Car" (1940 Packard) ad, 20–21
Society of Automotive Engineers (S. A. E.), Gleason, T. C., 244–246, 248, 250
"Sparkling Speedline Stripes" styling (1941 Packard), 53, 81, 112, 135, 152
"Speed-Stream" styling (1940 Packard), 10
"Standard of the World" slogan (1941 Cadillac), 147, 183
Steel Magazine, "Mirrors of Motordom," Allen, A. H., 70
Steidle, Jeff, 1–4, 42, 62–63, 88, 119
"Summer cooling," nameplate, 26
"Sunshine Turret Top," sunroof, option on 1941 Cadillac Fleetwood Series Sixty Special, owned by Zeiger, Dr. Richard, 151, 175–176, 180, 187–188
Supermarine P. T. (Patrol Torpedo) boat engines, by Packard, 136
Swaney, Jack, 67–68, 79–80, 137
Syracuse Herald American, 11

T

Temple, David W., 25–26, 85
Tenite dash (1940 Packard), 37
"That Packard Look," 89, 93

"Thermal shock," Bhatti, Mohinder S., Ph.D., 6, 159, 243

"Torpedo styling," 1941 Cadillac, 149–150

"Turn on the Cold!" (1940 Packard Weather Conditioner) ad, 2, 54, 68

V

Vacamatic transmission (1941 Chrysler Crown Imperial) ad, 224

Vandivert, R. M., 49–50

Venetian blinds (1941 Cadillac), 180–181

Verschoor, Ronald, 1941 Cadillac Fleetwood Series Sixty Special photos, 147, 175–188, 247–248

W

Waltman, Kevin *(PackardInfo.com)*, 12, 15–16, 33, 39–40, 65, 104, 108, 118, 138, 140–144

War Bonds, 145

War Production Board (WPB), 136

"Weather Conditioned Cars," Packard Service Letter (1940 Packard), 40

Weather Conditioner (1940 Packard), 108, 206

Weiss, Terry, owner of 1941 Packard Clipper Custom, air-conditioned, 78, 89–95, 128, 137

Wells, Bud, 121, 123, 129

Wenger, Terry, Sr., 190–193

Wilson, Chuck, 5

Wnuk, Paul, 1953 Oldsmobile air conditioning cool air distribution, 249

World War II, 119, 130, 136, 138

WPC News, 212–215

Y

Yost, L. Morgan, *Packard: A History of the Motor Car and the Company*, 19, 23, 117, 158, 217, 247

Z

Zeiger, Dr. Richard, 1941 Cadillac Fleetwood Series Sixty Special, air-conditioned, owner; photos by Verschoor, Ronald, 147, 151, 175–188, 247–248

Made in the USA
Coppell, TX
20 September 2023